ARCHI+SCAPE

一筑一景

文化・教育・住宅 Culture & Education & Residence

Dopress Books 度本图书 编

華中科技大學出版社
http://www.hustp.com
中国・武汉

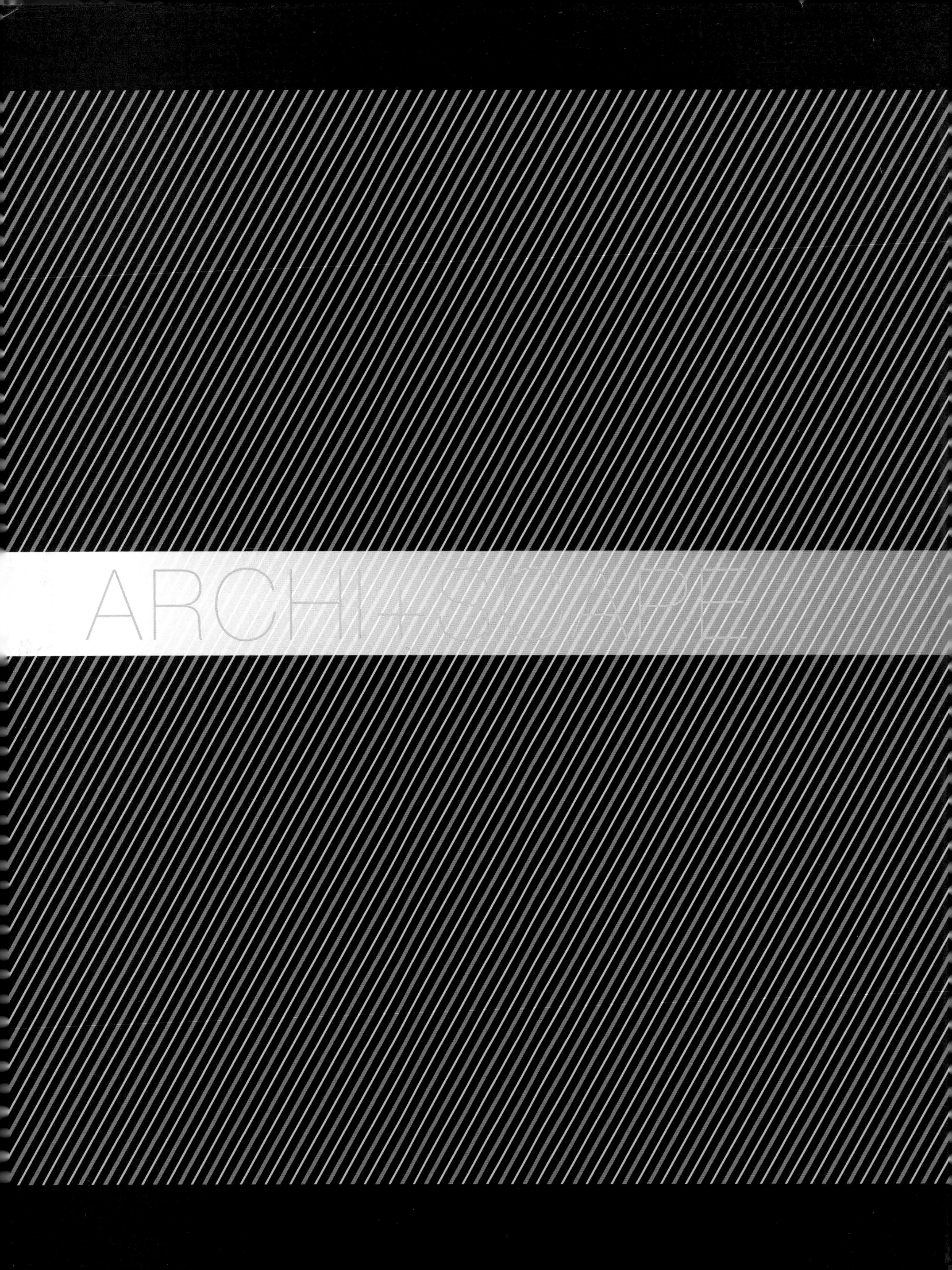
ARCHI+SCAPE

一筑一景

ARCHI+SCAPE

前言 FOREWORD

出色的建筑作品应是建筑形态、室外环境和室内空间的和谐统一、浑然天成，而实现这一理想模式的最佳途经，即将各部分空间进行一体化设计，在建筑与景观，整体与细节的关联中找到功能、技术与艺术的平衡。

《一筑一景》（ARCHI+SCAPE）系列图书创造性地将建筑（Architecture）与景观（Landscape）结为一体，形象地揭示了建筑与环境不可分割、互为依存的紧密关系。本系列书共有两册，书中精选的近100例近期优秀作品基于"建筑与环境相融"这一主题，很好地反映了当今世界建筑与环境设计的主流思潮与最新趋势，融入其中的新鲜创意与独特视角将大幅提升相关领域建筑师、设计师的灵感与技能。

本册书着重展示了文化、教育与住宅方面出色的建筑与环境设计作品，其中包括音乐厅、剧院、人文景区、画廊、展示中心、世博会建筑、博物馆、幼儿园、校园、住宅区等近50个精彩案例。

Remarkable architectural works reflect the unity of exterior environment, architectural morphology and interior space. The best way to realize this ideal is to initiate the integration of design as well as striking balance between function, technology and art in relation of architecture and landscape, the whole and part.

The book title *ARCHI+SCAPE* creatively combines architecture, landscape and interior, vividly revealing the inseparable and interdependent relationship between them. Focusing on the theme of architecture and environment in harmony, the almost 100 recent masterpieces selected worldwide well reflect the leading concepts and current trend of architecture and landscape design in today's world. Fresh ideas and unique views substantially promote the specialization of architects in this field and bring infinite inspiration to all designers and readers.

This book mainly shows the outstanding architectural works about culture, education and residence. There are almost 50 exciting projects featured in the book including concert halls, theaters, cultural attractions, art galleries, exhibition centers, museums, kindergartens, campus, residential areas, etc.

Cultural Zone 文化区

Exhibition & EXPO 展示与博览会

Museum 博物馆

目录 CONTENTS
ARCHI+ SCAPE

Kindergarten & School 幼儿园与学校

Residential Zone 住宅区

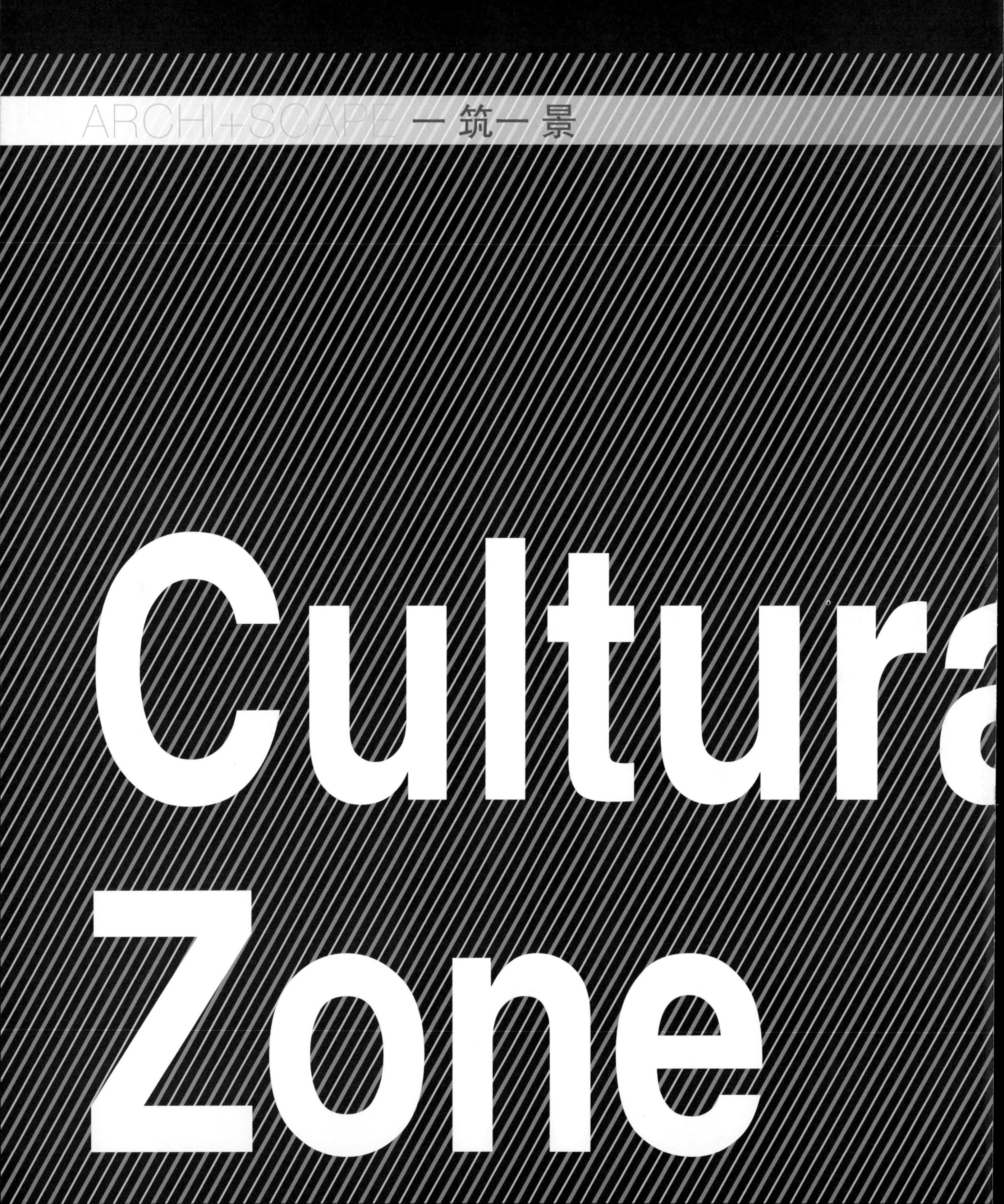
ARCHI+SCAPE 一筑一景
Cultura
Zone

文化区

NO.1 006-057

The Culture Yard

文化广场

Location / 地点:
Elsinore, Denmark
Date of Completion / 竣工时间:
2010
Area / 占地面积:
17,000 m^2
Architecture / 建筑设计:
AART Architects
Landscape / 景观设计:
AART Architects
Photography / 摄影:
Adam Mørk
Client / 客户:
Elsinore Municipality

Elsinore's old shipbuilding yard has been transformed into a 17,000 m^2 cultural and knowledge centre, including concert halls, showrooms, conference rooms, a dockyard museum and a public library. The Culture Yard symbolizes Elsinore's transformation from an old industrial town to a modern cultural hub. In this way, the yard is designed as a hinge between the past and present, reinforcing the identity of the local community, but at the same time expressing an international attitude, reinforcing the relation between the local and global community.

The contrast between past and present permeates the Culture Yard. For instance, the original concrete skeleton with armoured steel has been reinforced, but left exposed as a reference to the area's industrial past. The historic context has thus been the main structural idea in the design process, ensuring the keen observer would discover a chapter of history in every corner of the yard and every peeling of the wall. In other words, if you want to understand what Elsinore really is, what the intangible blur between past and present feels like, this is the place to visit. Thanks to the architectural features such as wrought iron stairs and concrete elements, interacting with modern glass structures and interior designs, the contrast between the days of yore and the present becomes evident. It is the Culture Yard's way of playing with the field of tension between old and new, making the notion of past versus present, the industrial society versus the information society, constantly present. Particularly striking, when viewed from the seafront and Kronborg Castle, is the multifaceted façade. Like a fragmented, yet strongly coherent structure, the enormous glass and steel façade challenges the historic site and stares unflinchingly across the Sound – the strait that separates Denmark and Sweden. The transparent façade also reinforces the relation between inside and outside, as you can peak in from street level and enjoy the magnificent sea view and view of Kronborg Castle from every floor of the building especially from the glass cave which in a dramatic gesture protrudes out of the building above the main entrance.

In this way, the façade encloses the yard in a distinctive atmosphere, as the dazzling and dramatic play of lines generates a sense of spaciousness. Although the façade is made of hundreds of lines and triangles, it appears as one big volume, generating a sense of place and time. The volume also takes the environment into account, since the façade not only functions as an aesthetic and spatial architectural feature, but also as a climate shield, reducing the energy demand for cooling and heating of the building.

埃尔西诺的造船厂旧址如今被改造成了一座17000m^2的文化中心，这里设有多个音乐厅、展厅、会议室、1个船坞博物馆和1个公共图书馆。该文化中心反映了埃尔西诺正由一个老工业城市向现代文化会所转变。因此，它被设计成衔接传统与未来的纽带，在强化社区特色的同时强调国际化，增强了本地和国际社会之间的联系。

新旧元素的对比贯穿于整个文化中心。例如，旧址遗留的混凝土夹杂装甲钢的构架被进一步加固，并毫不掩饰地暴露在外，“叙述”着该区域的工业背景。历史元素在设计过程中扮演了主要角色，使敏锐的观察者能够从场地的每一个角落和墙壁的每一处剥落中品味这里的历史。换句话说，如果你想真正了解埃尔西诺，了解其历史，那么这里就是最理想的场所。铁艺楼梯、混凝土的构造与现代化的玻璃结构、室内设计相协调。从海滨和卡隆堡宫望去，该项目最引人注目的就是建筑多面外墙。庞大的外墙采用玻璃和钢铁构造，呈现出零碎而又紧密相连的结构，矗立在这块历史悠久的场地上，“凝视”着分隔丹麦和瑞典的松德（Sound）海峡。透明的多面外墙还加强了内部和外部之间的联系，人们可以从各个楼层内部——特别是大厅上方突出楼体的玻璃洞穴处——欣赏壮丽的海景和卡隆堡宫。

建筑外墙使该场地洋溢着独特的氛围，其惊艳而生动的建筑线条则使空间更显宽敞。虽然外墙由数以百计的线条和三角形构成，它却是一个巨大的整体，给人带来视觉冲击感。外墙的设计也考虑到环境因素，因为它不仅是一栋可供欣赏的特色建筑，还承担着调节室内温度的功能，可以减少建筑物对供暖和制冷的能源需求。

Materials: Steel, Concrete, Glass, Wrought iron, Aluminum sheets.

KULTURVÆRFTET
HAVLIT.DK

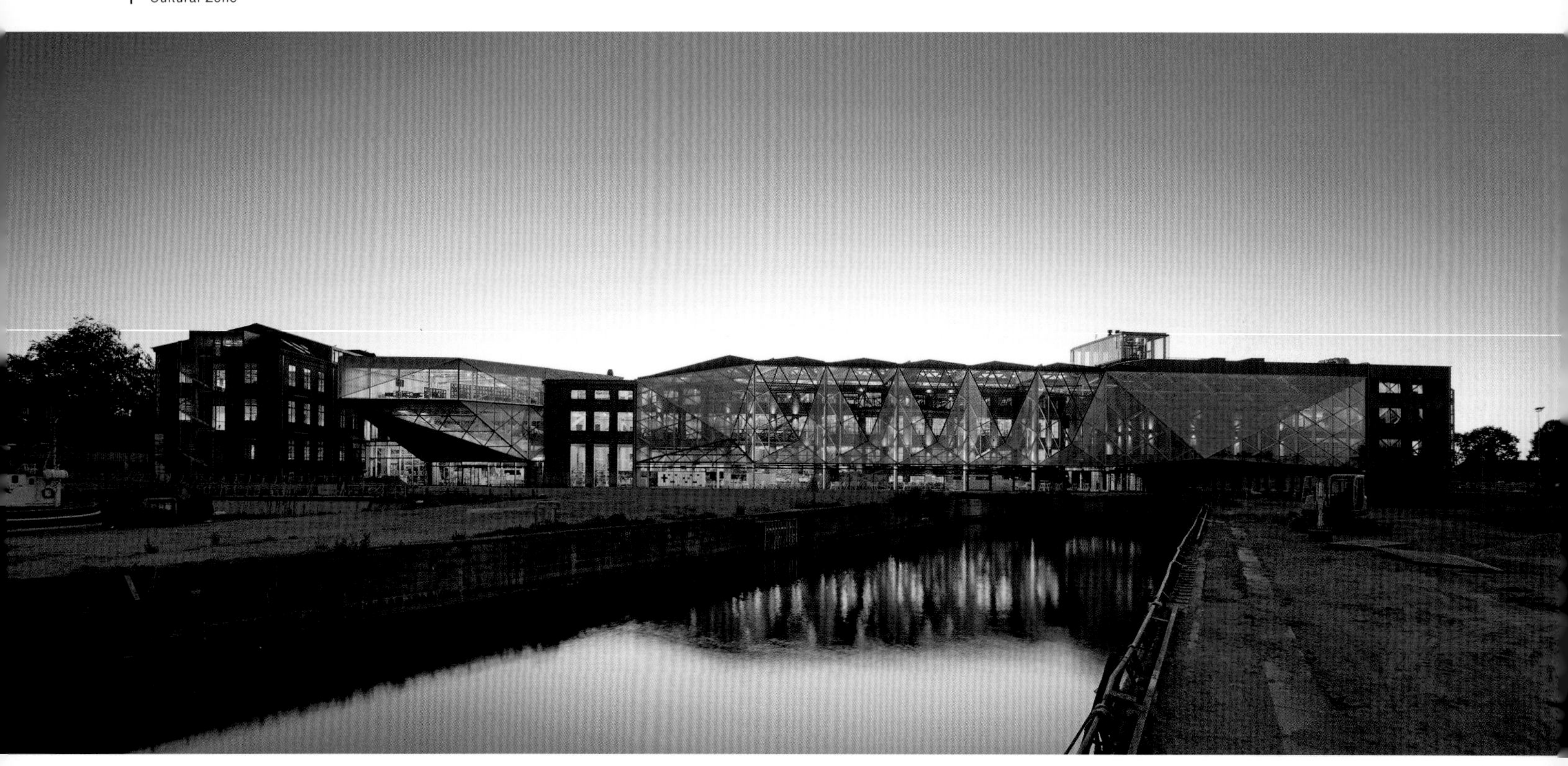

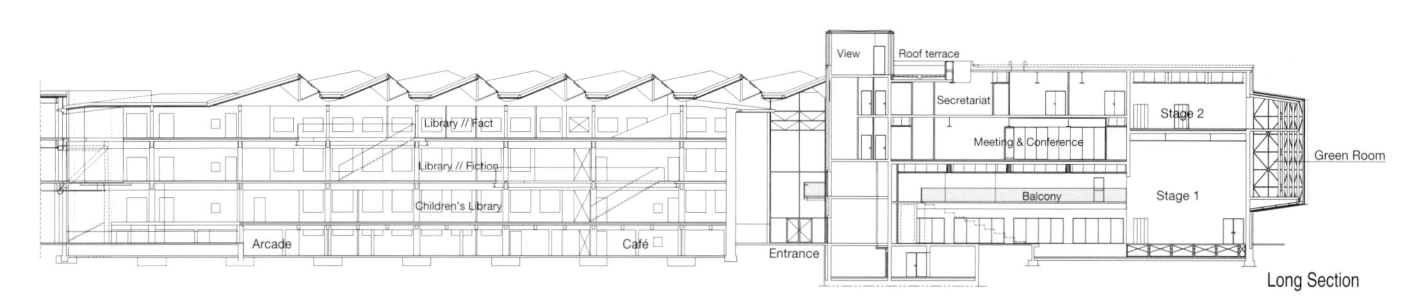

Long Section

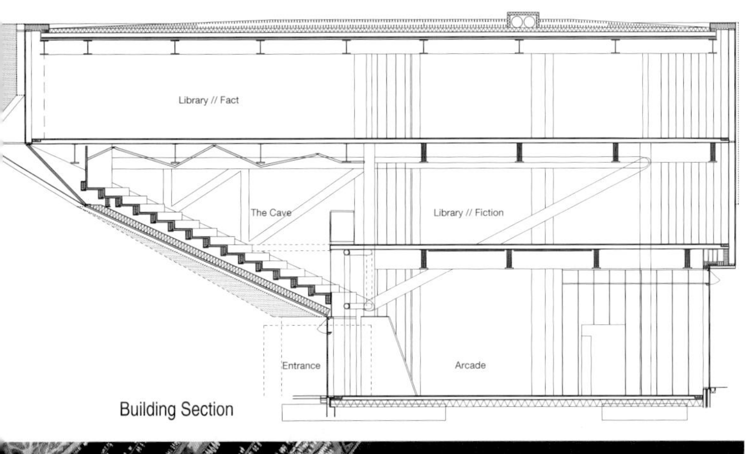

Building Section

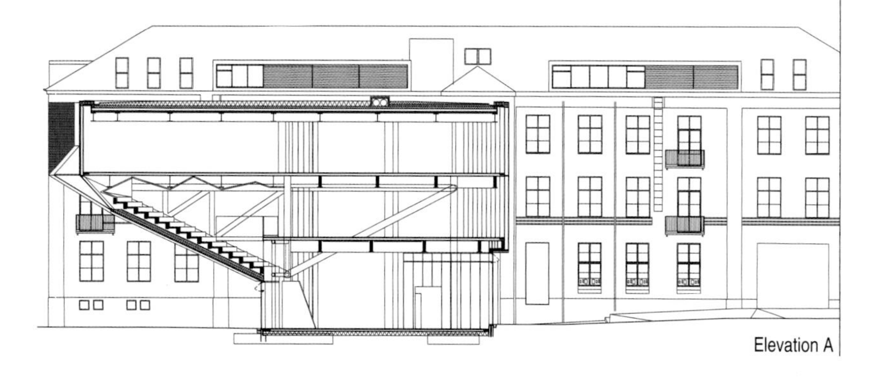

Elevation A

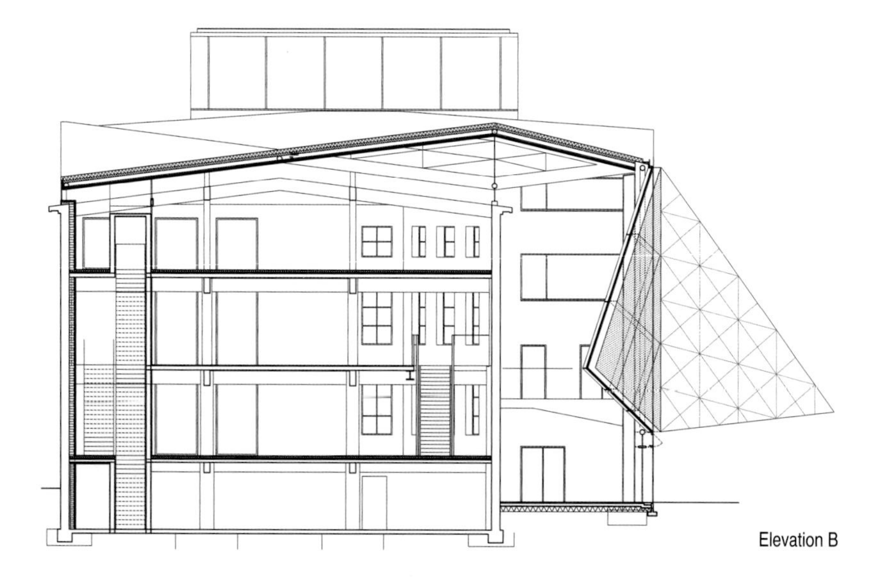

Elevation B

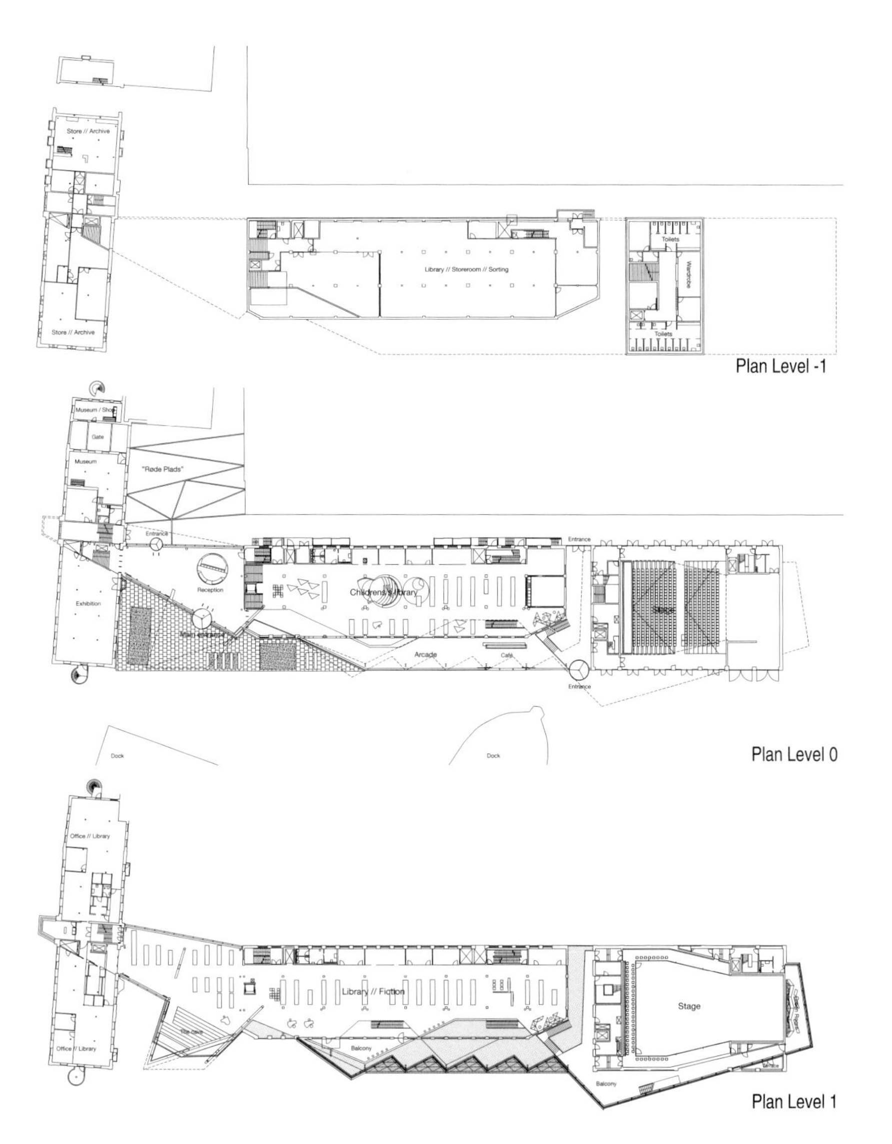

Store // Archive
Library // Storeroom // Sorting
Toilets
Wardrobe
Plan Level -1
"Røde Plads"
Entrance
Reception
Exhibition
Children's Library
Stage
Arcade
Café
Dock
Plan Level 0
Office // Library
Library // Fiction
Stage
Balcony
Plan Level 1

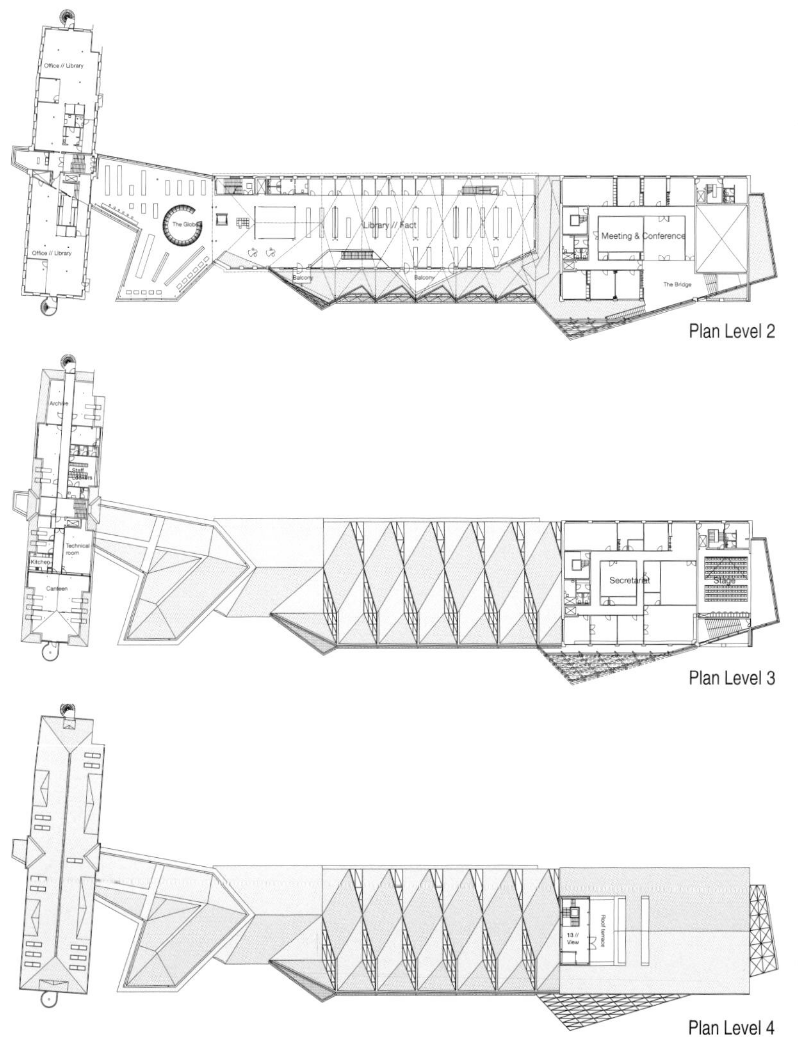
Office // Library
The Globe
Library // Fact
Meeting & Conference
Balcony
The Bridge
Plan Level 2
Archive
Technical room
Kitchen
Canteen
Secretariat
Stage
Plan Level 3
13 // View
Plan Level 4

BYGNING 12

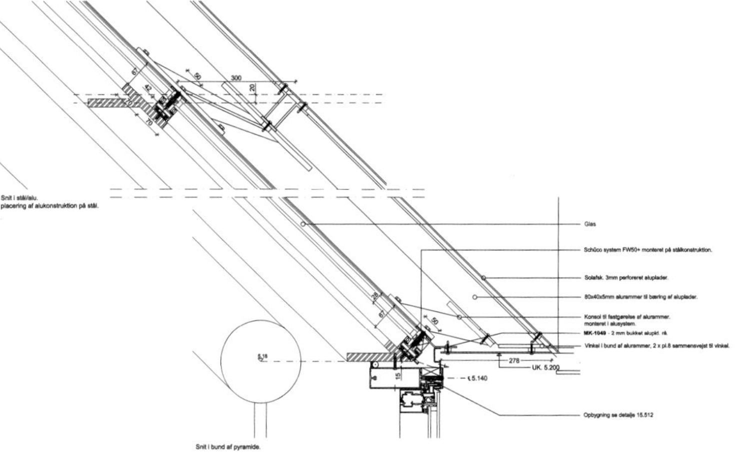
Snit i stål/alu.
placering af alukonstruktion på stål.
Glas
Schüco system FW50+ monteret på stålkonstruktion.
Solafsk. 3mm perforeret aluplader.
80x40x5mm alurammer til bæring af aluplader.
Konsol til fastgørelse af alurammer.
monteret i alusystem.
MK-1049 - 2 mm bukket alupld. rå.
Vinkel i bund af alurammer, 2 x pl.8 sammensvejst til vinkel.
278
UK. 5.200
5.140
15
Opbygning se detalje 15.512
Snit i bund af pyramide.

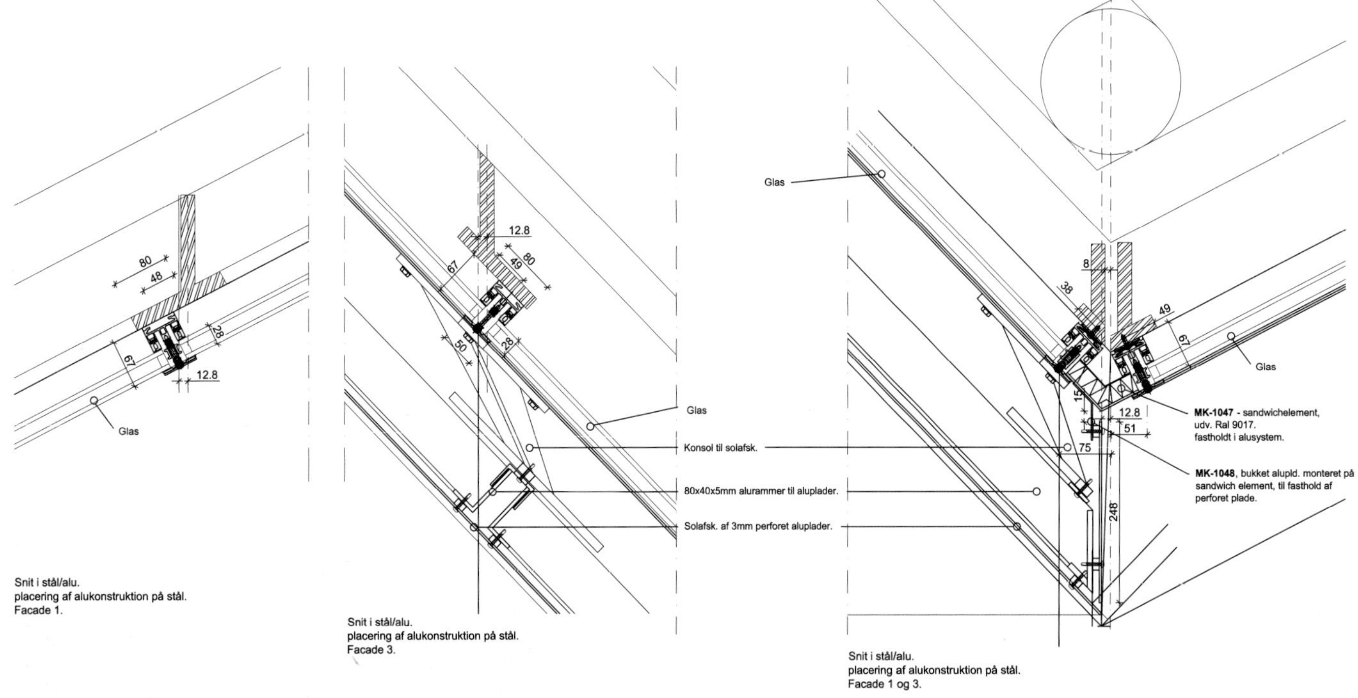
80
48
67
28
12.8
Glas
Snit i stål/alu.
placering af alukonstruktion på stål.
Facade 1.
12.8
67
49
80
50
28
Glas
Konsol til solafsk.
80x40x5mm alurammer til aluplader.
Solafsk. af 3mm perforet aluplader.
Snit i stål/alu.
placering af alukonstruktion på stål.
Facade 3.
Glas
8
38
49
67
Glas
15
12.8
51
75
248
MK-1047 - sandwichelement,
udv. Ral 9017.
fastholdt i alusystem.
MK-1048, bukket alupld. monteret på
sandwich element, til fasthold af
perforet plade.
Snit i stål/alu.
placering af alukonstruktion på stål.
Facade 1 og 3.

Details

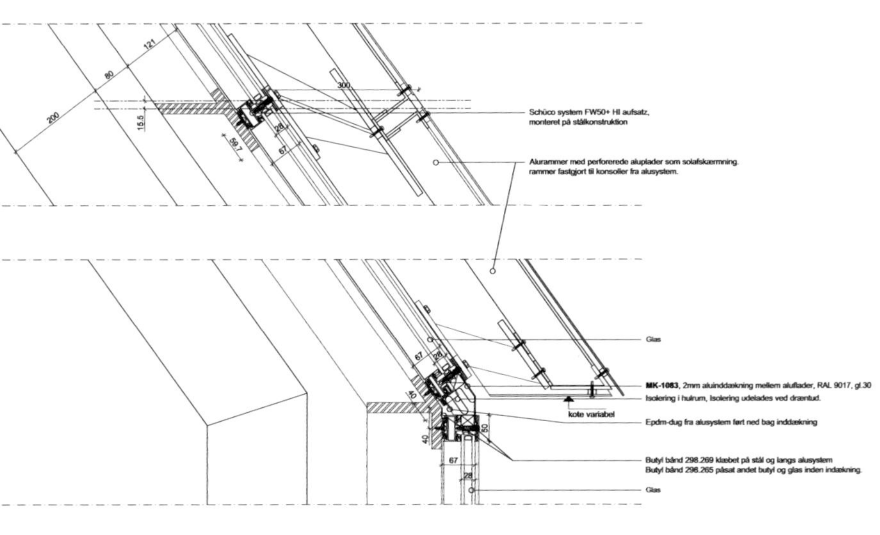

Schüco system FW50+ HI aufsatz,
monteret på stålkonstruktion
Alurammer med perforerede aluplader som solafskærmning.
rammer fastgjort til konsoller fra alusystem.
Glas
MK-1083, 2mm aluinddækning mellem aluflader, RAL 9017, gl.30
Isolering i hulrum, Isolering udelades ved drænud.
kote variabel
Epdm-dug fra alusystem ført ned bag inddækning
Butyl bånd 298.269 klæbet på stål og langs alusystem
Butyl bånd 298.265 påsat andet butyl og glas inden indækning.
Glas

HORISONTEN

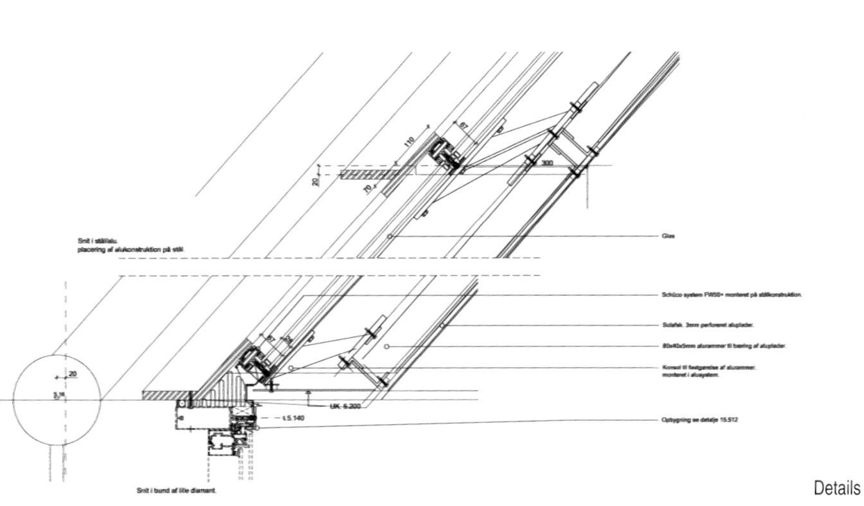

Snit i stållalu.
placering af alukonstruktion på stål.
Glas
Schüco system FW50+ monteret på stålkonstruktion.
Solafsk. 3mm perforeret aluplader.
Konsol til fastgørelse af alurammer.
monteret i alusystem.
UK. 6.200
Opbygning se detalje 15.512
Snit i bund af lille diamant.

Details

Details

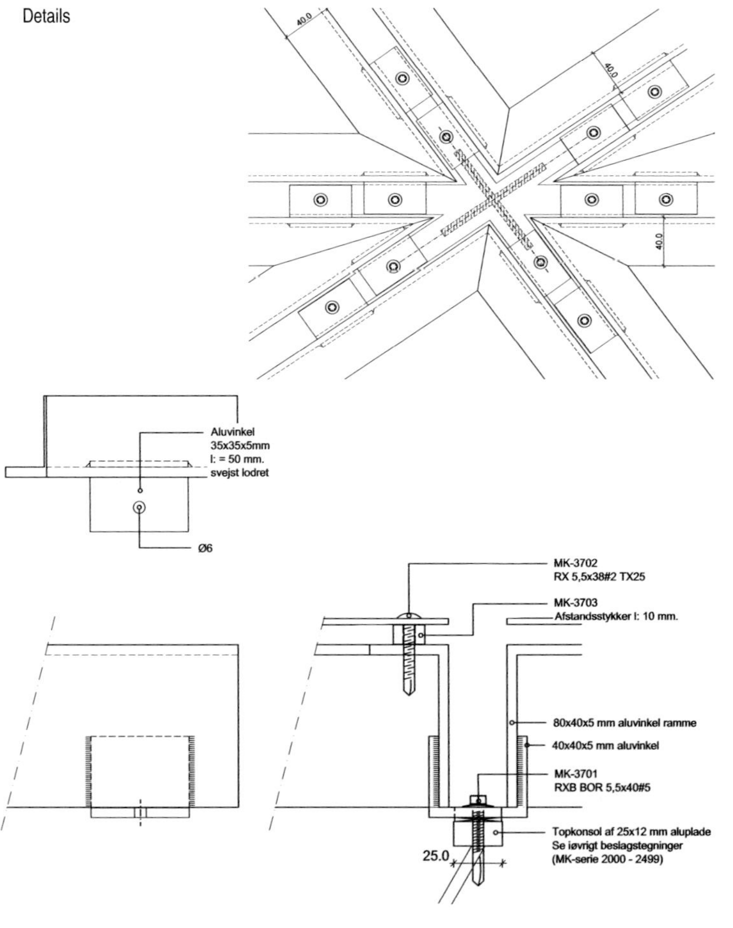
40.0
40.0
40.0
Aluvinkel
35x35x5mm
l: = 50 mm.
svejst lodret
Ø6
MK-3702
RX 5,5x38#2 TX25
MK-3703
Afstandsstykker l: 10 mm.
80x40x5 mm aluvinkel ramme
40x40x5 mm aluvinkel
MK-3701
RXB BOR 5,5x40#5
Topkonsol af 25x12 mm aluplade
Se iøvrigt beslagstegninger
(MK-serie 2000 - 2499)
25.0

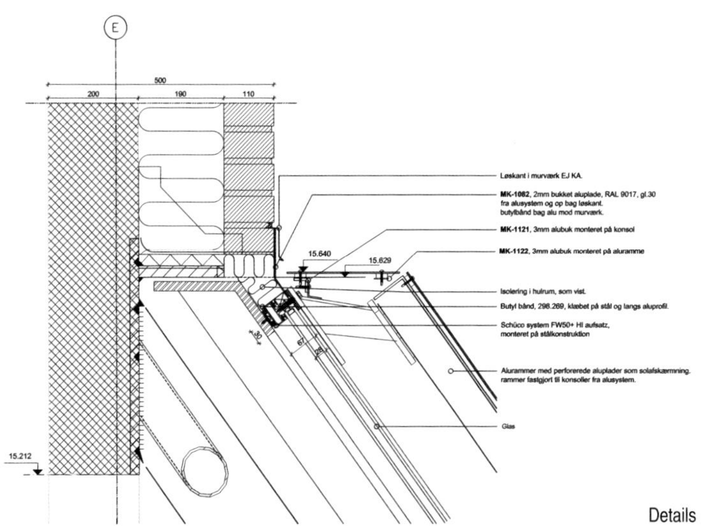
E
500
200
190
110
Løskant i murværk EJ KA.
MK-1082, 2mm bukket aluplade, RAL 9017, gl.30
fra alusystem og op bag løskant.
butylbånd bag alu mod murværk.
MK-1121, 3mm alubuk monteret på konsol
MK-1122, 3mm alubuk monteret på aluramme
15.640
15.629
Isolering i hulrum, som vist.
Butyl bånd, 298.269, klæbet på stål og langs aluprofil.
Schüco system FW50+ HI aufsatz,
monteret på stålkonstruktion
67
28
Alurammer med perforerede aluplader som solafskærmning.
rammer fastgjort til konsoller fra alusystem.
Glas
15.212

Details

Harpa-Reykjavik Concert Hall & Conference Centre

哈帕・雷克雅未克音乐厅与会议中心

Location / 地点:
Reykjavik, Iceland
Date of Completion / 竣工时间:
2011
Area / 占地面积:
28,000 m²
Architecture / 建筑设计:
Henning Larsen Architects and Batteriid Architects
Landscape / 景观设计:
Landslag efh
Photography / 摄影:
Nic Lehoux
Client / 客户:
Austurnhofn TR – East Harbour Project Ltd.

Harpa - Reykjavik Concert Hall and Conference Centre gathers inspiration from the northern lights and the dramatic Icelandic scenery.

Situated on the border between land and sea, the Concert Hall stands out as a large, radiant sculpture reflecting both sky and harbour space as well as the vibrant life of the city. The spectacular facades have been designed in close collaboration between Henning Larsen Architects, the Danish-Icelandic artist Olafur Eliasson and the engineering companies Rambøll and Art Engineering GmbH from Germany.

The Concert Hall of 28,000 m² is situated in a solitary spot with a clear view of the enormous sea and the mountains surrounding Reykjavik. The building features an arrival and foyer area in the front of the building, four halls in the middle and a backstage area with offices, administration, rehearsal hall and changing room in the back of the building. The three large halls are placed next to each other with public access on the south side and backstage access from the north. The fourth floor is a multifunctional hall with room for more intimate shows and banquets.

Seen from the foyer, the halls form a mountain-like massif that is similar to basalt rock on the coast, forming a stark contrast with the expressive and open facade. At the core of the rock, the largest hall of the building is the main concert hall, revealing its interior as a red-hot centre of force.

The project is designed in collaboration with the local architectural company, Batteríið Architects. And Nic Lehoux took the beautiful photos after it was completed.

哈帕・雷克雅未克音乐厅与会议中心位于冰岛首都雷克雅未克，其设计灵感来源于北极光和冰岛迷人的景色。

该音乐厅位于海陆交界处，这个大型的建筑物拔地而起，其光彩夺目的雕塑式设计映射着天空和海港，体现了充满活力的都市生活。宏伟的的外墙由Henning Larsen事务所、丹麦/冰岛艺术家Olafur Eliasson以及来自德国的Rambøll和Art Engineering GmbH工程公司倾力合作完成。

哈帕・雷克雅未克音乐厅占地2.8hm²，位于一个独立的场地内，它拥有开阔的视野，广袤的大海和雷克雅未克周围的山景。这栋建筑的前方设有接待室和休息厅，中部有四个大厅，后方设有办公室、行政室、排练厅和更衣室。三个大型音乐厅依次排列，公共入口位于南侧，后台入口位于北侧。大楼的第四层是一个多功能厅，可以举办演出和宴会。

从休息厅望去，音乐厅呈山峦状，如同海边的玄武岩，与富于表现力且奔放的外观设计形成鲜明对比。位于“岩石”核心的是这里最大也是最主要的音乐厅，红色调的内部设计使其成为一个激情四射的中央空间。

该项目是在当地建筑公司Batteriid Architects事务所的协作下完成的。项目竣工后由Nic Lehoux负责摄影。

Materials: Armoured glass;
Stainless steel;
Concrete.

Facade South

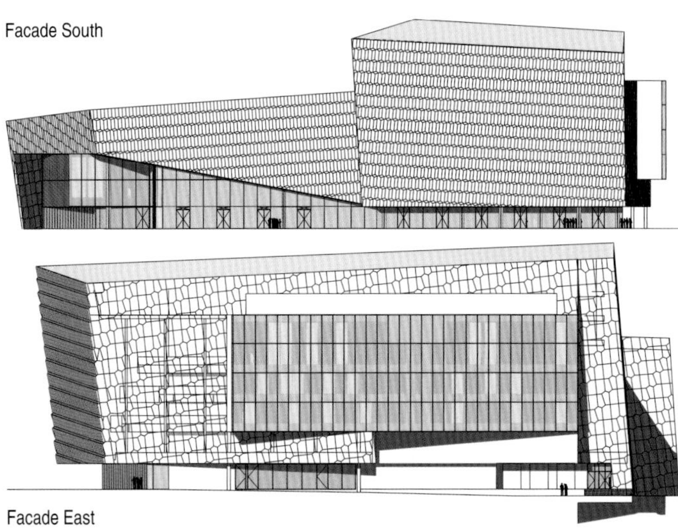

Facade East

Facade West A
Facade North
Facade West B

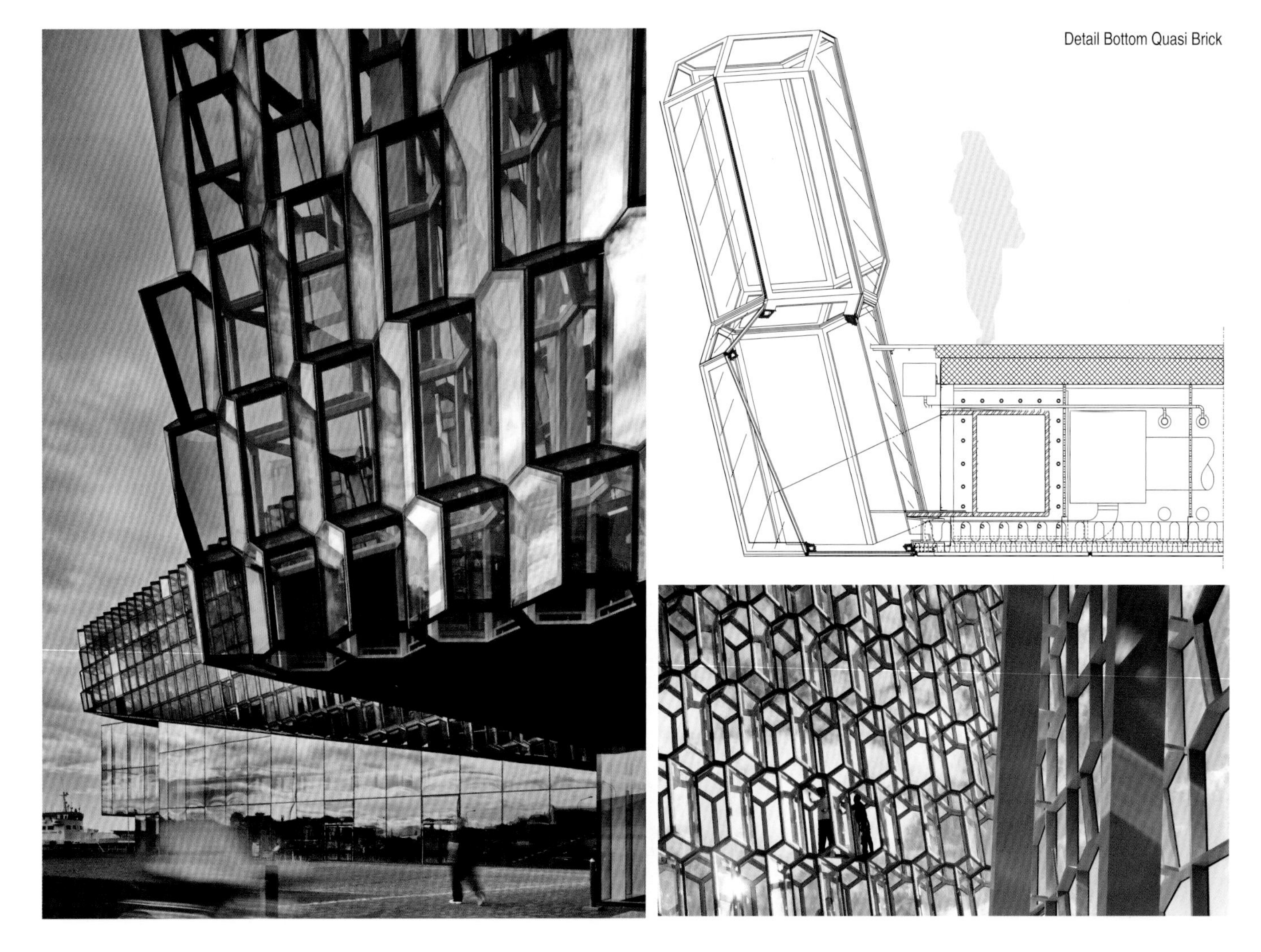
Detail Bottom Quasi Brick

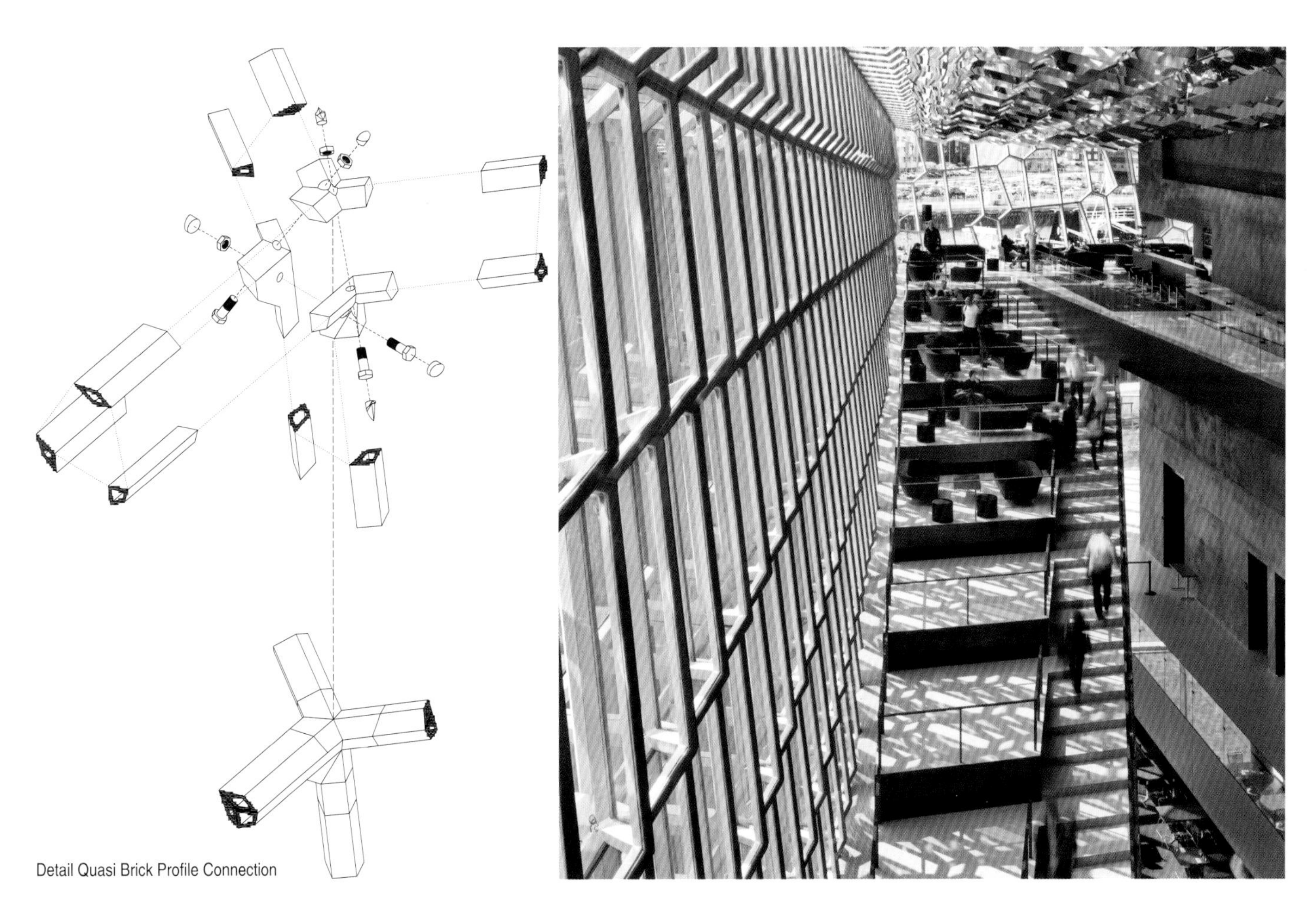
Detail Quasi Brick Profile Connection

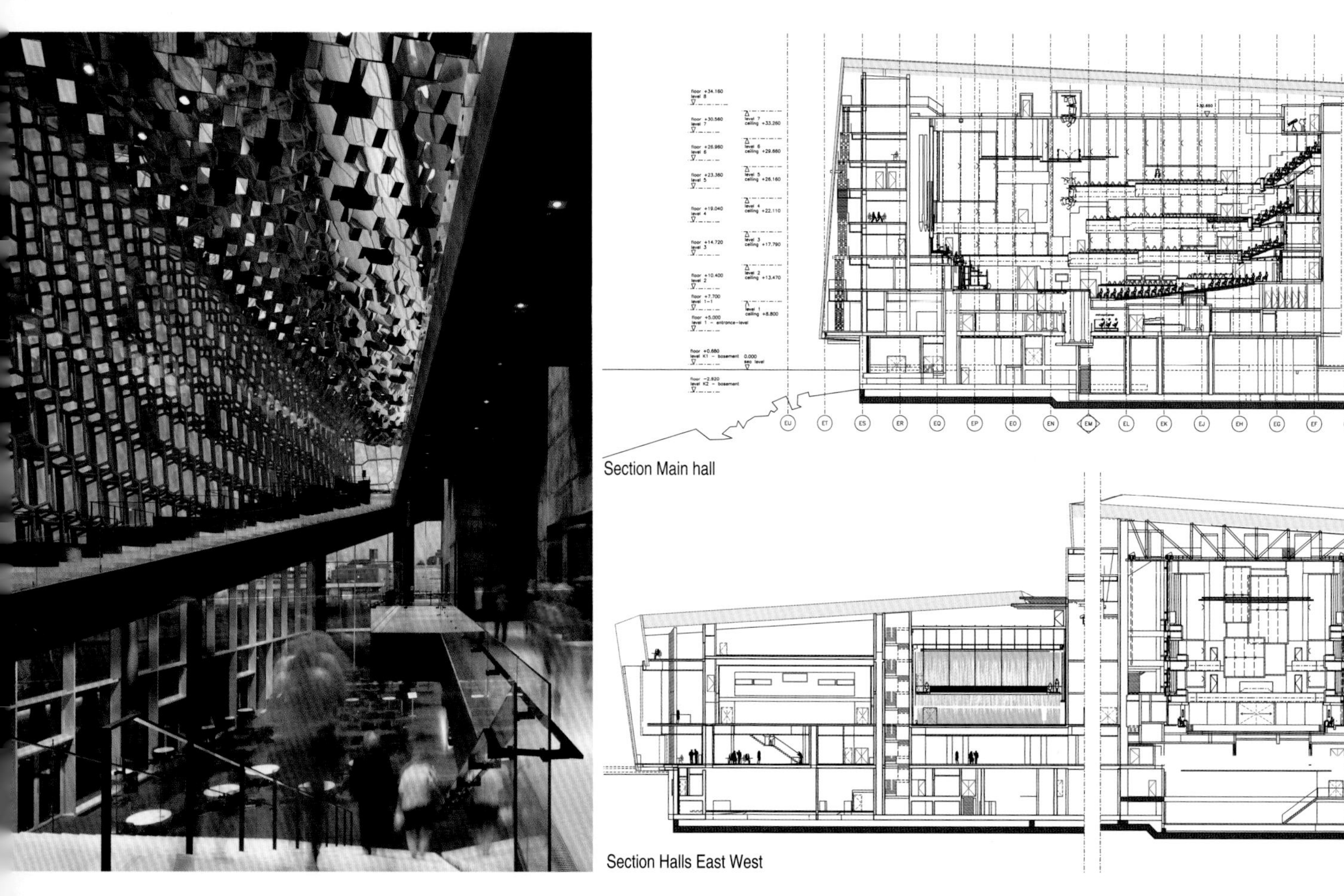

Section Main hall

Section Halls East West

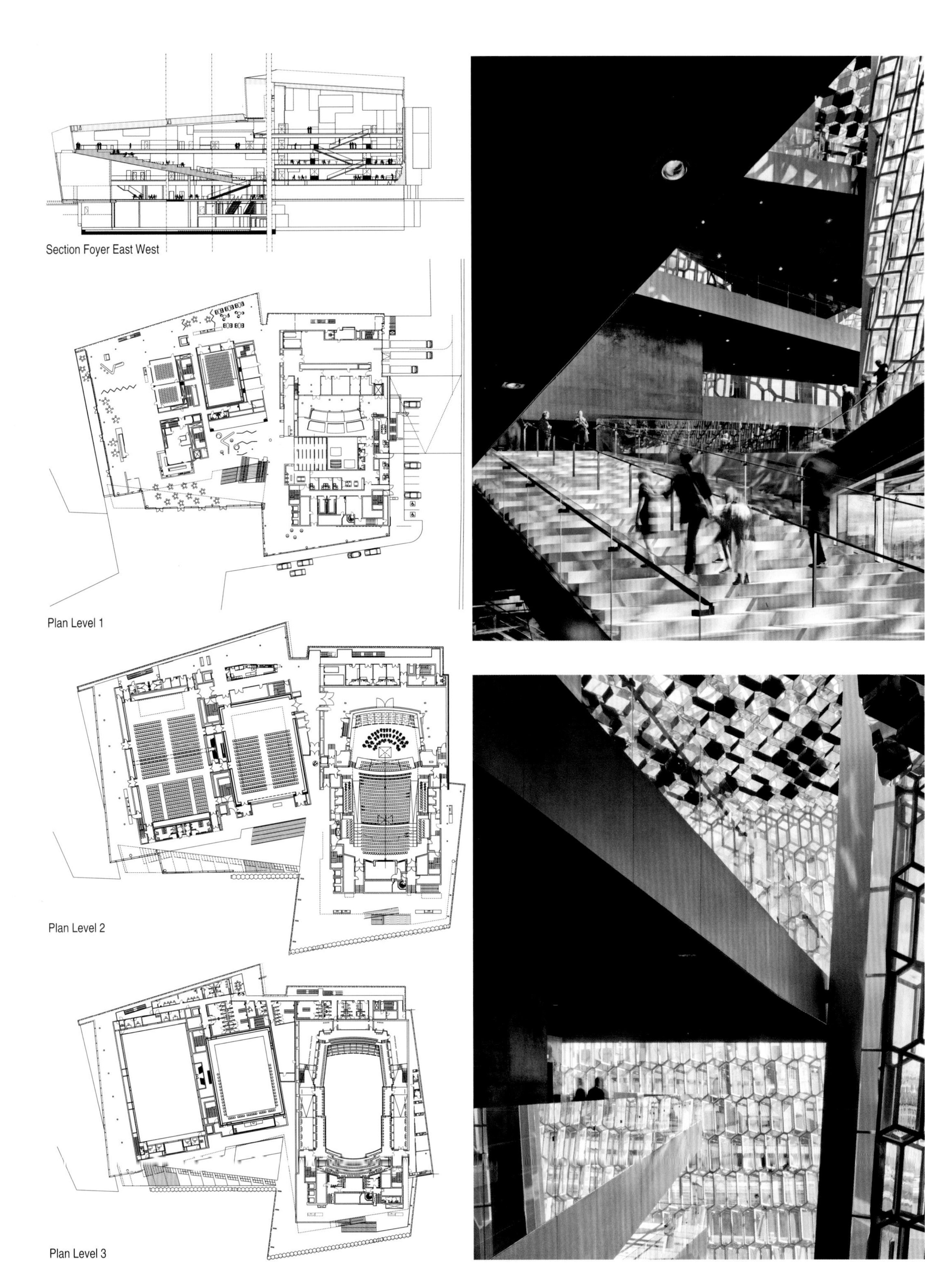
Section Foyer East West

Plan Level 1

Plan Level 2

Plan Level 3

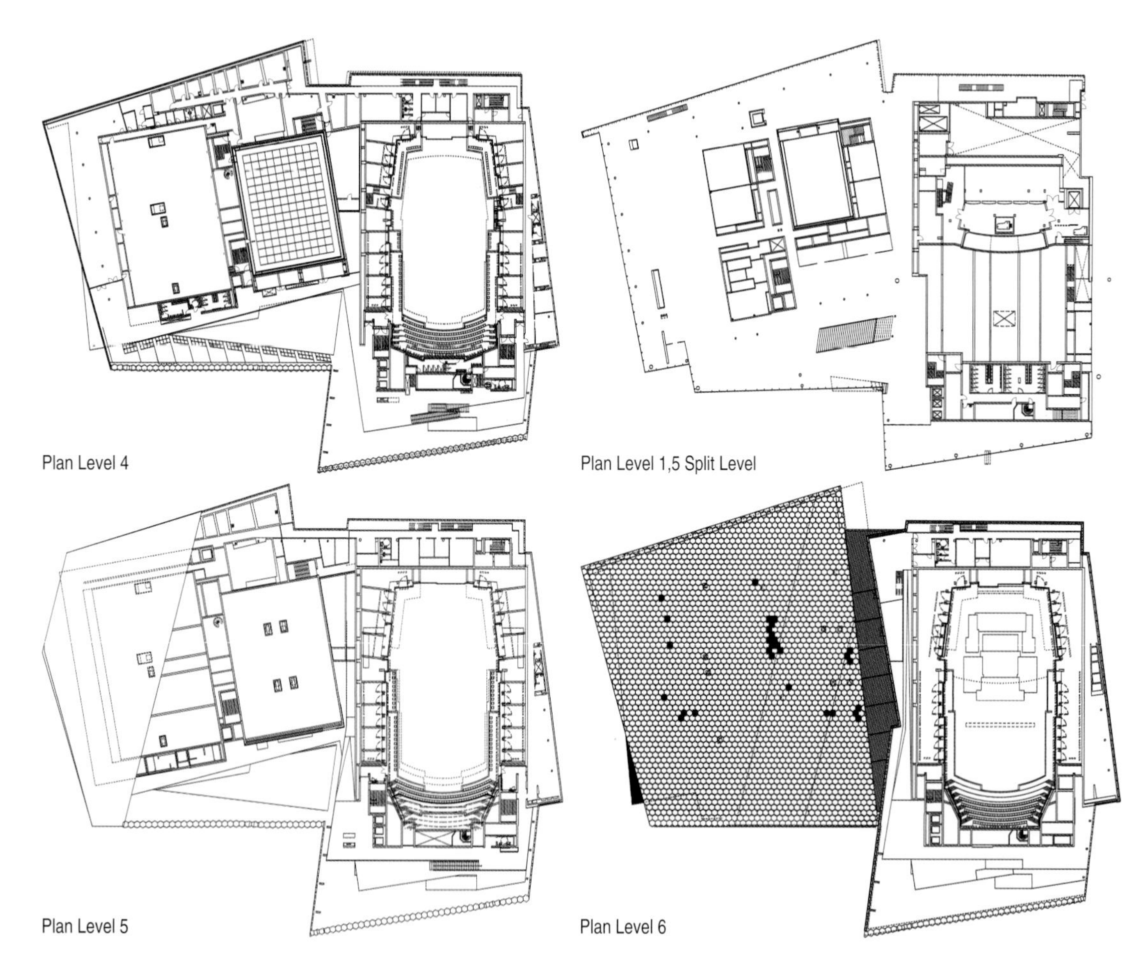

Plan Level 4

Plan Level 1,5 Split Level

Plan Level 5

Plan Level 6

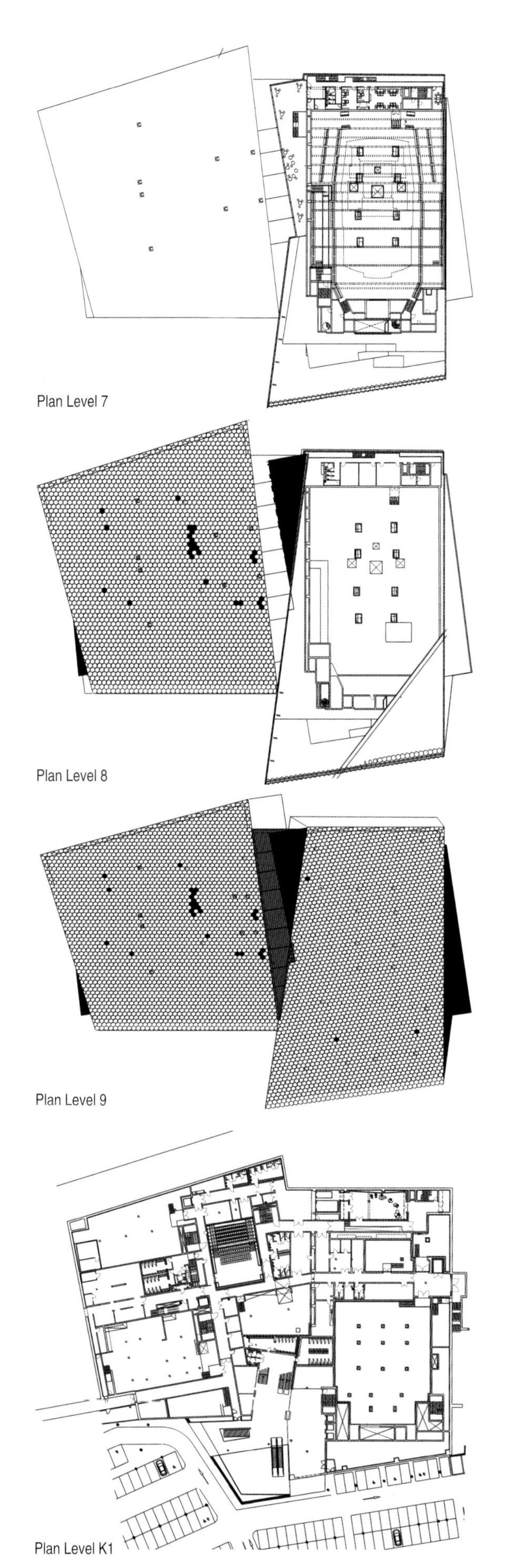
Plan Level 7

Plan Level 8

Plan Level 9

Plan Level K1

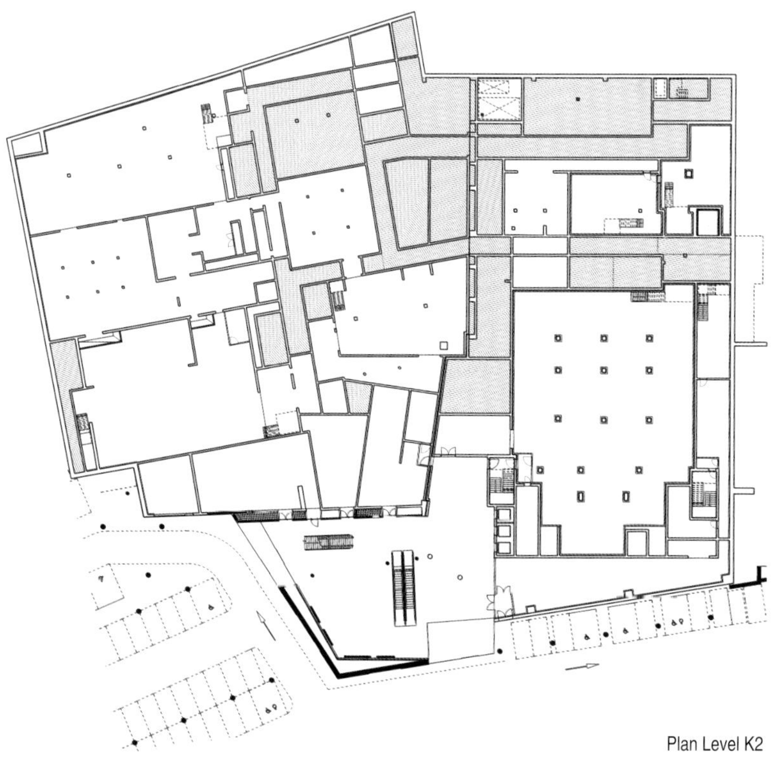
Plan Level K2

NO99 Straw Theatre

NO99秸秆剧院

NO99 Straw Theatre is an object standing on the verge of being a pure functional container on one hand, and an art installation on the other. The Straw Theatre is built on the occasion of Tallinn being the European Capital of Culture, to house a special summer season programme of theatre NO99, lasting from May to October 2011. Thus it is a temporary building, operating for half a year, built for a specific purpose, programme and location.

The Straw Theatre is built in central Tallinn, on top of the former Skoone bastion, one of the best preserved baroque fortifications of Tallinn. At the beginning of the 20th century, the bastion worked as a public garden, and during the Soviet era it was more or less a restricted recreational area for the Soviet navy with a wooden summer theatre and a park on top. With the summer theatre having been burnt down and the Soviet troops gone, for the last 20 years the bastion has remained a closed and neglected spot in the centre of town with real estate controversies and several failed large-scale development plans. In such a context, the Straw Theatre is an attempt to acknowledge and temporarily reactivate the location, test its potential and bring it back to use, doing all this with equally due respect to all historical layers of the site. The rectangular main volume of the theatre is situated exactly on the same spot as the navy summer theatre, and one descending flight of stairs of the latter is used as a covered walkway and entrance area to the Straw Theatre. The building is surrounded by various outdoor recreational functions including an oversized chess board, table tennis, swings, and a baking oven, all with a non-commercial and pleasantly low-key feel.

The dramatic appeal of the building stems from its contextual setting on the site and its black, uncompromisingly mute main volume contrasting with a descending tail with an articulate angular roof. And of course one cannot escape the effect of the material – uncovered straw bales, spraypainted black. The Straw Theatre is a unique occasion where straw has been used for a large public building and adjusted to a refined architectural form. For reinforcement purposes, the straw walls have been secured with trusses, which is a type of construction previously unused. As the building is temporary, it has not been insulated as normal straw constructions would require but has been kept open to experience the raw tactile qualities of the material and accentuate the symbolic level of the life cycle of this sustainable material.

Location / 地点:
Tallinn, Estonia
Date of Completion / 竣工时间:
2011
Area / 占地面积:
440 m^2
Architecture / 建筑设计:
Salto AB
Interior Design / 室内设计:
Salto AB
Photography / 摄影:
Martin Siplane, with exceptions by Paul Aguraiuja, Karli Luik

NO99秸秆剧院既是一座实用性建筑，又是一项艺术性构造。该剧院的建立是为了庆祝塔林被授予“欧洲文化之都”的年度称号，这里计划举办一项特殊的夏季活动，活动时间是2011年5月到2011年10月。因此，它是一个临时建筑，需持续运行半年，是为特定地区及用途而建造的。

剧院建于塔林市中心，地处Skoone堡垒的顶部。堡垒是塔林保存最完好的巴洛克式防御工程之一。在20世纪初，这个堡垒曾被用作公共花园，后来在前苏联时期，它几乎成为前苏联海军的限定娱乐区域，当时在堡垒顶部还有公园和一个木质的夏季剧院。随着夏季剧院被烧毁，在过去的20年里，这个位于市中心的堡垒一直处于封闭和搁置状态，成为一块存在争议的土地，针对它的几次大型发展计划均以失败告终。在这样的背景下，秸秆剧院项目旨在证明并暂时激活该地区的功能用途，让这里重新被利用起来。设计师在建设时要适当尊重场地的历史元素。新剧院的主体呈矩形结构，恰好位于海军夏季剧院的原址，而原址遗留下来的一段下行阶梯则被用作铺设的走道和新剧院的入口。建筑的四周设有各类户外休闲设施，包括一个大型棋盘、乒乓球台、秋千和一个烤箱，一切是那么亲切而质朴，丝毫没有商业化的气息。

该建筑独特的吸引力在于其场地的历史背景，以及它那黑色、静谧的主体结构和一节节由棱角分明的屋顶组成的下降式尾状结构。当然，建筑的材料是格外引人注目的，项目采用黑色喷漆的秸秆砌块，并将其裸露在外。该项目为秸秆这一材料在大型公共建筑和优雅建筑形态中的应用提供了机会。秸秆墙身由构架支撑以达到加固的目的，这种手法还是首次应用在建筑设计中。由于该建筑是临时性的，它没有像通常秸秆建筑那样进行隔热处理，而是采用开放式设计，让人体验原生材料的质地，并强调了这种可持续材料的生命周期。

Materials: Straw Bales; Laminated Wood.

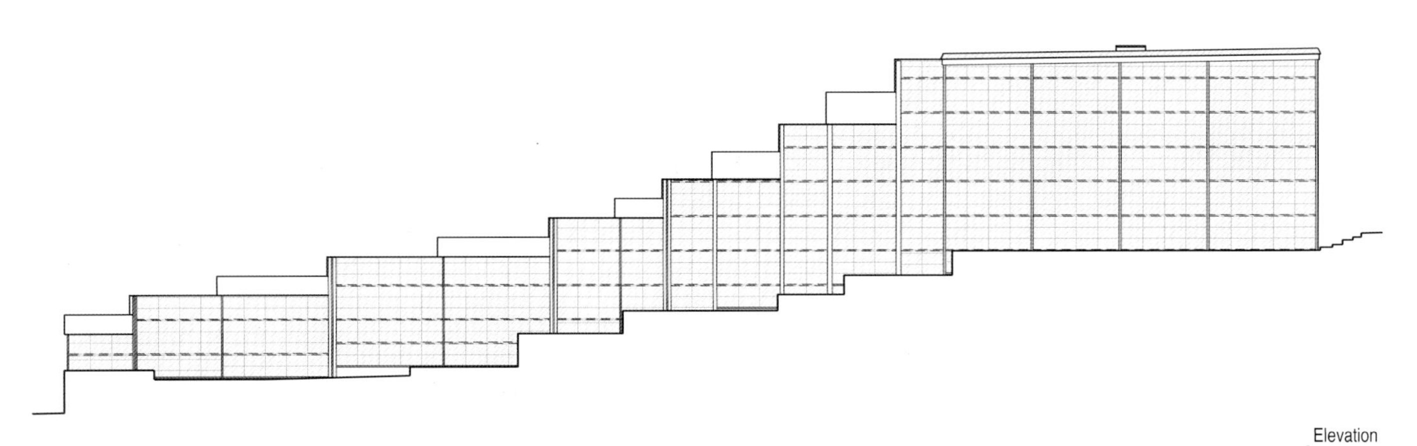

Elevation

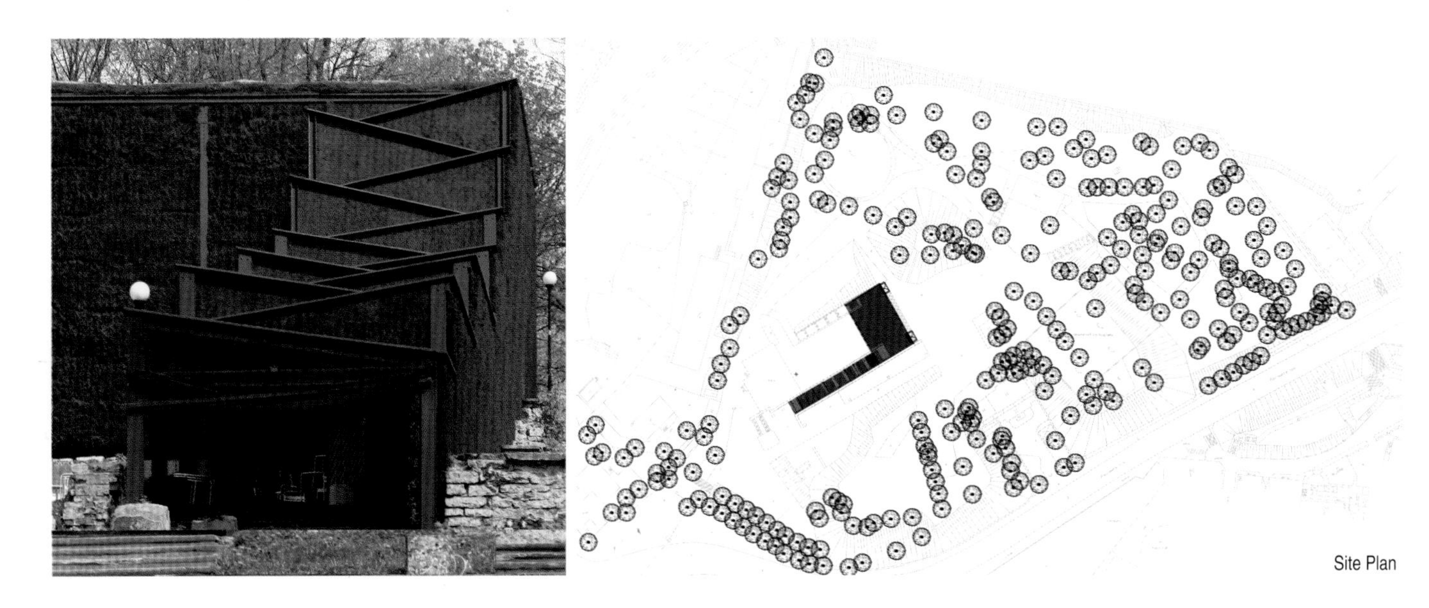

Site Plan

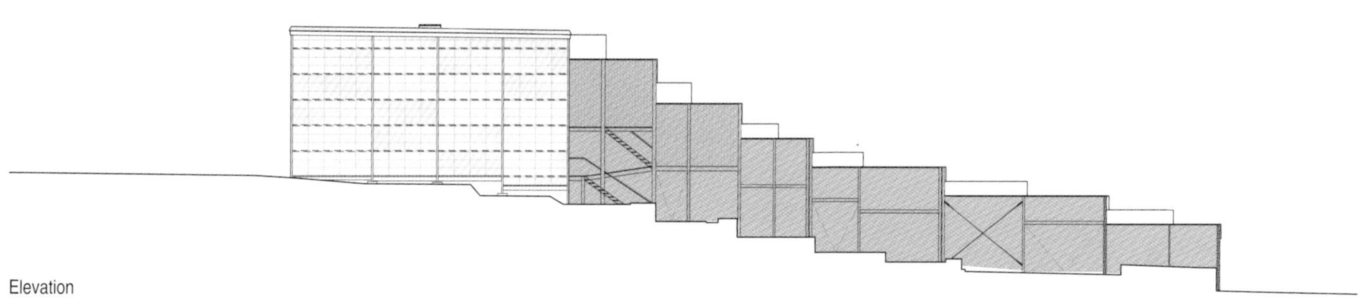

Elevation

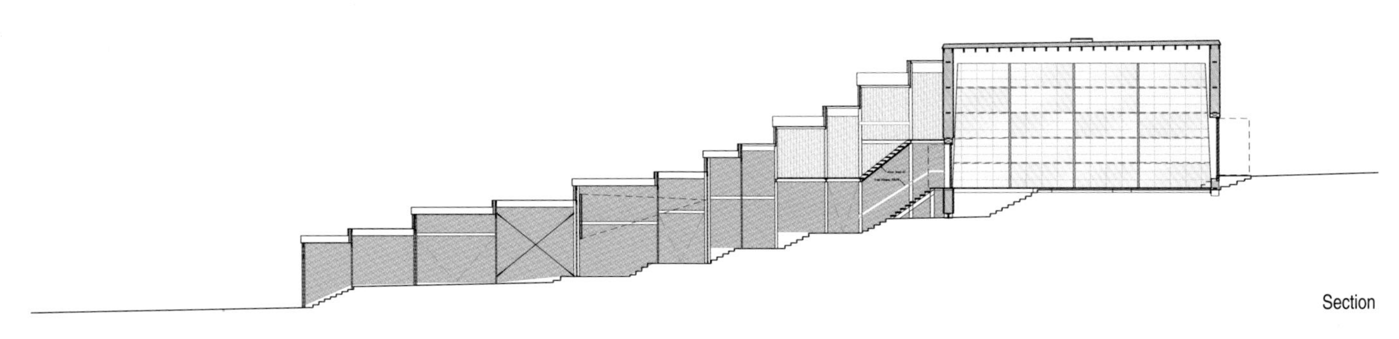

Section

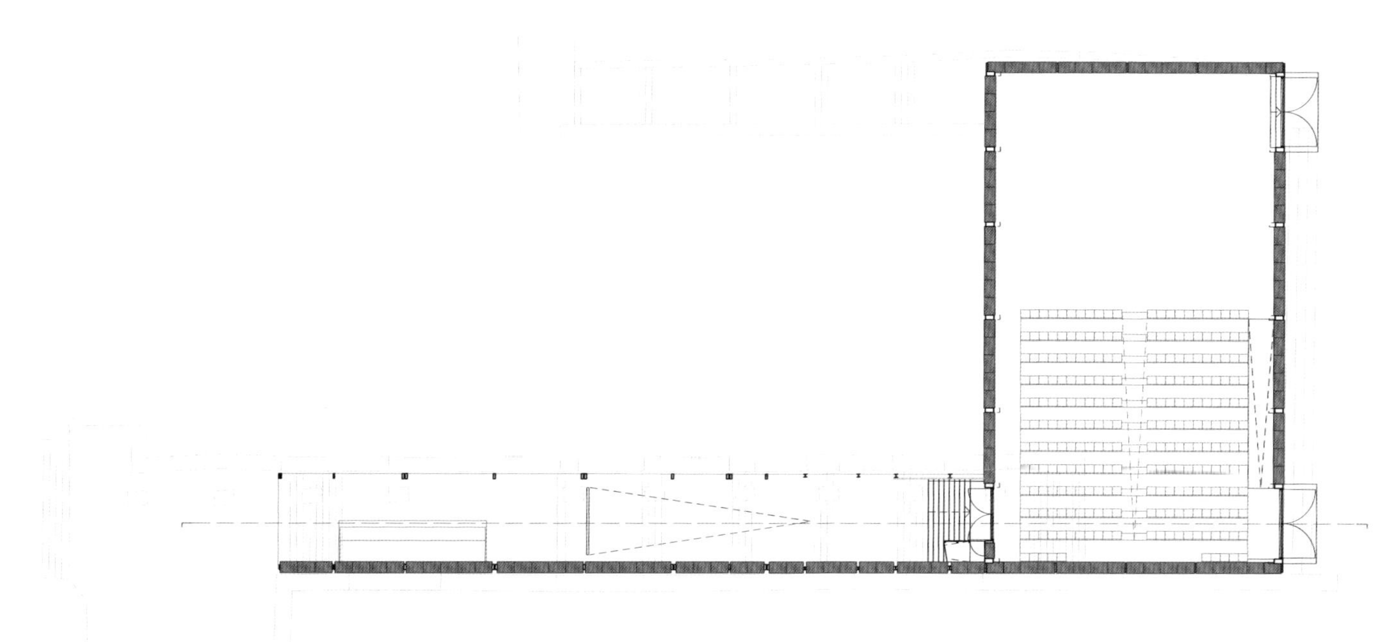

Plan

Milstein Hall in USA

美国米尔斯坦大厅

Milstein Hall is the first new building in over 100 years for the renowned College of Architecture, Art and Planning (AAP) at Cornell University in Ithaca, New York. Rather than creating a new free-standing building it is an addition to the AAP buildings creating a unified complex with continuous levels of indoor and outdoor interconnected spaces. Milstein Hall provides an additional 4,366 m² adding much-needed space for studios, gallery space, critique space and the first auditorium solely dedicated to the AAP. The existing AAP was housed in four separate buildings at the northern edge of the campus, forming the "back-side" of the Arts Quad-detached from its atmosphere yet disengaged from the Falls Creek Gorge to the north. Despite the proliferation of architectural styles, the four buildings share a single typology – the linear, corridor building – segregating the AAP's activities behind a labyrinth of security codes and dead ends. Milstein Hall offered an opportunity to simultaneously rethink the neglected north side of the Arts Quad and provided a space where the AAP's interdisciplinary potential could finally be exploited.

Milstein Hall, a box – modern architecture's typology par excellence – connects the second levels of Sibley and Rand Halls with a large horizontal plate. The interior space provides a typology entirely absent from the campus: a vast horizontal expanse that provides 2,323 m² of studio space and flexibility over time. Enclosed by floor-to-ceiling glass and a green roof with 41 skylights, this "upper plate" cantilevers almost 15 m over University Avenue to establish a relationship with the Foundry, a third existing AAP facility. The cantilever is supported by five exposed hybrid trusses designed to balance structural efficiency and maintain open circulation within the large open plan. Lifted, the building leaves the figure-ground of the AAP campus intact. Beneath the lifted box, a half-submerged "lower plate" contains a cluster of public programs – a 253-seat auditorium, gallery space and a 465 m² circular critique spaces – that serve the entire AAP. The roof of the lower plate rises to form a dome, a single manipulation that simultaneously supports the raked auditorium seating, becomes the stairs leading up to the studio plate above, and is the artificial ground for an array of custom fabricated exterior seating pods. The materiality of the lower level, constructed of exposed cast-in-place concrete, adds a contrast to the upper plate's glass and steel character. However they create frameworks of raw spaces to serve as a pedagogical platform for the AAP to generate new interaction driven by the students' and faculty's ambitions and explorations.

Location / 地点:
New York, USA
Date of Completion / 竣工时间:
2011
Area / 占地面积:
4,400 m²
Architecture / 建筑设计:
OMA
Interior Design / 室内设计:
OMA
Landscape / 景观设计:
Scape Landscape Architecture PLLC
Photography / 摄影:
Philippe Ruault, Cornell University, Frans Parthesius, Jack Pottle
Client / 客户:
Cornell University, College of Architecture, Art and Planning (AAP)

米尔斯坦大厅是康奈尔大学（位于纽约州伊萨卡市的著名高校）艺术与规划学院（AAP）百年来的第一栋新建筑。米尔斯坦大厅没有被设计为新的独立式建筑，而是作为AAP建筑群的扩展，呈现统一的形态，室内和室外的各层空间浑然一体。米尔斯坦大厅拥有4366m²的设计室、展厅、评论区和一个AAP专用会场。AAP目前包括校园北部边缘的四栋独立大楼，共同构成"艺术方庭"（Arts Quad）的"背面"，并与北方的秋溪峡谷相分离。尽管四栋大楼的建筑风格各有不同，但是它们有了一个共同的特性，即线性的回廊式的建筑。米尔斯坦大厅将使"艺术方庭"北部重现生机，令AAP的跨学科潜能得到真正发挥。

米尔斯坦大厅采用现代建筑典型的盒状结构，通过一个巨大的横板式结构与Sibley Hall和Rand Hall的第二层相连。盒状建筑的内部设计可谓是独一无二的：这里是一片水平的广阔区域，设有2323m²的设计室，富有弹性的空间可在不同时期灵活运用。该结构被落地窗和包含41个天窗的绿色屋顶所围绕。这个悬挑建筑由5个可见的混合桁架支撑，桁架不但起到平衡结构的作用，还让周围开阔的空间畅通无阻。悬挑的盒状建筑下方设有一个半下沉空间，内设一系列公共设施——一个可容纳253人的会场、画廊和一个465m²为AAP的各系院所共用的圆形评论区。下层结构的屋顶为圆形，圆顶的屋顶里是通往上层工作室的梯道。下层结构由暴露的现浇混凝土建成，与上层由玻璃和钢铁构成的建筑风格形成对比。然而，上、下两部分都设立了未经处理的空间构架，并以此作为AAP的教学平台，有助于学生和教师的互动。

Green Roof: Sedum; **Façade:** Glass, Turkish Marble; **Ceiling:** Perforated Aluminum Panels; **Dome:** Plywood, MDO board; **Elevator (Vertical Room):** Plywood Panels; **Auditorium:** Fixed and Loose Seats, Boardroom Seats, Curtains; **Lighting:** Lutron Control System

GREEN ROOF
STUDIO
STUDIO
ACCESS
AUDITORIUM
CRIT / GALLERY
FACULTY PLAZA
EXHIBITION
SERVICE / STORAGE

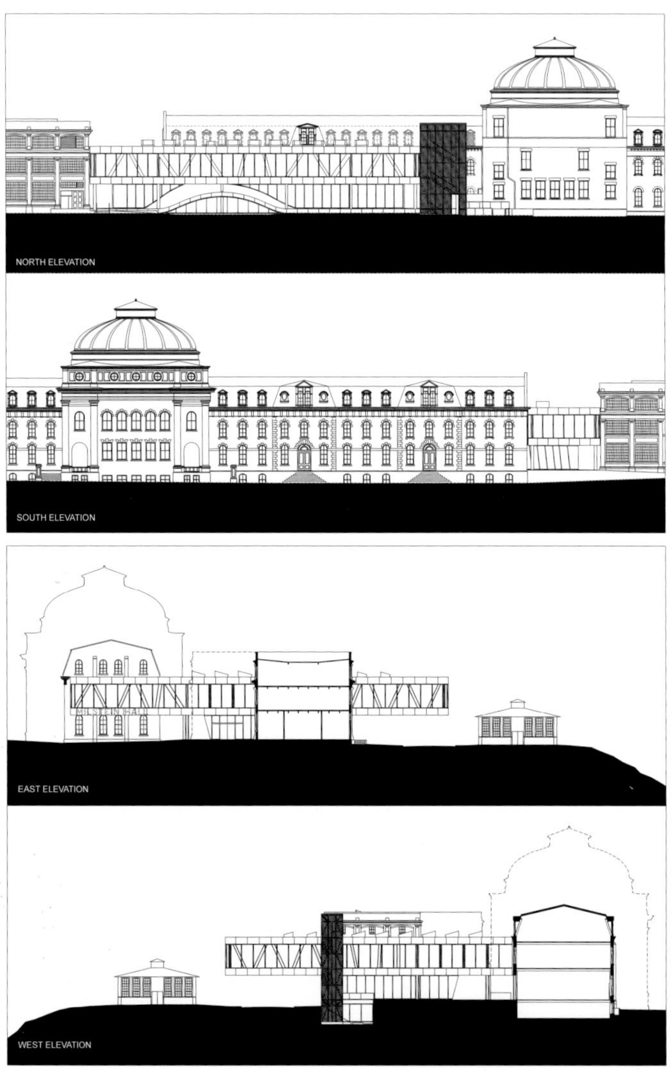
NORTH ELEVATION
SOUTH ELEVATION
EAST ELEVATION
WEST ELEVATION

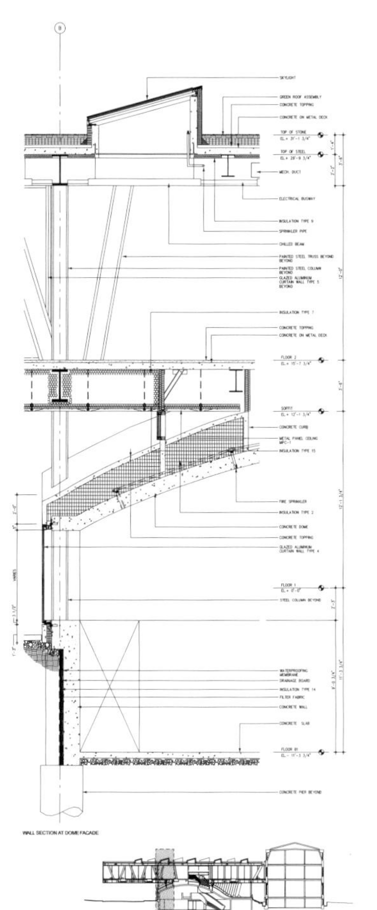

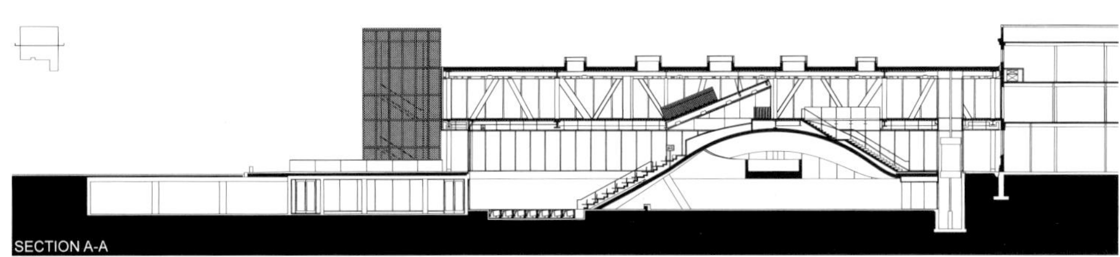
SECTION A-A

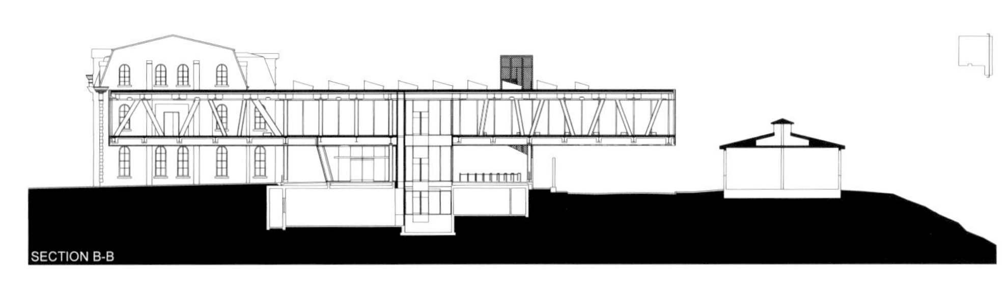
SECTION B-B

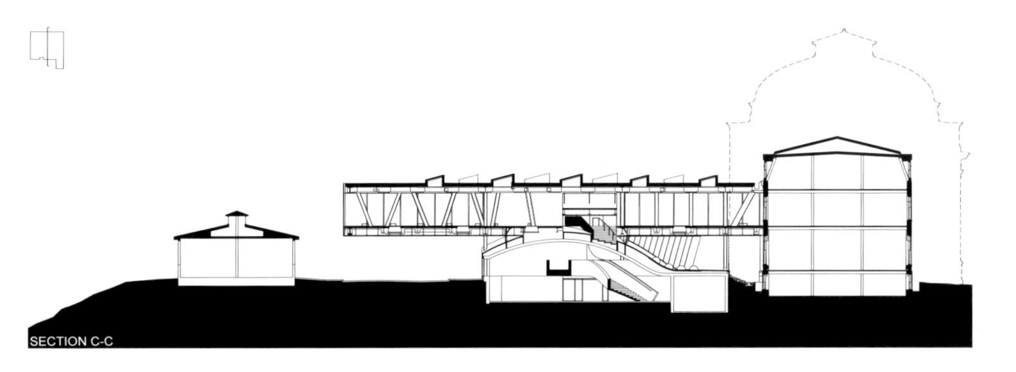
SECTION C-C

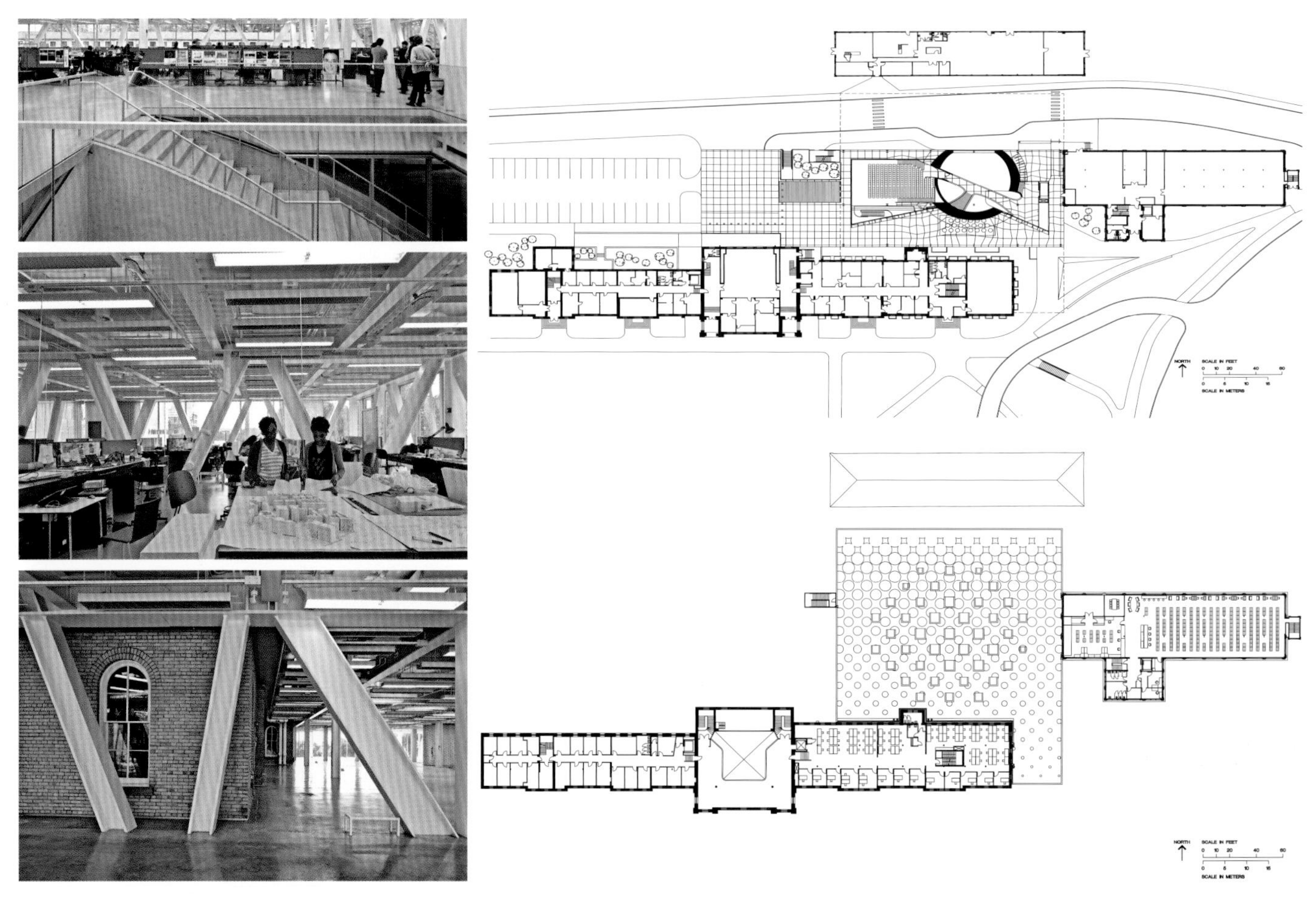

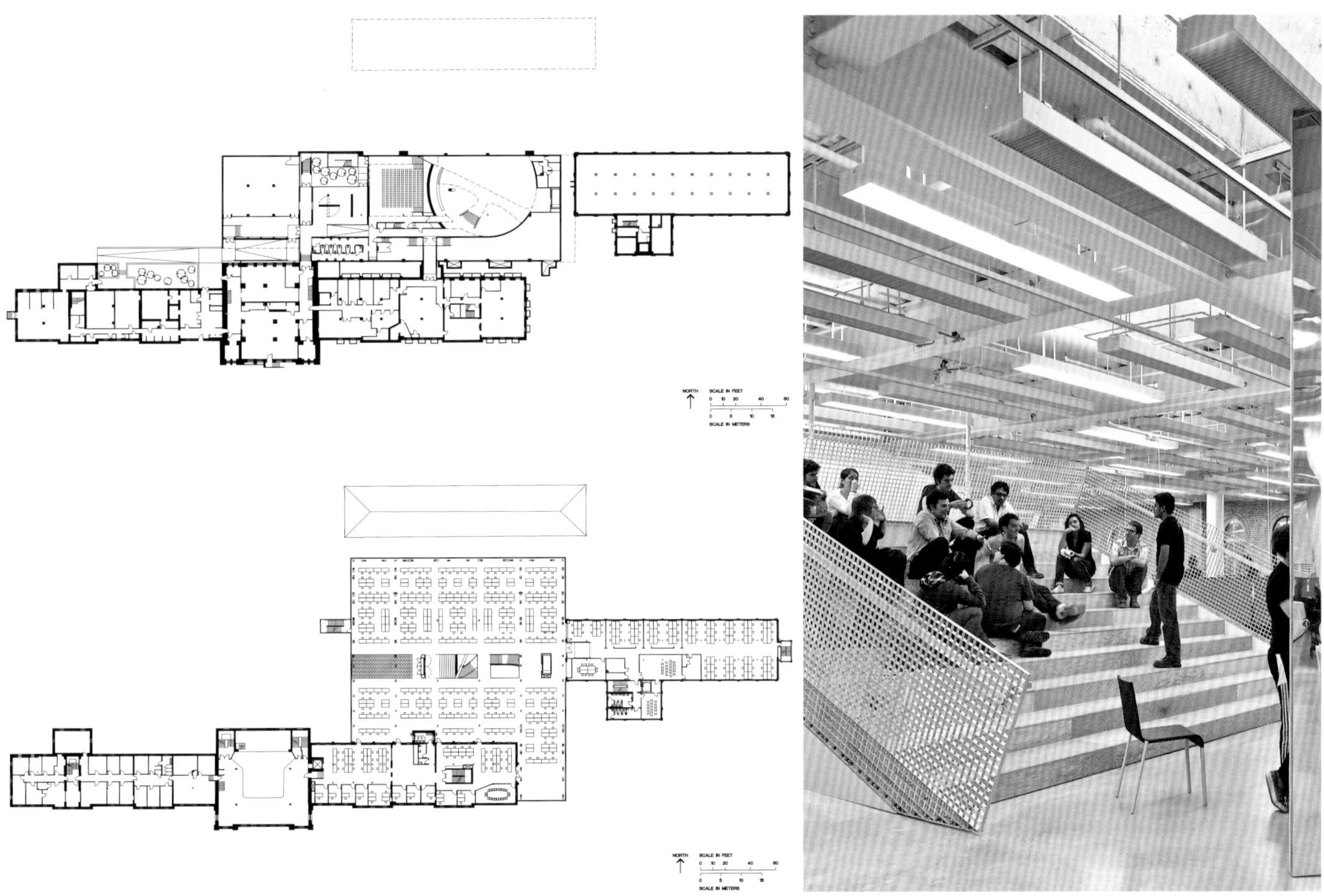
NORTH
SCALE IN FEET
0 10 20 40 60
0 5 10 15
SCALE IN METERS
NORTH
SCALE IN FEET
0 10 20 40 60
0 5 10 15
SCALE IN METERS

Garzoni Villa Garden & Collodi Butterfly House

加尔佐尼别墅花园及科洛迪蝴蝶馆

The Garzoni villa and garden complex is made up of the Villa Garzoni (built between the mid-sixteenth and the seventeenth centuries), that embodies two important architectural styles, that is the city mansion and the country villa, the Palazzina d'Estate (first decade of the eighteenth century), the Garden (medieval in origin) that has the flamboyant and theatrical character typical of Italian Baroque throughout the eighteenth century.

The restoration was preceded by research, searches of the archives, testing, taking of samples, diagnostic investigations, with regard both to the buildings and their component parts, and to the geo-morphological aspects, in order to acquire an accurate picture of the degradation and the instability of the architectural parts, and of the garden with its vegetation and all its decorations (sculptures, paintings, etc.). After the construction of an extensive and careful diagnostic picture, the next step was to prepare a plan that would guarantee the preservation of and respect for this complex. Hence the restoration involved the consolidation and functional renewal of the buildings, the façades, on which the original colours and decorations were restored, and the garden, in which all the species of trees were catalogued and planted in the positions in which the originals were located, restoring the terracotta statues adorning it. Along with the restoration work on what is already there, the project for the enhancement and functional renewal of the complex involved the creation of a new building, the Collodi Butterfly House, an innovative greenhouse building dedicated to the life, cataloguing and displaying of a wide range of species of butterflies, and containing a significant amount of vegetation suitable for creating the right habitat for the life-cycles of thousands of butterflies. The "ancillary" buildings have also been restored for use as a ticket office, coffee shop and bookshop, a museum for temporary and permanent displays and areas for teaching and for holding conferences.

Today the complex houses displays and equipment and, together with the historic Parco di Pinocchio, forms the backbone of the entire set-up of Collodi.

加尔佐尼别墅花园是一个复合建筑空间，在其构成部分中，加尔佐尼别墅始建于16世纪中叶，同时具备了城市豪宅和乡村别墅两种风格；Palazzina d' Estate建于18世纪初；加尔佐尼花园起源于中世纪，其华丽和戏剧性的设计彰显了意大利18世纪典型的巴洛克建筑风格。

整修前的准备工作包括对建筑及其组成部分和地理形态进行研究、勘测、测试、取样和调查，旨在准确地定位出退化和不稳定的建筑部分。有了以上广泛而细致的信息后，设计师制订了详细计划，以确保这些复合建筑得到保存和维护。整修项目时，首先对建筑的外墙进行巩固和功能性重建，恢复其表面的原有色彩和装饰；其次，园林中所有树种被分门别类地进行移植；点缀在园林里的赤陶雕像也得到修复。这个以整修建筑和更新功能为主旨的项目不但重修了现有设施，还设立了一个新的建筑——科洛蒂（Collodi）蝴蝶园。这是一个创新型温室，它作为演绎生命的舞台，收纳并展示着种类繁多的蝴蝶，同时种植大量的植被，为成千上万的蝴蝶提供适宜的栖息地。附属建筑也都被翻新，并分别被用作售票处、咖啡馆、书店、临时和长期展览用的博物馆、教学区和会议区。

如今，这群复合建筑内部设置了展览的空间和设备，与历史景点皮诺曹主题公园一同构成了科洛蒂地区的整体规划。

Location / 地点:
Collodi, Italy
Date of Completion / 竣工时间:
2008
Area / 占地面积:
21,000 m^2
Landscape / 景观设计:
Emilio Faroldi Associati, Gurrieri Associati
Photography / 摄影:
Marco Buzzoni, Marco Introini, Goirani
Client / 客户:
Villa e Giardino Garzoni srl

Floor and Footpaths: Stabilised Gravel, Cocciopesto and Paving Stones;
Kerbs and Low Walls: Sandstone, Pebbles, Stone Slabs, River Pebbles, Rough-Hewn Stone and Mortar;
Furniture: Antique Statues in Stone and Terracotta, Stone Vases, Water Pools, Long and Short-Stemmed Hedging Stone Steps, Ramps in Stone and Gravel, Stone and Metal Gazebos and Stone Benches;
Plants: Acacia Julibrissin, Acacia Dealbata, Carpinus Betulus, Camphor, Cedrus Atlantica, Cedrus Deodara, Chamaecyparis, Cupressus Sempervirens, Fraxinus Excelsior, Aesculusus Hippocastanum, Yucca, Lingustrum Japonicum, Paulonia Imperialis, Phoenix Dactylifera, Quercus Ilex, Taxus Baccata, Washingtonia, Chamaerops Humilis and Chamaerops Excelsa.

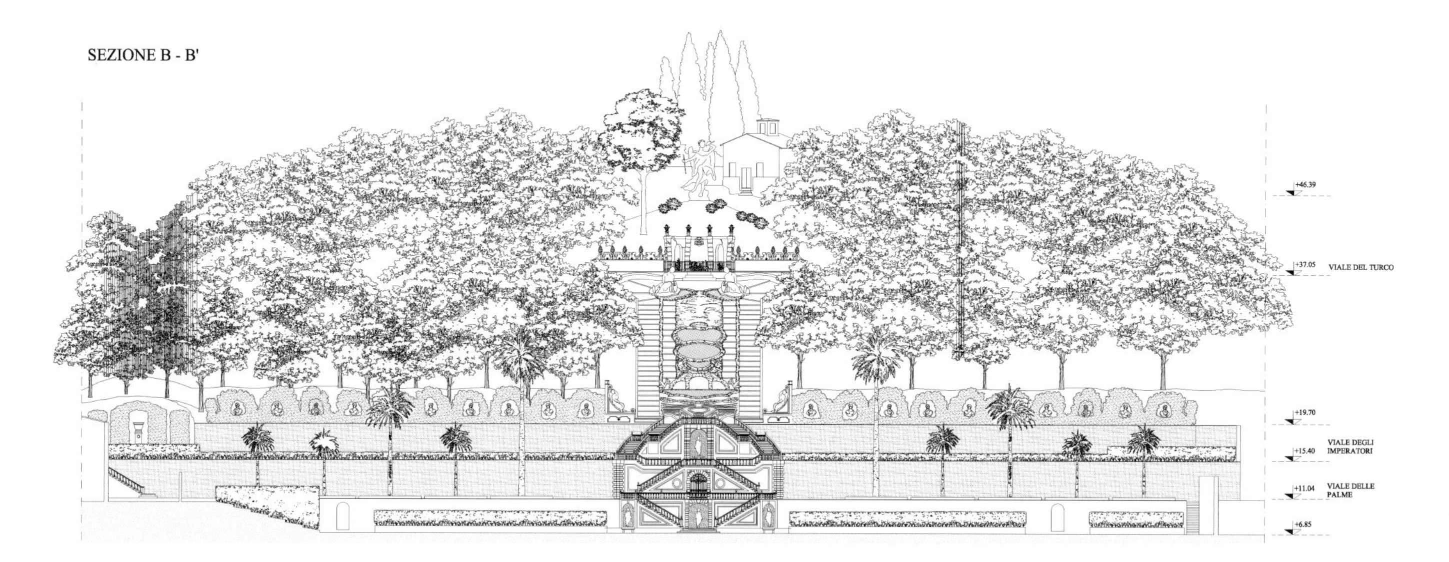
SEZIONE B - B'
+46.39
+37.05
VIALE DEL TURCO
+19.70
VIALE DEGLI IMPERATORI
+15.40
+11.04
VIALE DELLE PALME
+6.85

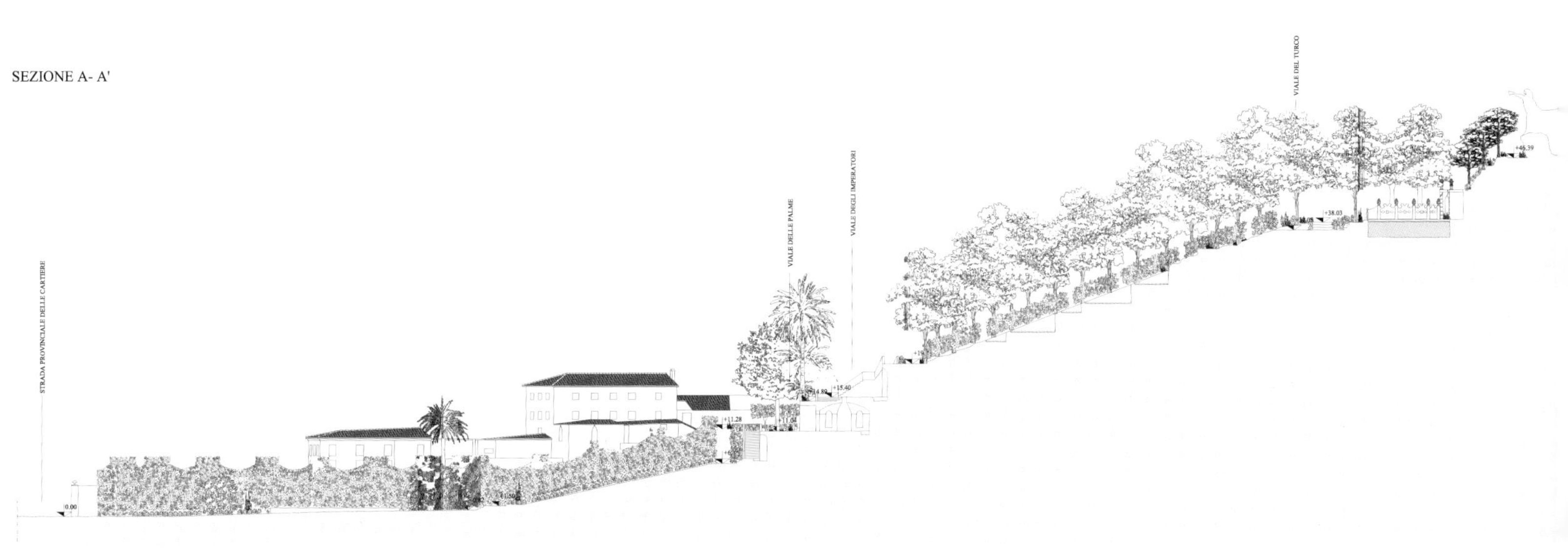
SEZIONE A- A'
STRADA PROVINCIALE DELLE CARTIERE
VIALE DELLE PALME
VIALE DEGLI IMPERATORI
VIALE DEL TURCO
+11.28
+15.40

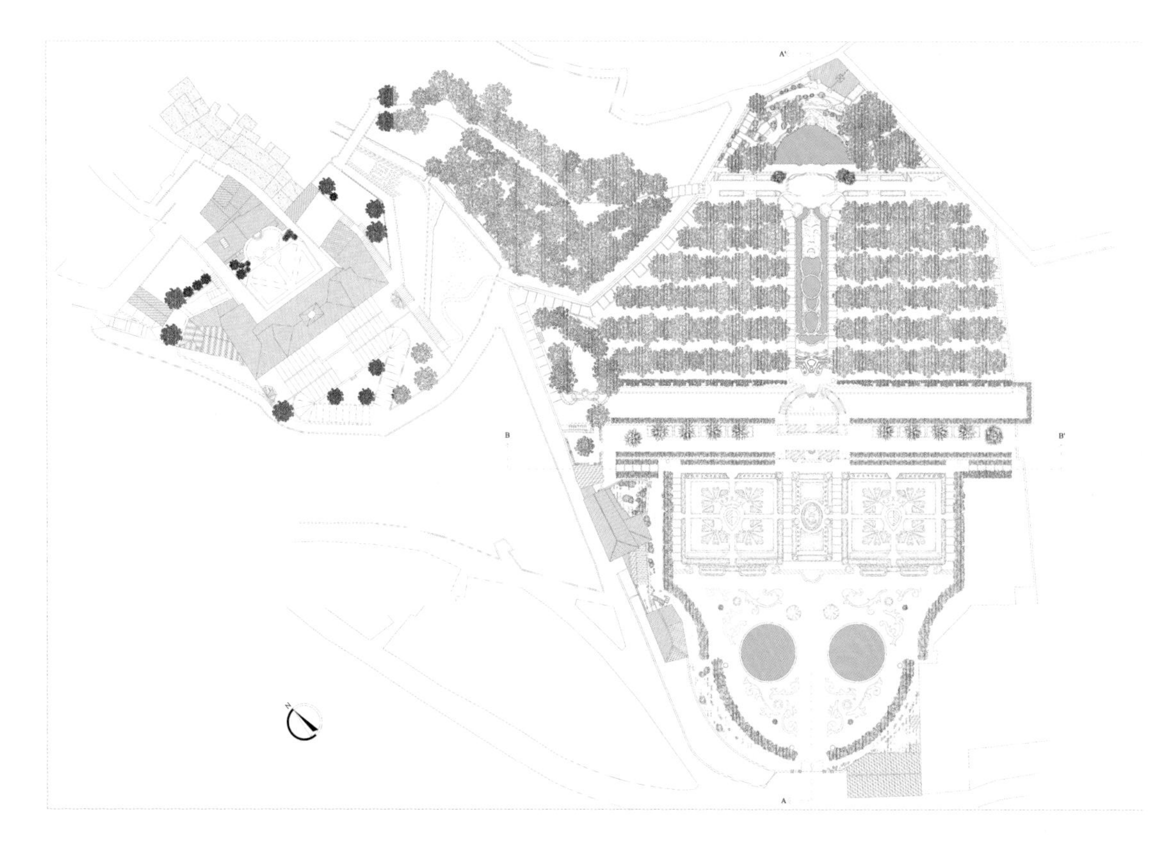

Elevation

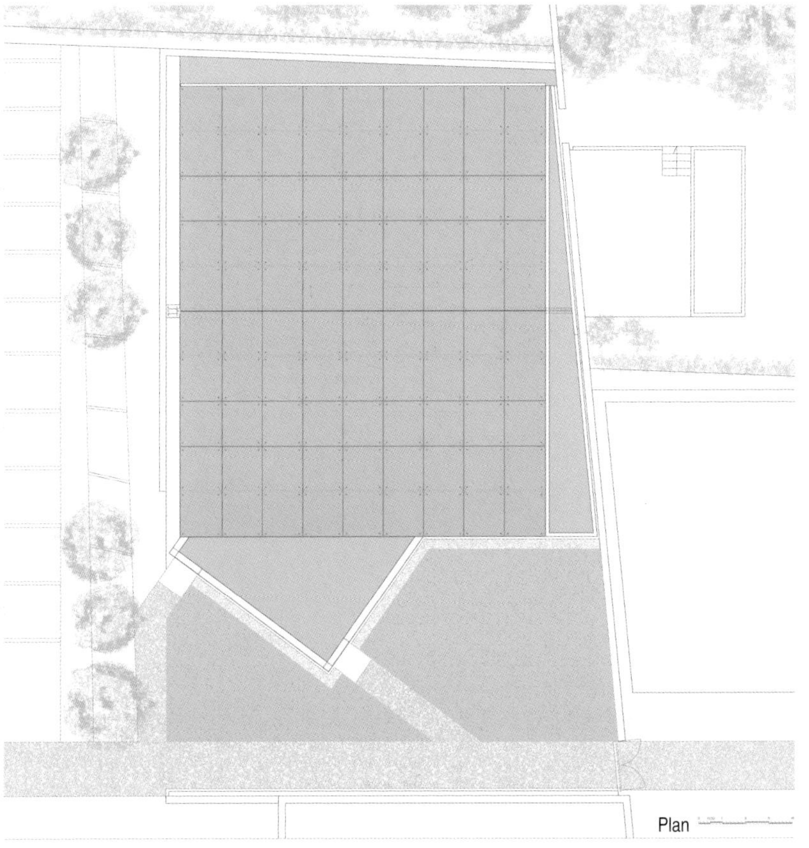
Plan

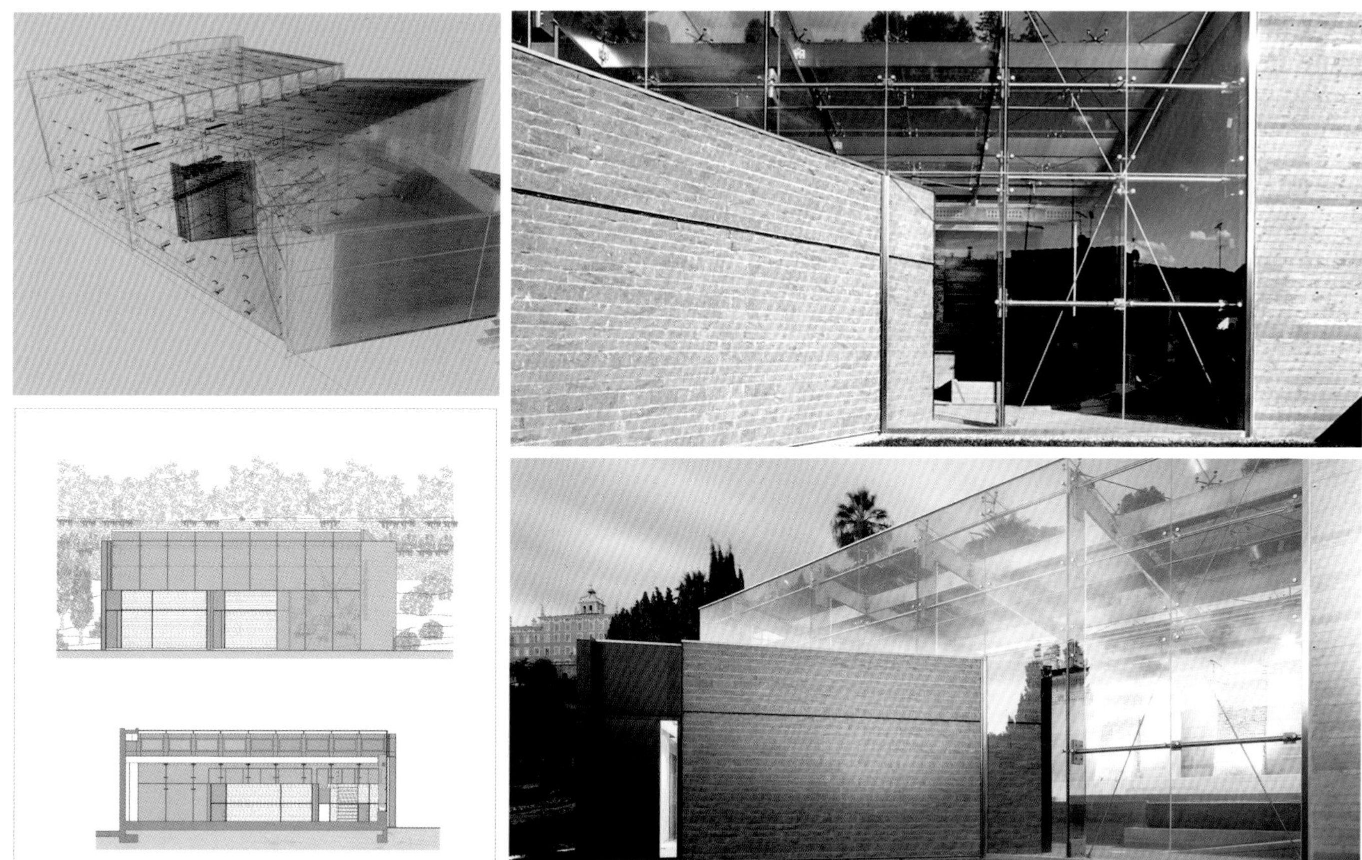

Elevation

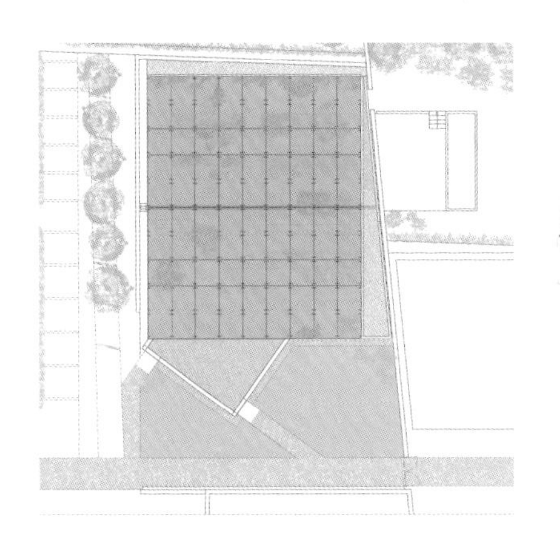

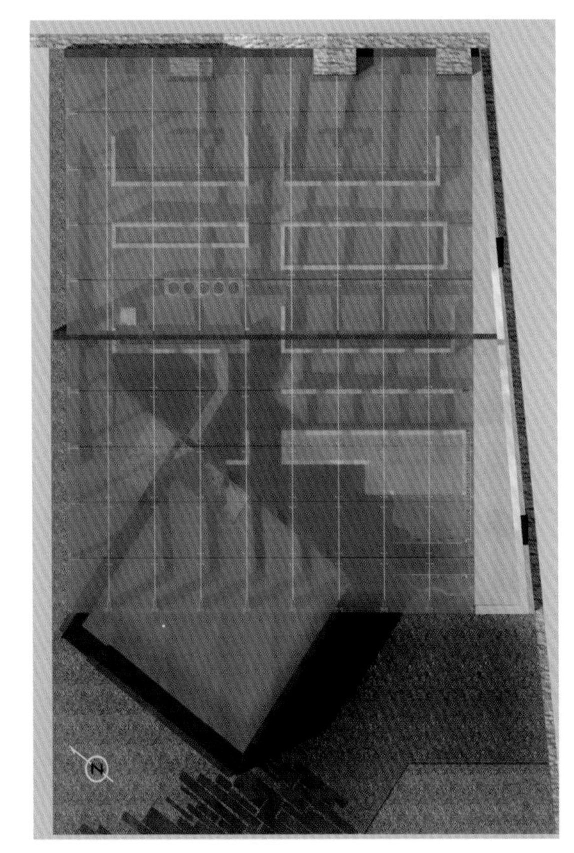

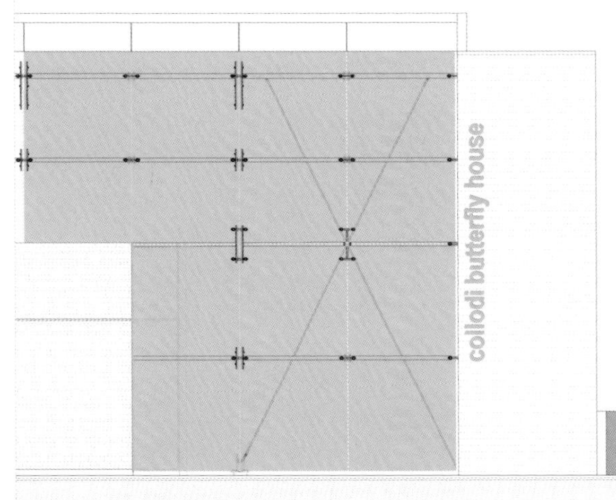

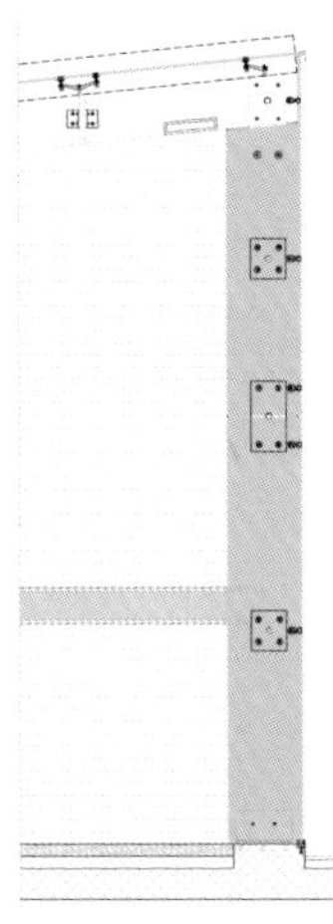

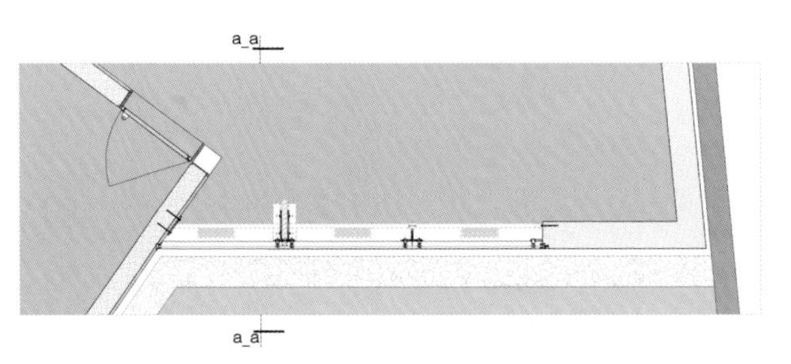

Detail

Jewish Reconstructionist Congregation Synagogue

重建的犹太教教堂

The new LEED Platinum certified synagogue for the Jewish Reconstructionist Congregation replaces the old building at the edge of a residential area, across from a city park and the tracks of the Skokie Swift commuter train.

The synagogue's design reflects the vision and goals of the congregation. The design criteria were developed through goal setting workshops with the JRC board and building committee. The consensus opinion was that the building should symbolically reflect the values of Judaism and Reconstructionism. Evanston's zoning ordinance, limiting building height and lot coverage, impacted the final building program and design solution. The congregation originally identified 3,902 m^2 of dedicated space to serve their needs. The design met these needs in 2,936 m^2 of flexible and convertible space. The use of spaces from week to week, hour to hour, was traced over the building levels to find the best balance of all needs. The project demolished the existing 1,988 m^2 synagogue and constructed a new 2,936 m^2 facility on the same site. The new building has three floors containing the Congregation's offices, early childhood program, and chapel on the first floor; their education offices, classrooms and library on the second floor; and the sanctuary, social hall and kitchen on the third floor. The new facility includes spaces for worship and social events, as well as classrooms for Early Childhood, Religious, and Adult Education programs. The building includes space for the Congregation's staff, a large teaching kitchen, the youth group, arts and crafts, and a library with a media center and a language laboratory.

The new building is built on the foundations of the old. Local demolition rubble is placed in wire cages to create "gabion" walls to retain the edges of gardens and children's playgrounds. The memorial trees that shade the existing building were cut down and reconstituted as paneling on the Ceremonial door in order to preserve the memory of those associated with their planting and care. The Congregation has placed, throughout the building, their collected words – lyrics, testaments, calls for protest – to be added to and to be enshrined in the building as a permanent testament of the Congregation's work.

重新修建的犹太教教堂是“能源与环境设计先锋奖”（LEED）的白金认证项目，它取代了原来的老建筑，位于住宅小区一侧，对面是城市公园和Skokie Swift通勤列车轨道。

教堂的设计体现了会众的设想和目标。设计的标准是通过与犹太教重建派会众（JRC）和建设委员会的多次研讨发展而来的。大家的一致观点是让这栋建筑反映犹太教和重建派的价值观。由于埃文斯顿市的土地规划条例对建筑的高度和占地面积作出了限制，该项目的最终建设计划和设计方案也受到了影响。教会方最初划分出3902m^2的空间，专门用于这项计划。而项目的设计则仅用了2936m^2的灵活与可转换空间就满足了他们的需求。该项目拆除了原来1988m^2的犹太教堂，又在这里重新建立起一座2936m^2的建筑。新大楼共有三个楼层，其中一楼设立会堂办公室、幼儿教育处和教堂，二楼设立教育办公室、教室和图书馆，三楼设立圣堂、交谊厅和厨房。这项新设施提供了信仰和社交活动场所、幼儿教室以及宗教、成人教育项目。

新的会堂建立在原建筑的旧址上。拆毁过程中产生的碎石被放置在铁丝笼中，并组成“石笼”墙，“石笼”墙围绕在花园和儿童游乐场的边缘。遮掩建筑的树木被砍伐掉，成为礼堂大门的镶板，这样可以为树木的栽培者带来一段回忆。楼房遍布着教堂会众的箴言集——包括赞美诗、圣约、祷告文，它们被融入建筑之中，作为教堂会众工作的结晶而世代相传。

Location / 地点:
Evanston, USA
Date of Completion / 竣工时间:
2011
Area / 占地面积:
32,000 m^2
Architecture / 建筑设计:
Ross Barney Architects
Interior Design / 室内设计:
Ross Barney Architects
Landscape / 景观设计:
Ross Barney Architects
Photography / 摄影:
Ross Barney Architects
Client / 客户:
Jewish Reconstructionist Congregation

Skylights: Solatube;
Solar Lighting: Se'lux;
Glass: PPG Industries;
Carpet: Interface.

JEWISH RECONSTRUCTIONIST CONGREGATION

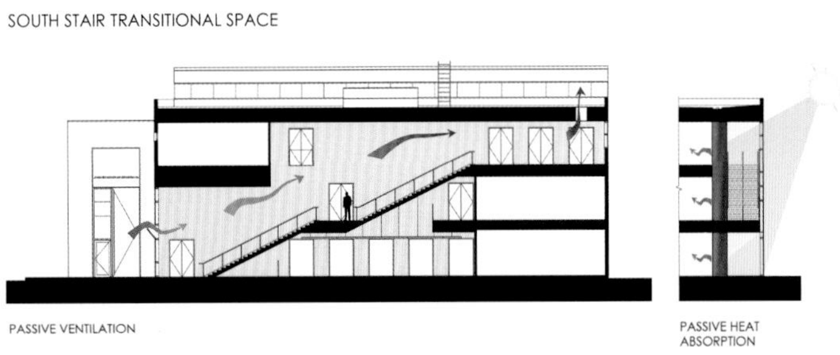
SOUTH STAIR TRANSITIONAL SPACE
PASSIVE VENTILATION
PASSIVE HEAT ABSORPTION

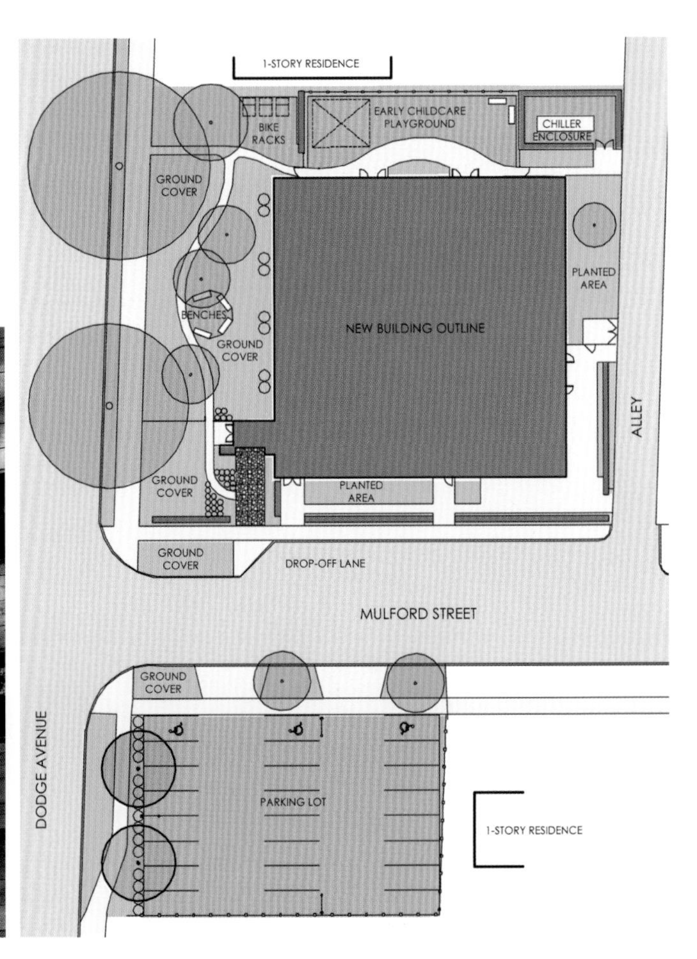
1-STORY RESIDENCE
BIKE RACKS
EARLY CHILDCARE PLAYGROUND
CHILLER ENCLOSURE
GROUND COVER
PLANTED AREA
BENCHES
GROUND COVER
NEW BUILDING OUTLINE
ALLEY
GROUND COVER
PLANTED AREA
GROUND COVER
DROP-OFF LANE
MULFORD STREET
GROUND COVER
DODGE AVENUE
PARKING LOT
1-STORY RESIDENCE

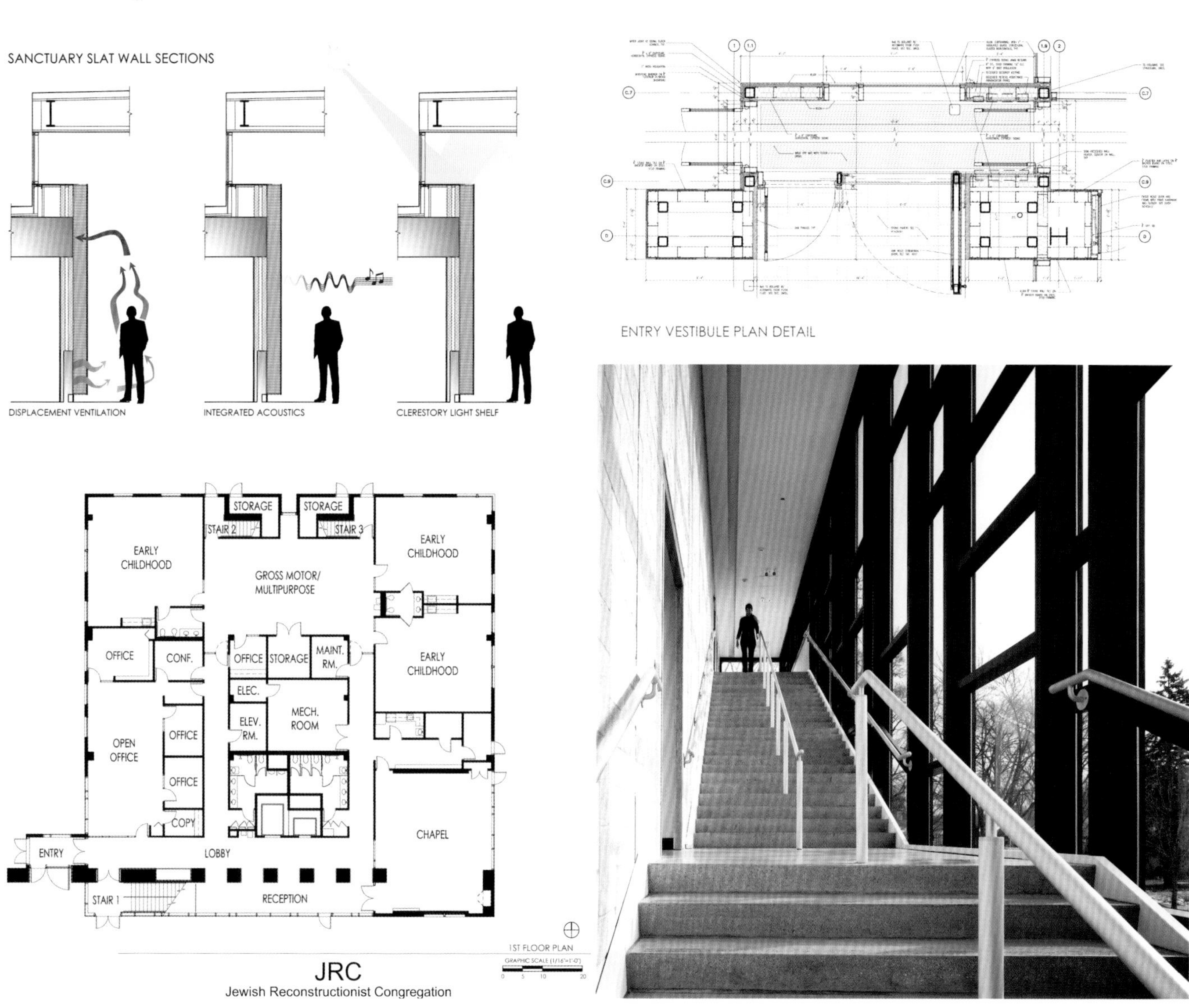
SANCTUARY SLAT WALL SECTIONS
DISPLACEMENT VENTILATION
INTEGRATED ACOUSTICS
CLERESTORY LIGHT SHELF
ENTRY VESTIBULE PLAN DETAIL
STORAGE
STORAGE
STAIR 2
STAIR 3
EARLY CHILDHOOD
GROSS MOTOR/ MULTIPURPOSE
EARLY CHILDHOOD
OFFICE
CONF.
OFFICE
STORAGE
MAINT. RM.
EARLY CHILDHOOD
ELEC.
ELEV. RM.
MECH. ROOM
OPEN OFFICE
OFFICE
OFFICE
COPY
CHAPEL
ENTRY
LOBBY
STAIR 1
RECEPTION
1ST FLOOR PLAN
JRC
Jewish Reconstructionist Congregation

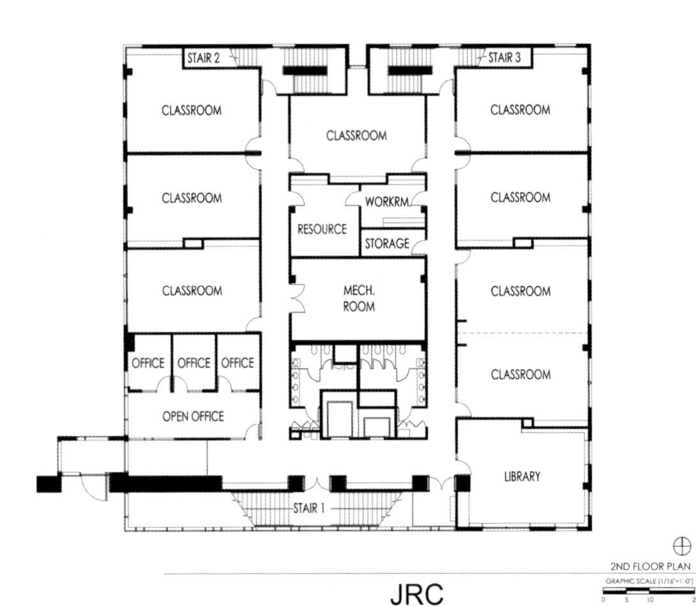

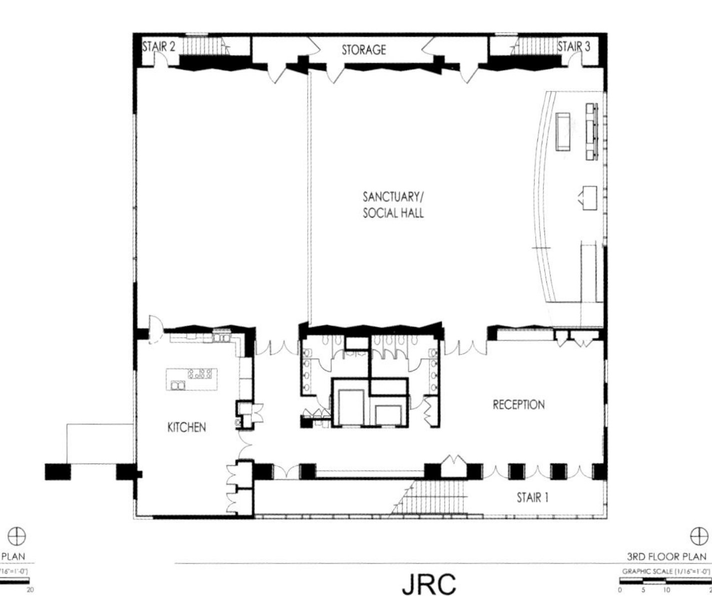

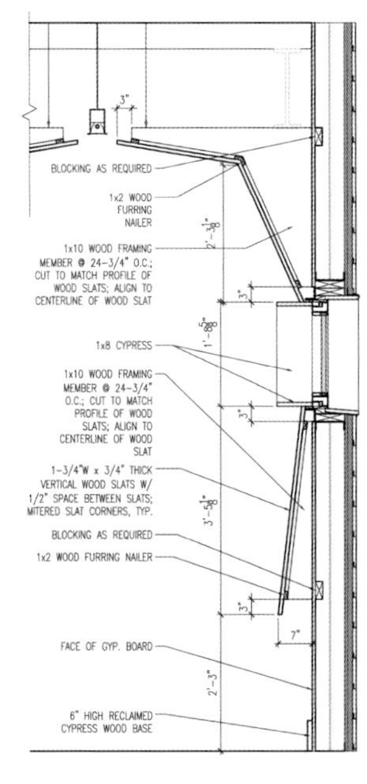

CHAPEL WOOD SLAT WALL & CEILING SECTION

WALL SECTION

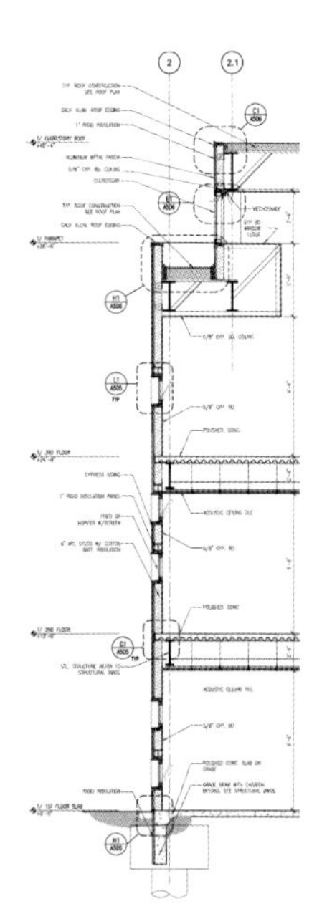

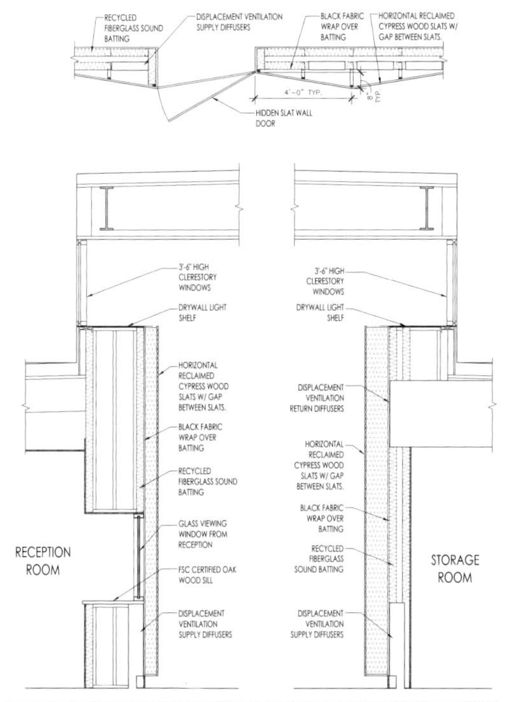

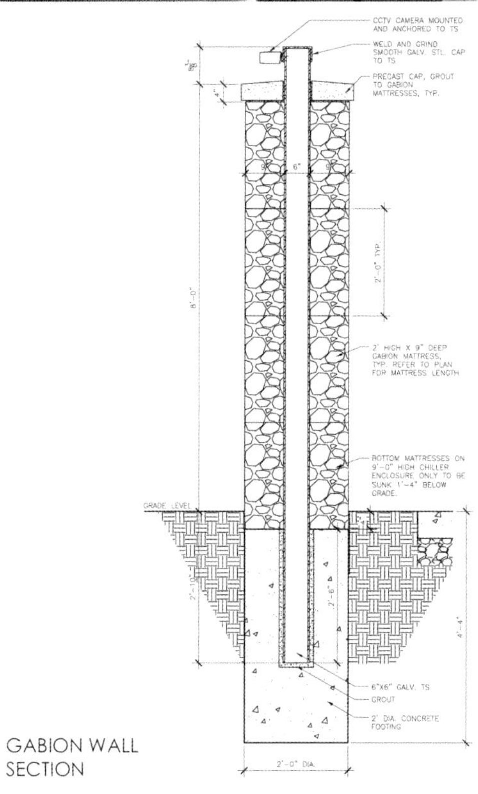

GABION WALL SECTION

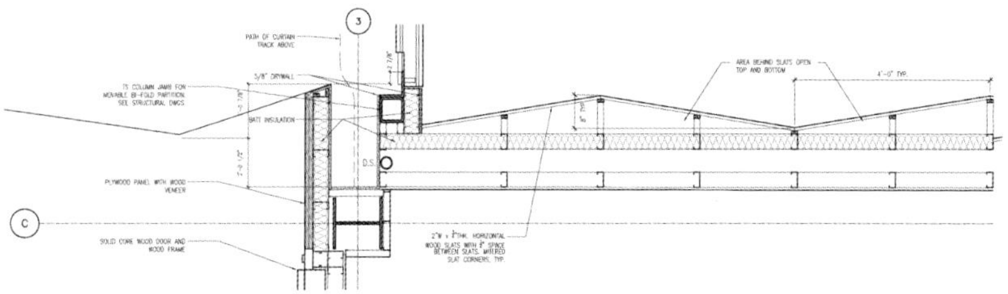

WOOD SLAT WALL PLAN DETAIL

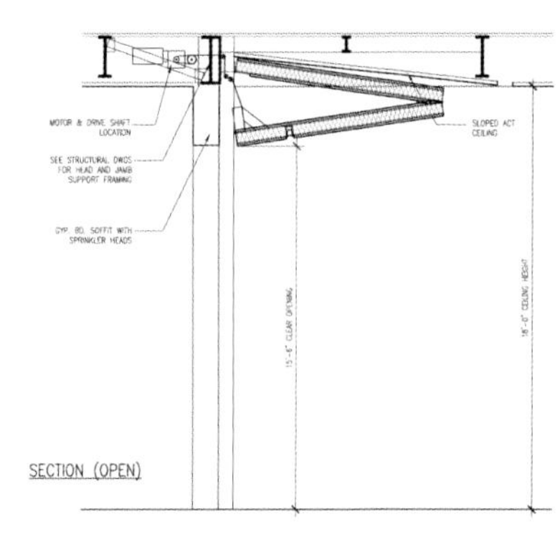

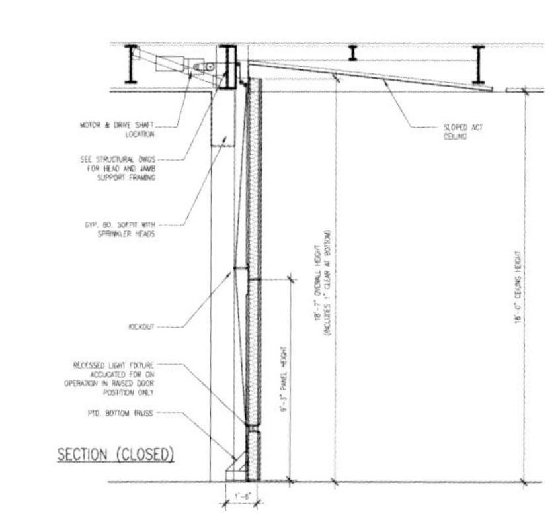

SANCTUARY BI-FOLD DOOR SECTIONS

ARCHI+SCAPE 一筑一景
Exhibit
& EXP

展示与博览会

NO.2 058-103

Triple V Gallery

三V画廊

Designed as a permanent show gallery and tourist information center for China's largest developer Vanke, MOD's dramatic design for the Triple V Gallery has become an icon along the Dong Jiang Bay coastline. Despite its obvious sculptural qualities, the building's DNA evolved rationally from a careful analysis of key contextual and programmatic perimeters – resulting in the Triple V Gallery's triangulated floor plan as well as the 3 soaring edges that have come to define its form.

The client's program called for 3 main spaces: a tourist information center, a show gallery and a lounge for discussion. Requiring their own entrances, the tourist center and the show gallery are orientated to separate existing pedestrian pathways and can be operated independently. As an extension of the show gallery, the lounge area is where discussions are conducted. This space takes advantage of the panoramic views of the coastline and comprises a sculptural bar counter. The V elevations that form the main 3 walls originate from the program requirements, needing to create entryways as well as a view passage to the seaside. The strongest local influence was the physical context and being near the sea. This prompted architects to explore the use of corten steel on the facade and timber planks for the interiors.

They design a cantilevered edge condition so that the sharp edges of the V (in plan) would be light and free. Initially, the structural engineer specified really large columns at the edge, ruining the whole idea. Tectonically, the building responds to the coastal setting and is finished in weather-sensitive corten steel panels on its exterior and timber strips on the interior walls and ceiling for a more natural feel.

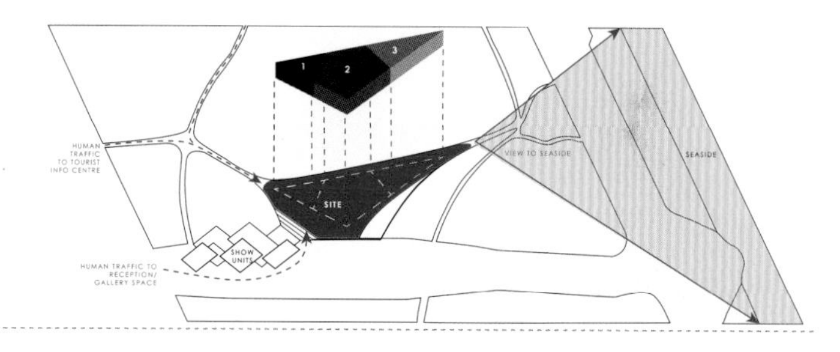

三V画廊是由Ministry of Design事务所（MOD）为中国最大的地产商——万科——设计的，它被长期用作展厅和旅游信息中心，现已成为东江湾海岸线的标志性建筑。建筑虽然具有很强的雕塑风格，却是理性地建立在一系列缜密分析的基础之上的。

万科要求建筑包含三个主要空间：旅游信息中心、展厅和聊天休息室。旅游中心和展厅拥有各自的入口，同时，两部分空间可以独立运作。休息室由展厅扩展而来，是一处供人们讨论交流的场所。这里可将海景尽收眼底，此外还设有一个雕塑吧台。

三面主墙构成了"V"形的外立面，该设计旨在满足功能的需求，创造出门廊以及通向海边的视觉通道。对设计造成最大影响的因素就是这里的物理环境和沿海气候。因此设计师在外墙上使用耐候钢，而在建筑内表面使用木材。

设计采用了悬臂式构造，使"V"形立面更为醒目。起初，结构工程师打算在建筑边缘设置很大的柱子，但这会破坏整个构想，因此结构工程师选择了其他解决办法。构造方面，为适应沿海环境，建筑的外墙面采用了耐候钢以适应气候条件，室内墙壁和顶棚则采用了木材，打造出自然舒适的环境。

Location / 地点:
Tianjin, China
Date of Completion / 竣工时间:
2011
Area / 占地面积:
750 m²
Architecture / 建筑设计:
Ministry of Design
Interior Design / 室内设计:
Ministry of Design
Photography / 摄影:
CI&A Photography

Architectural Cladding: Corten Steel Panels Pre-weathered;
Exterior: Customized Acrylic, Precast Concrete Brick in Grey, WPC Flooring in Grey, Metal Frame;
Facade: Curtain Wall, Double Glazed Glass Panes, Exterior Wall Paint;
Interior: Self-levelling Cement Flooring in Light Grey, Customized Cut Pile Carpet, Timber Parquet, White Oak.

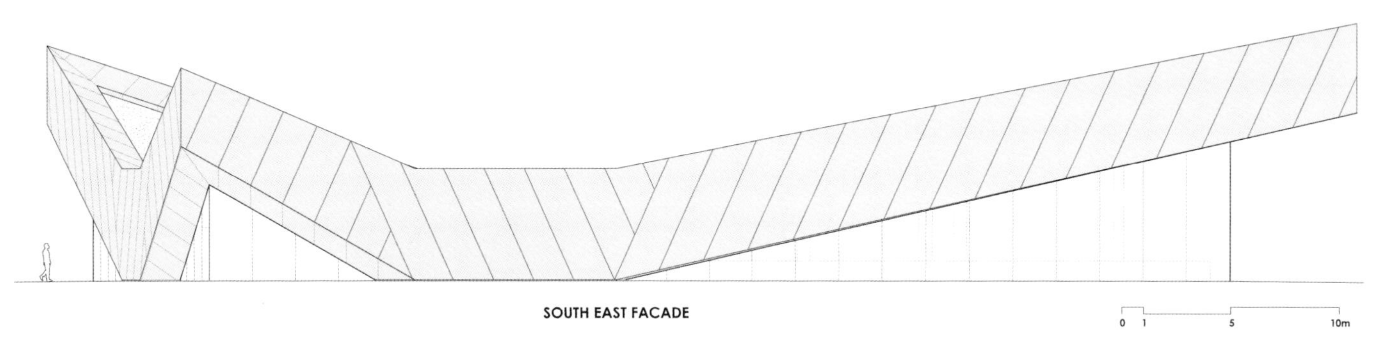

SOUTH EAST FACADE

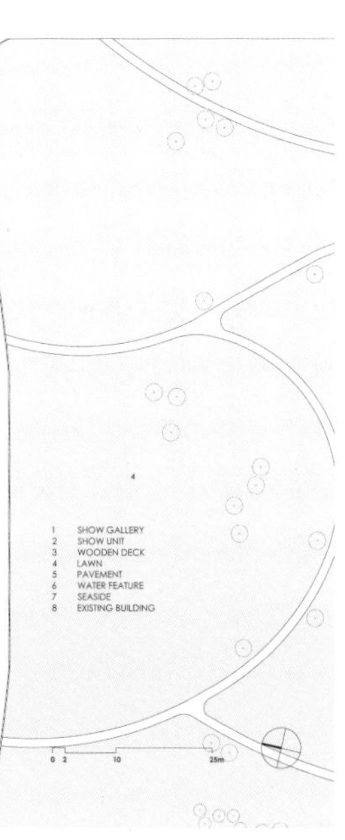

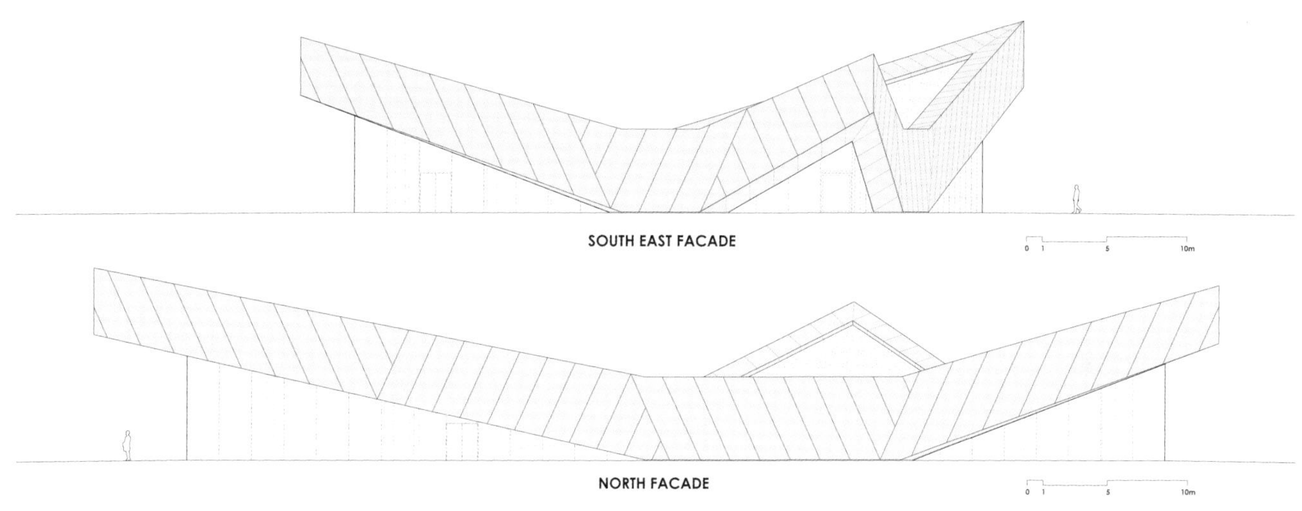

SOUTH EAST FACADE
0 1 5 10m
NORTH FACADE
0 1 5 10m

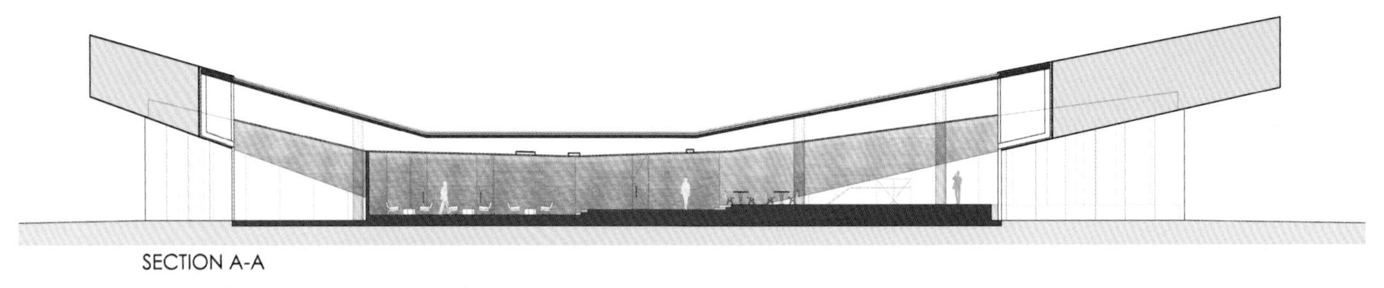

SECTION A-A

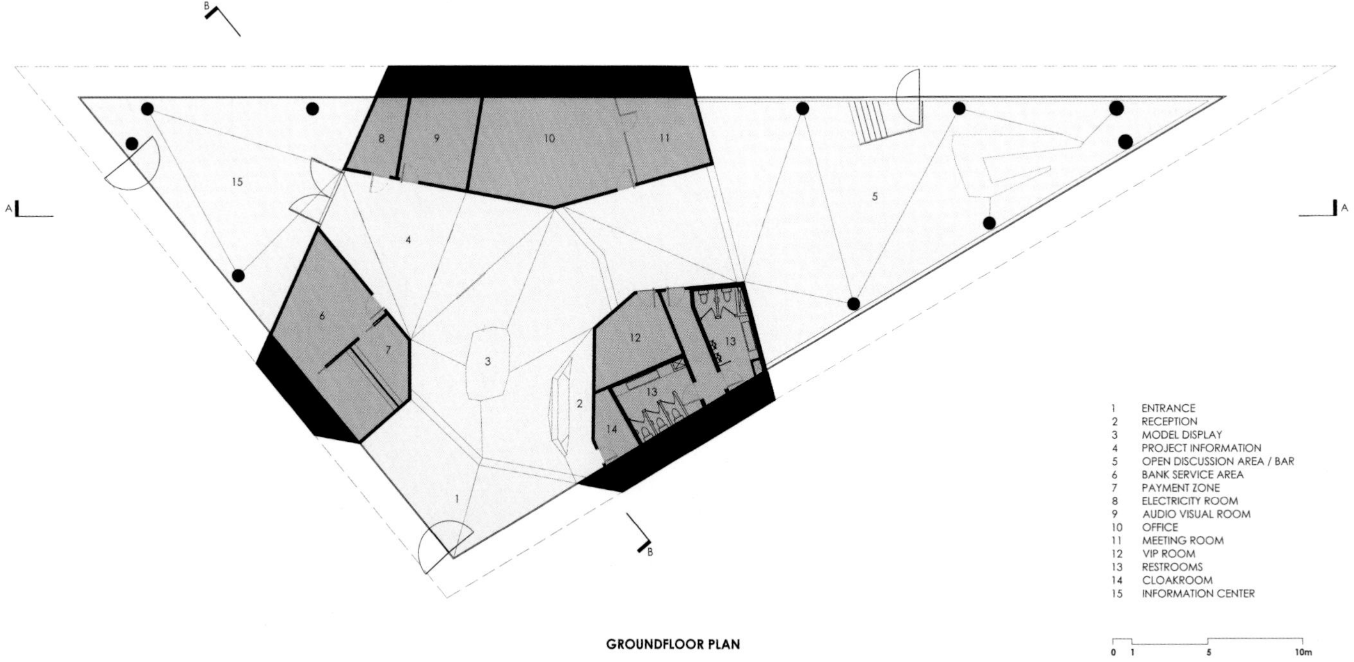

GROUNDFLOOR PLAN

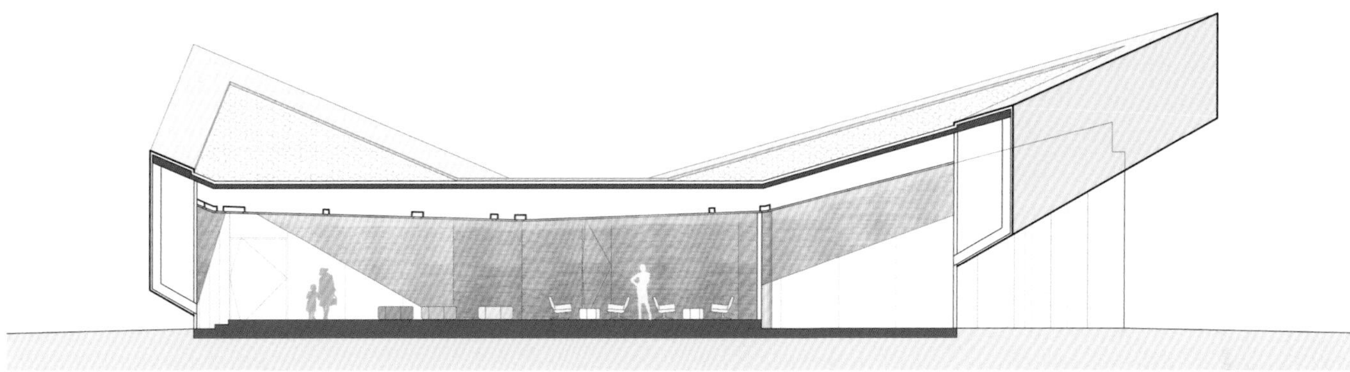

SECTION B-B

Arts Centre - Casa das Mudas

室内植物艺术展示中心

The construction of this Arts Centre aims to create an exhibiting space outside the limits of the capital. Given this specificity, it was unsettling to understand how a window or a door could be designed for a building of this scale and complexity in a rural environment, or how to create a room for exhibits that could simultaneously provide great spatial flexibility, when works of art are more and more unpredictable, and adding the question of being located on an ultra-remote island, that sometimes restricts the design of an architectural element to the size of the container. Considerably high mountains orographically define the district where the Arts Centre was built, with its urban area in the deepest valley. By implanting itself on the lineal peak of one of these mountains, that ends abruptly with the sea, it seeks to resign the "mountainous mass", acting as topography. Its interiority and complexity are only revealed when you are close to the building.

A mineral and sophisticated platform underneath the Casa das Mudas covers the whole museological complex. To sculpt this platform proposes a functional duality. Firstly the design of a viewpoint, time to pause and contemplate, recuperating the slow speed of the reading of the landscape, now partly broken by the creation of tunnels and motorway structures. Then, to provide a vast museological programme that attributed a new validity to the island. A vitreous patio organises and denounces the programme. In its interior the patios and pathways cut, burrow and link all the functions, going against the environment at certain times, vertically in relation to the mountain and horizontally to the sea. Two spatial characteristics stand out in the way they are worked and constructed. The exhibition rooms, of varying dimensions to allow greater amplitude of use, and which occur in series, reducing their design to the minimum, camouflaging all the technical infrastructure, in a way that the work of art is the core element of the visit. In opposition the auditorium presents a detailed and affirmative design in order to provide maximum values as a space for shows and to allow a wide variety of programming. Confronted with creating an Arts Centre, in the intensity of a rural landscape, with the insular and ancestral wish to observe the sea, this building was designed.

该艺术中心的建设旨在创造一个城市范围之外的艺术展示空间。鉴于项目的特殊性，如何在乡村地区为这种规模庞大、结构复杂的建筑设计窗户和门变得十分困难。由于艺术作品与日俱增的不可预测性，如何为展示场地设计一个灵活的空间也成为一个难题。由于项目场地位于极偏远的岛屿，建筑设计往往会受到场地大小的限制。高高的山脉凸显了艺术中心所在的区域，而市区则位于最深的山谷中。该中心建立在其中一座山的峰顶处，紧邻大海，只有当你靠近建筑时，才能发现它的内在性和复杂性。

本项目中石材铺装的结构复杂的平台覆盖着整个展示空间。塑造这个平台可满足双重功能：首先，它的结构改变一部分供人们驻足、思考、欣赏的观景点；其次，它能够容纳庞大的展示中心，为岛屿带来新的合理布局。一个玻璃天井使项目显得有序而通透。在内部，天井和通道连接了所有的功能性设施。该项目有两个别具特色的空间特性：各个展厅大小不一，有利于使用者更灵活地利用空间。其串联的结构将设计的复杂性降至最低，同时掩饰了所有的基础设施，使艺术作品成为展厅的核心元素。相反，礼堂则采用了细致的设计，发挥了展示空间的最大潜能，使其适用于举办多种文化活动。该项目设计满足了艺术中心的功能需求，凸显了乡村景观，也满足了人们观望大海的愿望。

Location / 地点:
Calheta, Portugal
Date of Completion / 竣工时间:
2004
Area / 占地面积:
12,000 m^2
Architecture / 建筑设计:
Paulo David Arquitecto, Paulo David
Landscape / 景观设计:
João Nunes, Proap
Photography / 摄影:
FG + SG, Fernando Guerra, Leonardo Finotti
Client / 客户:
Sociedade de Desenvolvimento da Ponta do Oeste

Interior Floor: Solid Oak;
Interior Walls: Concrete or Plaster, Painted White, Swisspearl Panels in Anthracite Colour, Stainless Steel and Frosted Glass for the WCs;
Interior Ceilings: Plaster Painted White;
Interior Handrails in Stairs: White Corian;
Interior Furniture: Vitra Eames Chais, Designed stools in Black Leather and Stainless Steel, White Corian Structures;
Interior Illumination: Concealed (Embedded) Fluorescent Lighting;
Exterior Walking Paths: Grounded Basalt stone and Asphalt Mixture;
Exterior Cladding: Sawn Basalt Stone, Glued to the Concrete Walls;
Exterior Handrails and Safeguards: Black Painted Steel;
Window Frames: Steel and Double Glass;
Exterior Illumination: Floor Washlights;
Plants: Endemic and Aromatic Plants on the Roof Top, Magnolia Tree.

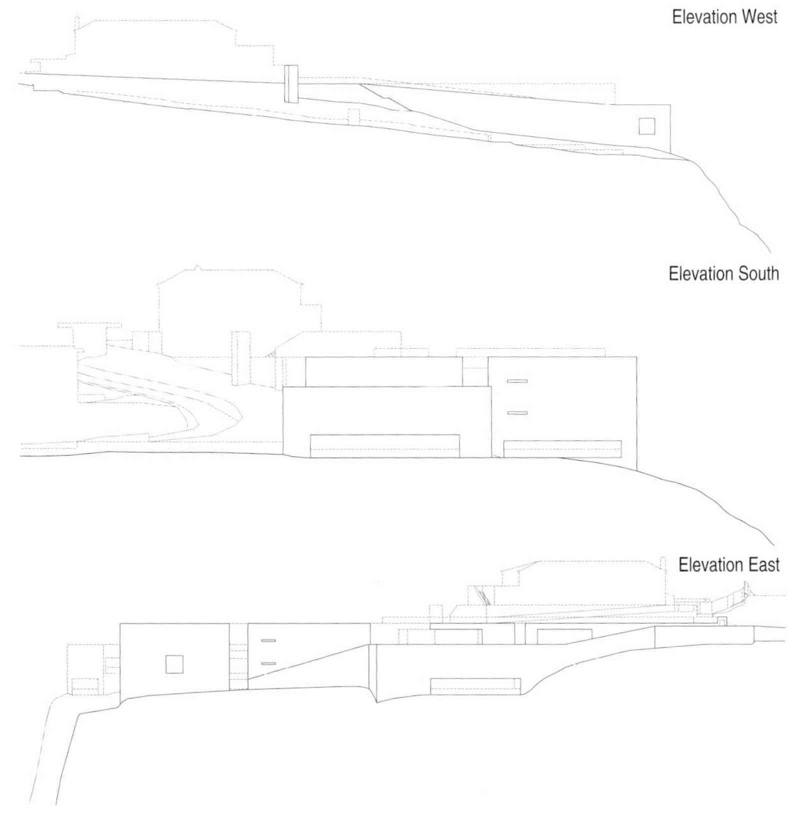
Elevation West
Elevation South
Elevation East

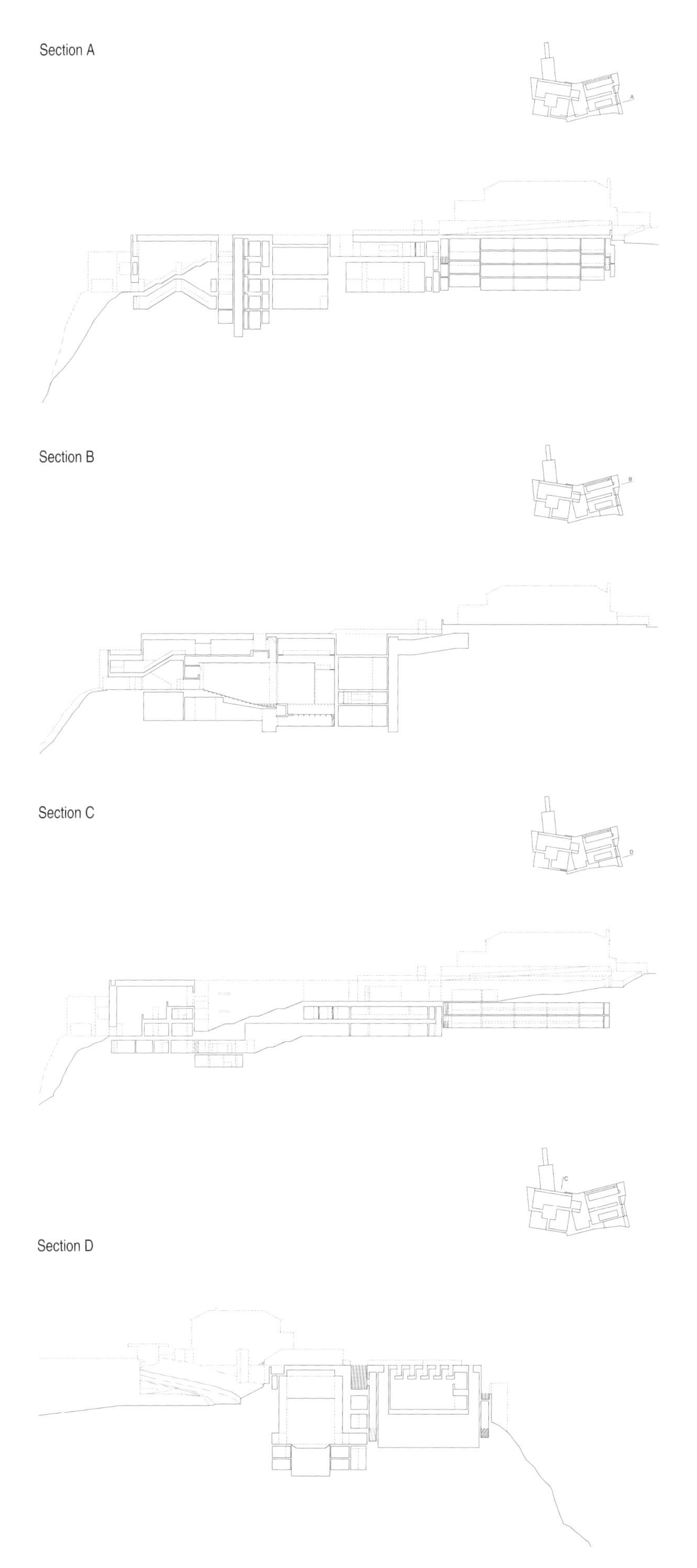
Section A
Section B
Section C
Section D

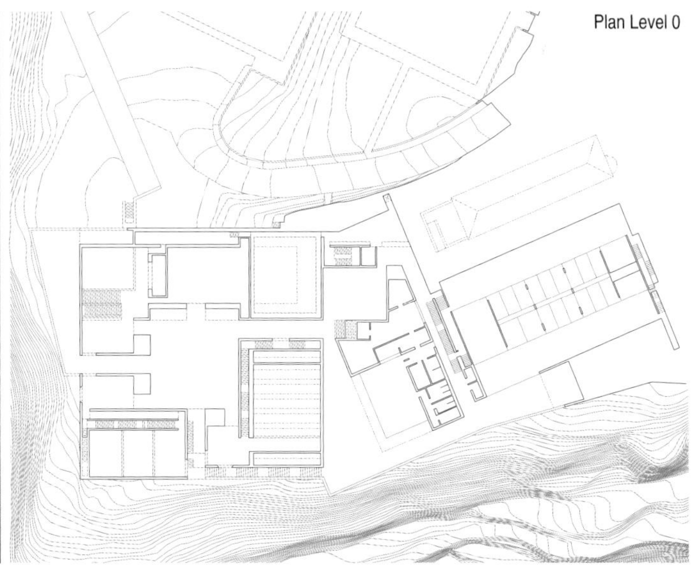

Plan Level 0

Plan Level -1

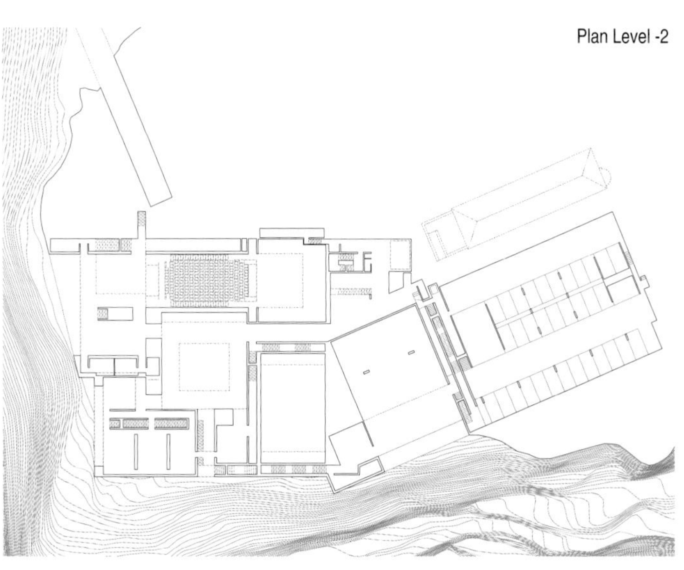

Plan Level -2

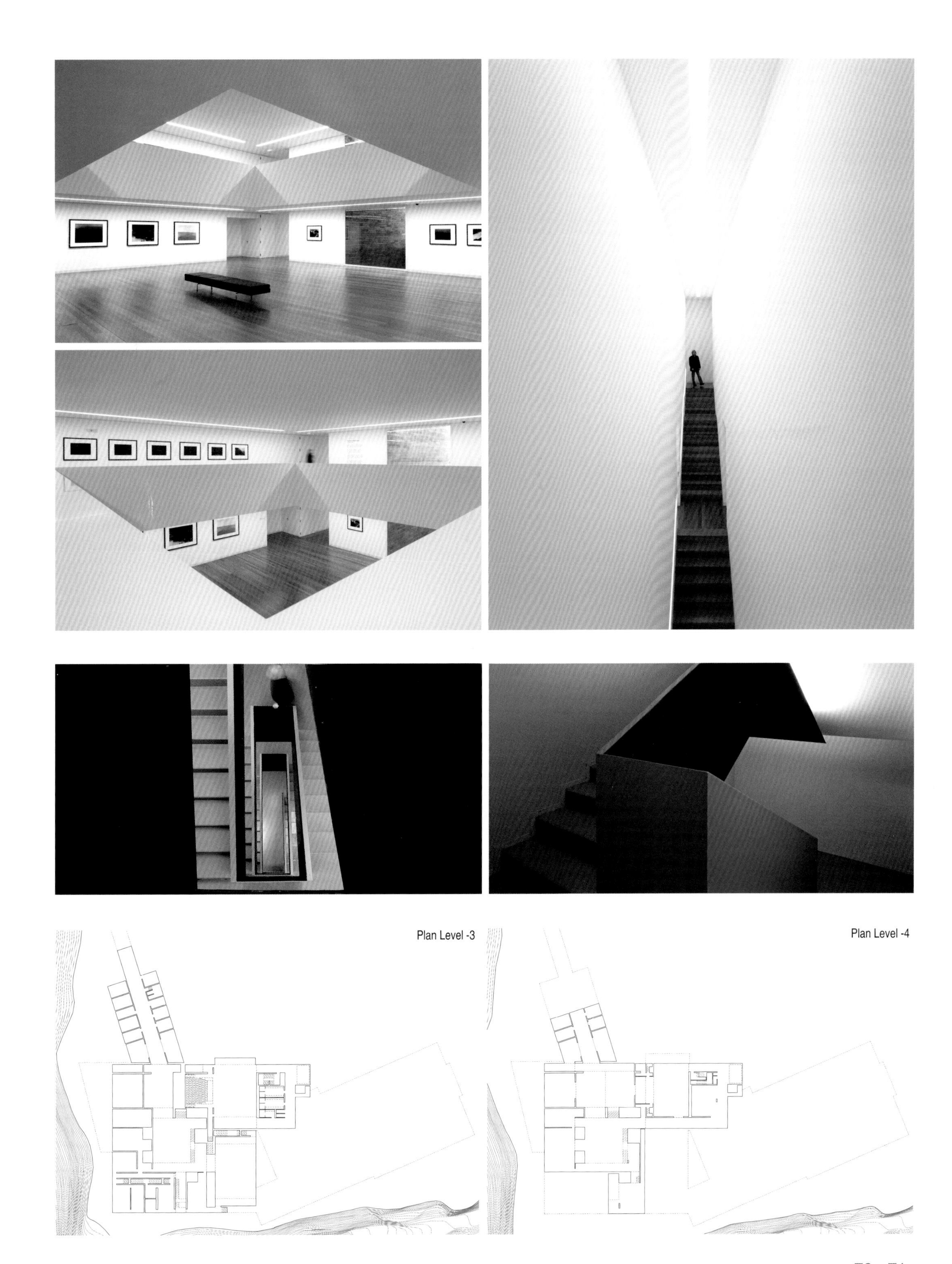
Plan Level -3
Plan Level -4

EXPO 2008 Main Building

2008年世博会建筑

This project for the main area of EXPO 2008, with its urban planning and the design of the building to house most of the exhibition areas, i.e. the international pavilions and those occupied by Spanish regions, was a considerable challenge in several respects.

Firstly, because EXPO 2008 is international, the specific BIE format had to be applied. This meant using the same construction concept when designing all the exhibition pavilions, and required the project to be seen as a single unit. This was an opportunity to provide Zaragoza with a first-rate building complex able to blend in with its natural and urban settings. It was also a chance to design the exhibition site so that once the EXPO was over it could be transformed with as little rebuilding as possible into a service and leisure area that could then be completed and consolidated as an interesting area of the city. Thirdly, the large roof not only gave the entire project a seamless appearance and image but also created an outstanding architectonic and urban identity. Finally, underpinning the entire project by the "Water and Sustainable Development" concept and theme was a driving force not only in the realm of ideas but also as regards practical, countable questions such as energy.

The criteria applied to the arrangement of the new-build blocks clearly differentiated between those facing north near the Rabal ring road and those facing south near the river Ebro. The northern blocks provided a barrier shielding the site from the noise of traffic and the strong north wind, whilst the public zones look straight towards the river Ebro with breath-taking views of the Basílica of Nuestra Señora del Pilar.

Another of the project's most important formal aspects was the use of organic forms inspired throughout by the fluidity of water associated with the same concept of searching for continuous, flowing spaces of a more amiable and interesting type in outdoor spaces rather reminiscent of the natural way water behaves in Nature.

该项目是为2008年世博会设计的主场区，是用于举行大部分展览的建筑，即国际馆和西班牙展馆区域，这个项目面临了相当大的挑战。

首先，由于2008年世博会是国际性的活动，项目需要符合国际展览的特定要求。这意味着所有展馆都要使用相同的建筑理念，并成为一个整体单元。本案设计的建筑群与自然和城市环境相融，为萨拉戈萨注入了新的生命力。其次，当世博会结束后，展览场地能通过最简单的重建工程转化为服务区和休闲区，成为城市里一片趣味盎然的区域。再次，巨大的屋顶不仅为整个项目带来无缝的外观形象，而且建立了一个显著的建筑和城市特征。最后，"水和可持续发展"的设计理念和主题不仅是影响该项目构思的主要因素，还是解决能源等实际问题的实用性措施。

新建筑群的布局清晰地将位于Rabal环路附近朝北的建筑和埃布罗河附近朝南的建筑区分开来。北部建筑构成一处屏障，使场地免受外部的交通噪声和强烈的北风的干扰，而公共区域直接面向埃布罗河，人们在这里可以观赏比拉尔圣母大教堂的壮丽景色。

该项目另一个重要的外观特征就是具有像水体一样流畅的线条，设计师通过连续、流畅的设计创造了一个具有趣味性的建筑，不禁让人联想起水在自然状态下的美景。

Location / 地点:
Zaragoza, Spain
Date of Completion / 竣工时间:
2008
Area / 占地面积:
250,000 m²
Architecture / 建筑设计:
ACXT
Interior Design / 室内设计:
ACXT
Landscape / 景观设计:
ACXT
Photography / 摄影:
Aitor Ortiz
Client / 客户:
Expo Agua Zaragoza 2008

Floor: In Situ Concrete, Concrete Paving Stone;
Walking Paths: Wood;
Furniture: Concrete Banks, Wood Banks, Stainless Steel Railings;
Plants: Grass, Sedum Vegetation (Green Roof) ;
Structure: In Situ Concrete, Post-Tensioned Concrete, Precast Concrete Slabs;
Others: Green Roof, Ponds and Artificial Waterfalls.

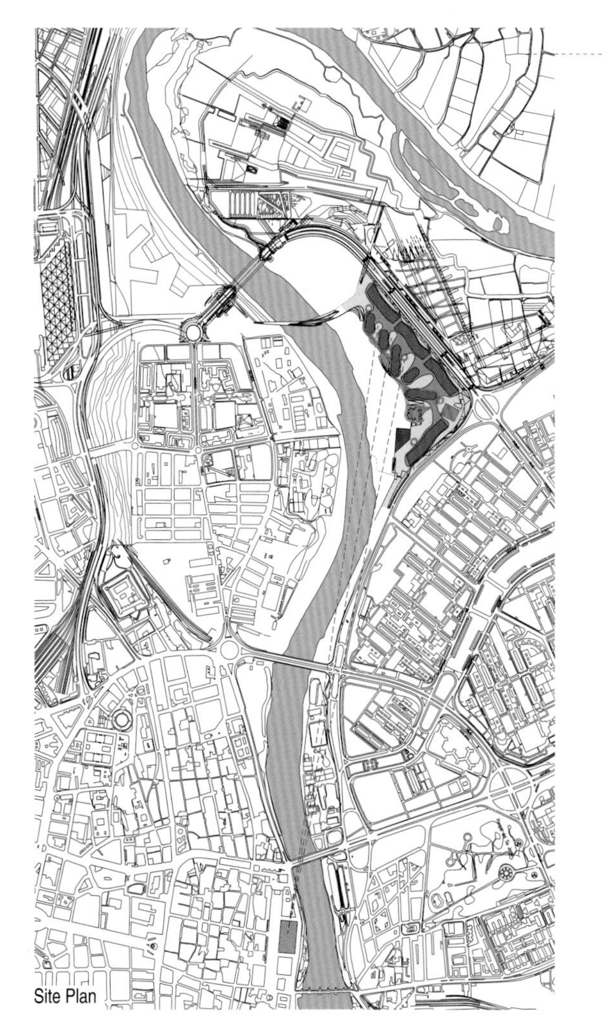

Site Plan

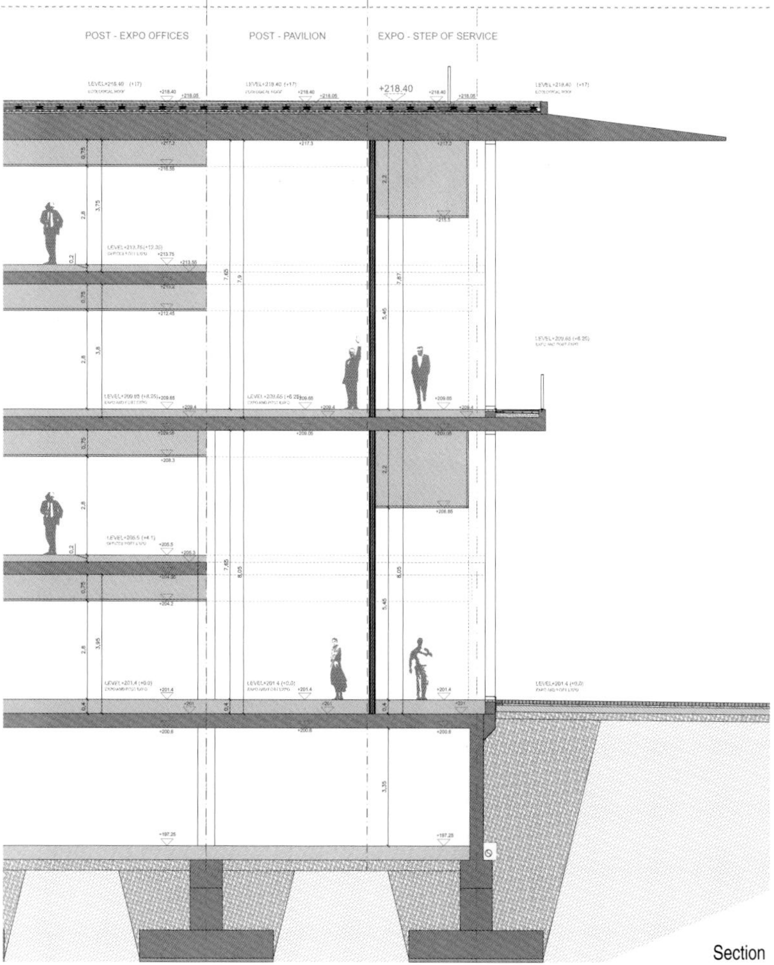

Section

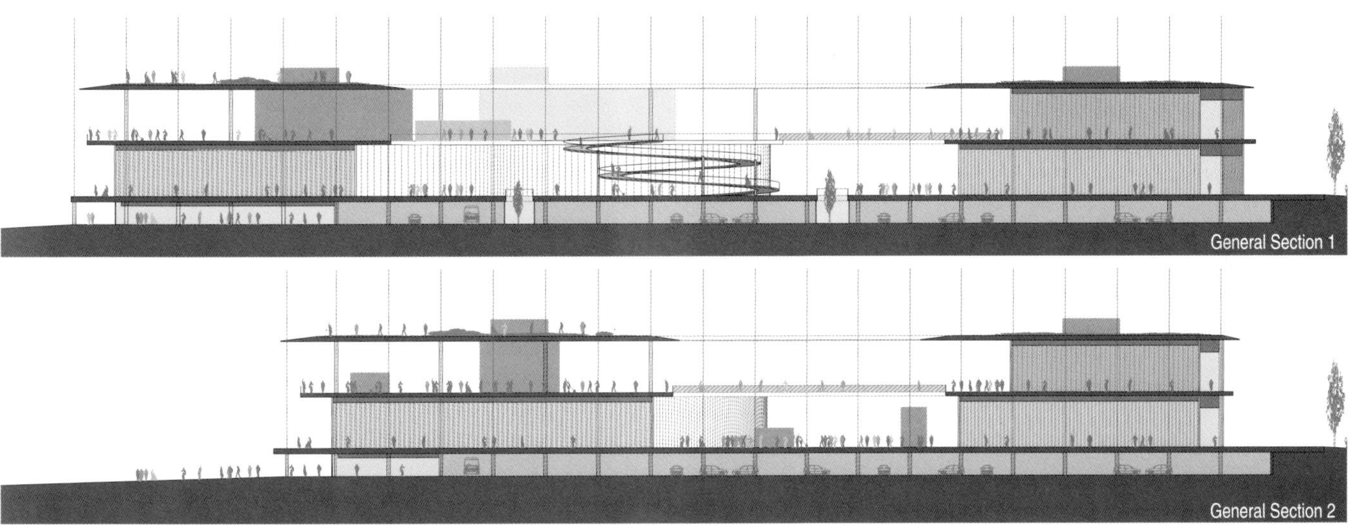

General Section 1

General Section 2

Japón

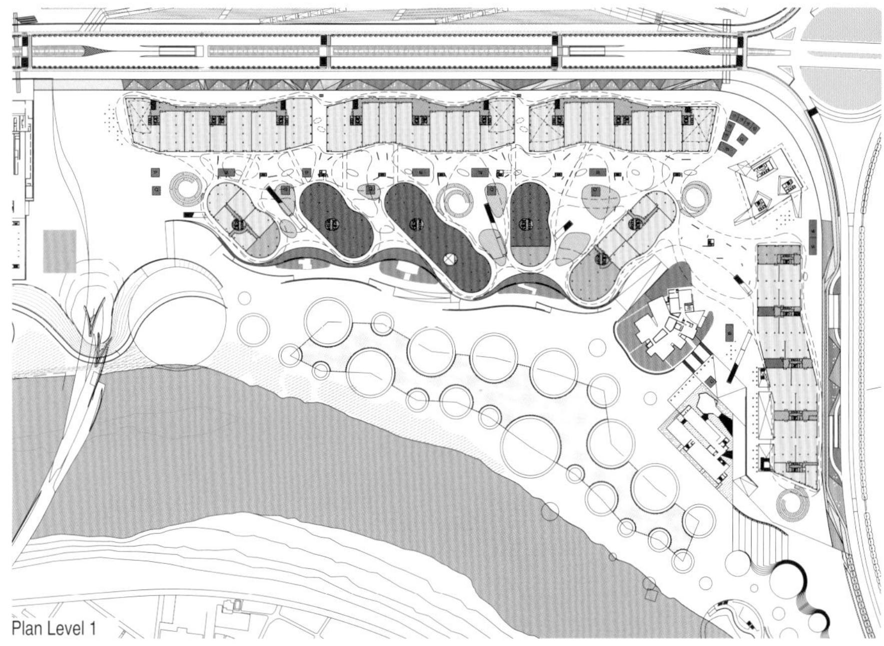

Plan Level 1

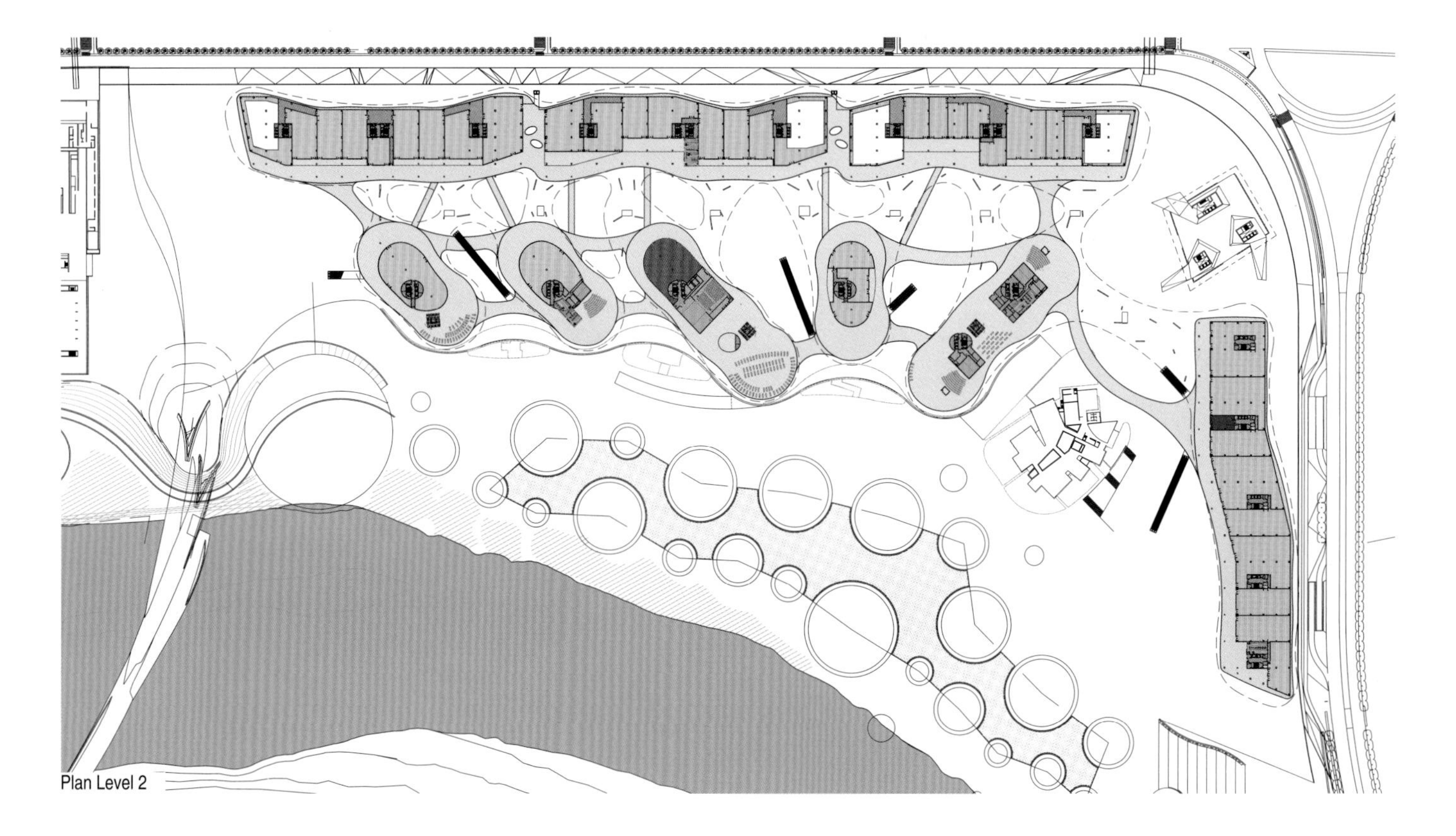

Plan Level 2

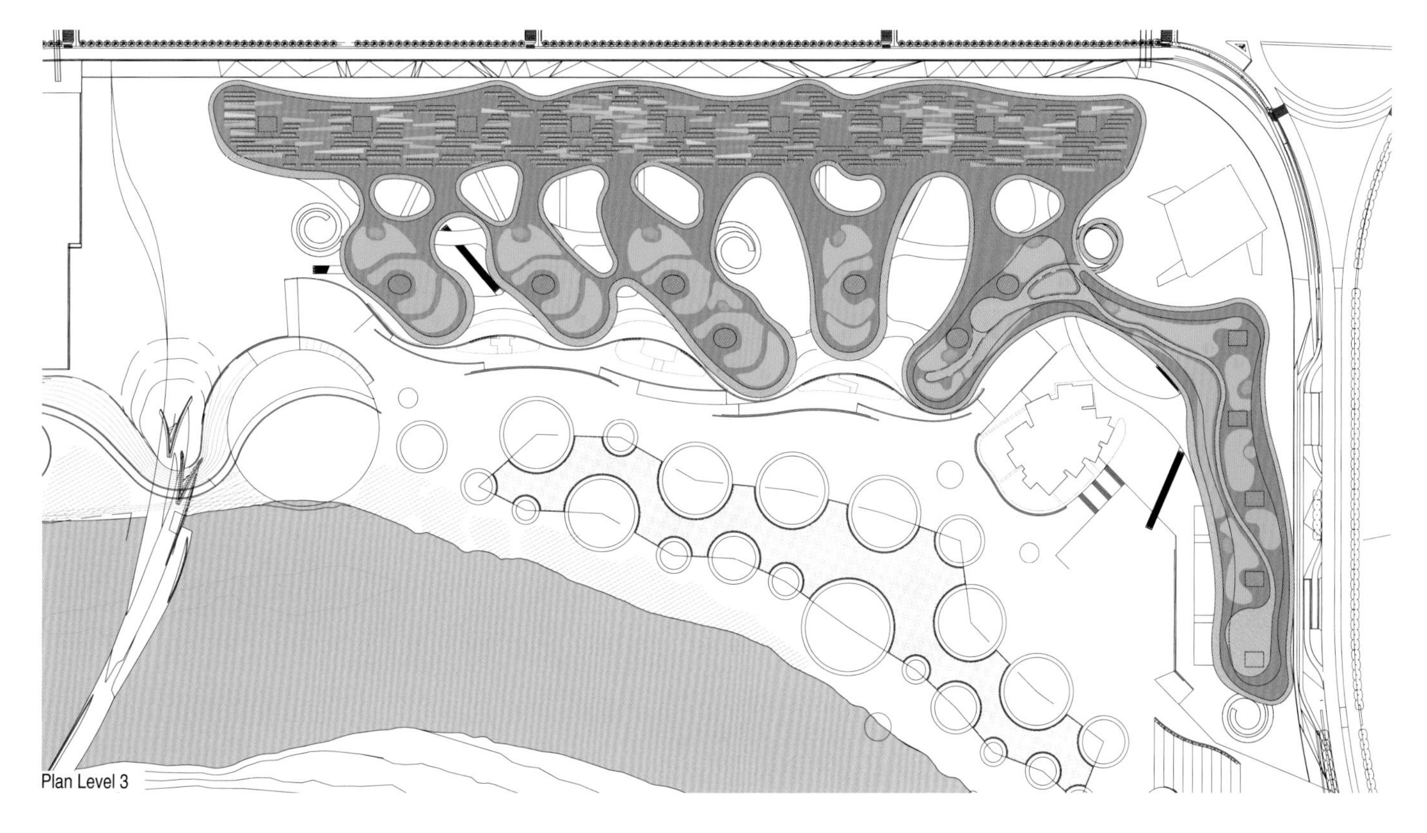

Plan Level 3

Incheon Tri-bowl

仁川Tri-bowl纪念馆

The Incheon Tri-bowl is started with the idea of being against general thoughts about architecture. It is composed of a curved floor with a flat roof instead of a flat floor with a curved roof of general architecture. This memorial hall is basically used for exhibitions and is designed to have a wide ceiling for those exhibitions. LED lamps, composing the ceiling, make different information available on those at any times. The structure is floating on a rectangular reflection pond and patrons will be entering it through a long bridge passing underneath the mass. The interior was finished with a transparent material and also a lightweight structure being separated from the shell body. The circulation of patrons is in the trace of a continuous cubic curve which is making multiple actions like circling, going up and down, etc. There are service spaces consisted of rooms for exhibition, performance and rest as well as office space. The exhibition and performance space can accept about 400 people and the patrons can see the exhibits hanging from the ceiling through the trace of a cubic curve. It is mainly a one-way curved surface from a free curve and a conic curve under the shell body. The exterior is finished with exposed concrete on the lower part of the shell while the upper part of the shell is finished with aluminum panels.

Incheon Tri-bowl is an uncommonly seen project that reverses the common understanding of an architectural space. Unlike the accustomed architecture space that applies small differences to the flat floor and ceiling, the Tri-bowl creates a free-curved floor under a flat ceiling. The building is used as a gallery space, in which the flat ceiling itself was planned from the beginning to be used as an exhibit, which leads architects to the solution of giving a radically shaping the floor of the building rather than the ceiling. This structure floats on a reflecting pond where there is a long bridge the visitors can enter under the extreme structure. The bridge continues inside the building and acts as the main circulation of the building. The programmatic space consists of an exhibition/theater space and service/relaxation space. The exhibition/theater space can accommodate up to 400 people, where visitors can also view the exhibition pieces of work that hang on the ceiling structure.

本项目的设计理念颠覆了建筑设计的一般思路。一般建筑通常采用平面基底和弧面屋顶，本项目却是弧面基底和平面屋顶的结构。这个纪念馆主要用于展览，因此顶棚的设计较为宽阔。LED灯具安装在顶棚中，使人们能在任何时间清楚地看到所需的信息。这个建筑仿佛漂浮在一个清澈如镜的矩形池塘上方，游客将沿着建筑下方一座长长的桥进入建筑内部。室内装潢采用了一种透明材料，并采用轻盈的结构。游客的通行路线沿着连续的曲线设置，可以满足人们在这里绕行、上下走动的需求。服务空间由展览区、演出大厅、休息室以及办公区构成。展览及表演空间可以容纳约400人，他们可以通过曲线路线观赏悬挂于顶棚的展品。建筑主体为一个单向弧面，由下方的自由曲线与圆锥曲面构成。建筑壳体下方的外表面以裸露的混凝土装饰，而壳体上部为铝板贴面。

仁川Tri-bowl纪念馆是一项罕见的项目，它的设计颠覆了人们对建筑空间的一贯认识。在一般的建筑空间中，平展的底面与顶棚差异很小，而Tri-bowl却在平面顶棚下方创造出一个无拘无束的曲线形楼底。该建筑作为一个展示空间，其平面布局就是为了满足展览需求的，这促使设计师优先考虑底面的建筑形式，而非顶棚。整个建筑仿佛漂浮在波光粼粼的池塘上方，池塘中有一座长长的桥，引导游客进入这个特色建筑的底层。这座桥在建筑内部继续延伸，并成为建筑的主体通行路线。项目空间由展览、剧院空间和服务（休息）空间构成。展览空间可容纳多达400人，游客在这里还可以欣赏悬挂于顶棚的展品。

Location / 地点:
Incheon, South Korea
Date of Completion / 竣工时间:
2010
Area / 占地面积:
2,869 m²
Architecture / 建筑设计:
iArc Architects
Interior Design / 室内设计:
iArc Architects
Landscape / 景观设计:
MIN Landscape
Photography / 摄影:
Youngchae Park
Client / 客户:
Incheon City

Materials: Exposed Concrete;
Anodizing Sheet Round Panel.

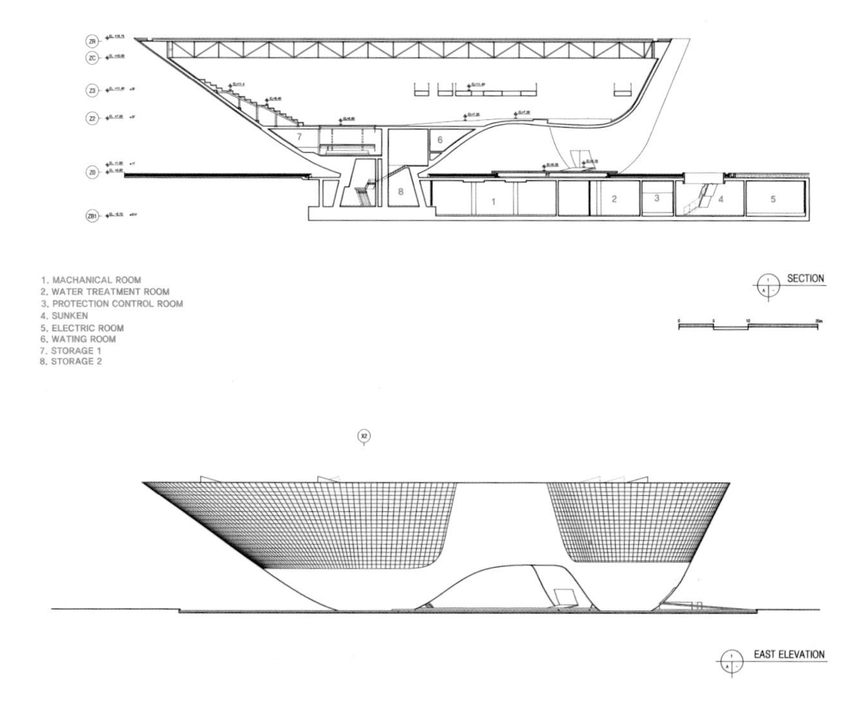
Key-Plan
1. MACHANICAL ROOM
2. WATER TREATMENT ROOM
3. PROTECTION CONTROL ROOM
4. SUNKEN
5. ELECTRIC ROOM
6. WATING ROOM
7. STORAGE 1
8. STORAGE 2
SECTION
EAST ELEVATION

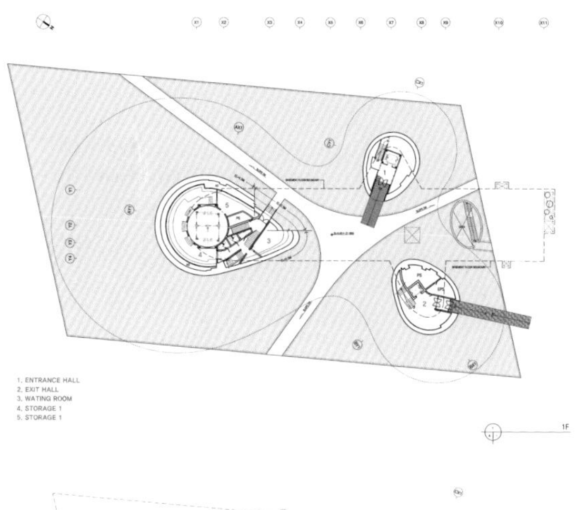

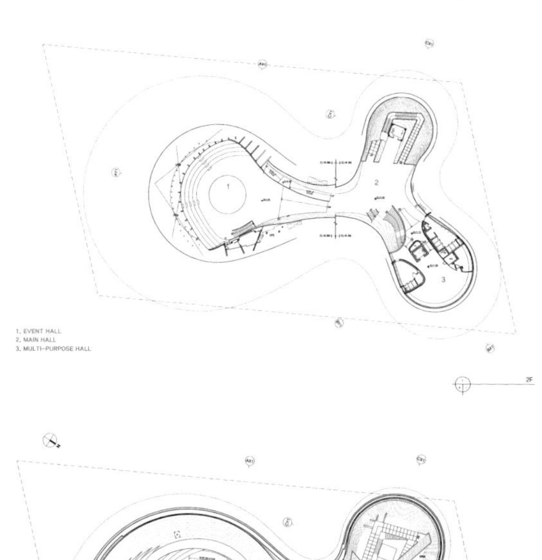

1. HALL 1
2. HALL 2

3F

Jodrell Bank Visitor Centre

焦德雷尔班克游客中心

Designed for the University of Manchester's Centre for Astrophysics the project includes two new Visitor and Exhibition Pavilions which have been built to inspire budding young scientists and showcase cutting-edge research 'as it happens'. Located directly adjacent to the Grade I listed Lovell Telescope the design of the buildings is inspired by this historic landmark and its setting in the surrounding landscape. The visitor centre forms phase one of a wider redevelopment of the visitor facilities, has a total floor space of 1,000 m^2, and comprises.

A Planet Pavilion entrance building including an orientation centre and a stylish glass-walled café with spectacular views of the Lovell Telescope. The building features an embossed artwork across the 55 m long front elevation representing their galaxy as captured by the Lovell Radio Telescope.

A Space Pavilion incorporating a multi-purpose event space and education pod as part of an interactive 'Live Science' exhibition space. The building opens out onto a large terrace giving visitors the opportunity to get up close to the Lovell Telescope to see and hear it work while it turns and rotates during the observations.

New landscaping was designed to guide visitors towards the telescope, which improved visitor footpaths in the Arboretum. The landscape works also include a 'Galaxy Maze' landscape installation within the Arboretum, designed by garden designer and TV presenter Chris Beardshaw.

Location / 地点:
Manchester, UK
Date of Completion / 竣工时间:
2011
Area / 占地面积:
945 m^2
Architecture / 建筑设计:
Feilden Clegg Bradley Studios
Exhibition / 展示设计:
Thomas Matthews
Landscape / 景观设计:
Grant Associates
Photography / 摄影:
Hufton & Crow
Client / 客户:
University of Manchester

该项目为曼彻斯特大学天体物理中心建造，包括两个新的展览馆，旨在展示年轻科学家最前沿的研究。该项目直接与一级保护建筑洛弗尔天文台相邻，其设计灵感来自于这个历史悠久的著名地标性建筑以及周围的景观。游客中心的重建是整个地块规划和发展的首要任务，其占地面积总计945m^2，包括以下几个区域.

首先看到的是行星馆。这里包括一个定位中心和一间时尚的咖啡馆。咖啡馆采用玻璃幕墙，人们可以在此欣赏壮观的洛弗尔望远镜。该建筑正立面的外墙采用了55m长的特色浮雕设计，象征着洛弗尔望远镜捕获到的银河系景象。

其次是一座太空馆。它拥有一个多功能活动空间和一处教育用的吊舱，后者是互动性“现场科学”展览空间的一部分。该建筑通向一个大型露台，让游客有机会近距离观看洛弗尔望远镜，看它如何旋转着巨大的肢体观测外太空。

另外，项目重新进行了环境美化，更好地将游客导向望远镜景点，同时改善了植物园的游行路线。环境美化还包括在植物园中建立一个“银河迷宫”，并由园林设计师兼电视节目主持人克里斯·比尔德肖负责设计。

Cladding: Composite Metal Panels;
Floor: Painted Concrete;
Walking Paths: Blacktop Macadam, Resin Bound Gravel;
Furniture: Purpose Made Reception and Shop Furniture, Coloured Black MDF with Clear Lacquer Finish;
Illumination: LED Lighting by Zumtobel;
Plants: Native Oak Trees, Beech Hedges, Grass Turfing and Seeding;
Others: "Galaxy Maze" Landscape Installation within Existing Arboretum with mix of Hedging and Decorative Planting.

Jodrell Bank Visitor Centre - **Phase One Masterplan**

1 Planet Pavilion (Admissions Building)
2 Space Pavilion (Exhibition / Event Building)
3 Lovell Telescope
4 Control Building
5 Visitor Footpaths
6 Visitor Carpark
7 Galaxy Maze Landscape Feature

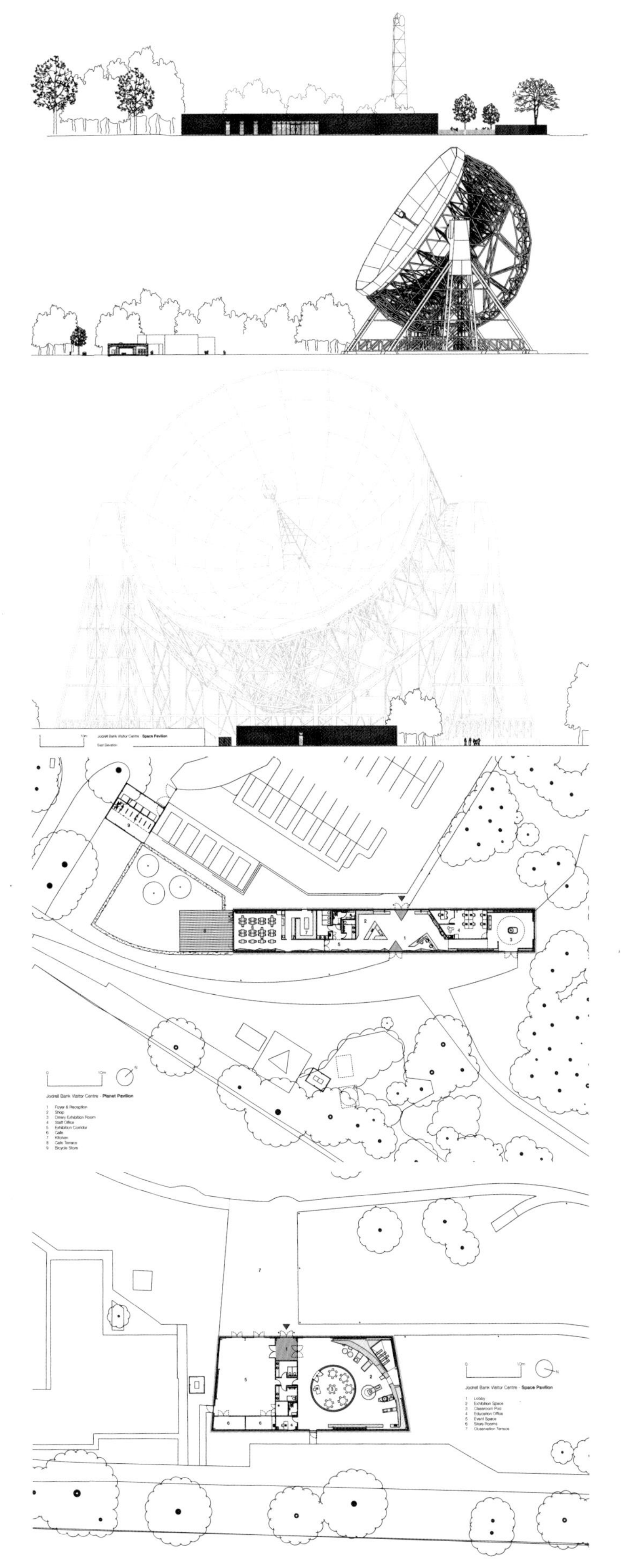
Jodrell Bank Visitor Centre - Space Pavilion
East Elevation
0
10m
N
Jodrell Bank Visitor Centre - Planet Pavilion
1 Foyer & Reception
2 Shop
3 Orrery Exhibition Room
4 Staff Office
5 Exhibition Corridor
6 Cafe
7 Kitchen
8 Cafe Terrace
9 Bicycle Store
0
10m
N
Jodrell Bank Visitor Centre - Space Pavilion
1 Lobby
2 Exhibition Space
3 Classroom Pod
4 Education Office
5 Event Space
6 Store Rooms
7 Observation Terrace

How the Telescope Works
Seeing the Invisible

Black Holes
Life of Stars

Orquideorama in Colombia

哥伦比亚的生态建筑

This project rises from a hexagonal geometry that allows for growth and flexibility in the floor plant. Besides, it works with the scale of the trees and some of their strategies: it concentrates all the networks in the trunk, mixes structure and organic force, and opens translucent foliage.

Assembling modules of seven hexagons, they can define a spatial, structural and bioclimatic pattern that allows repetition, controlled growth and flexibility, both to avoid touching the already existing trees and to adapt to a limited budget.

This pattern or module has a central hexagon that functions like a hollow trunk in which the technical network (structural, electrical, and water collection), the organic network, a hot air exit, and an access to rain and humidity for the internal gardens, concentrate. These trunks were understood as converging vortexes, and therefore the geometry of the wood lining displays a concentrated torsion force through a hyperbolic paraboloid. They tried to combine organic and structural forces.

The intelligence of the pattern amplifies through repetition, but the spatial and bioclimatic qualities are already present in each part. You could say the qualities of the pattern emerge from inside out.

The space they propose isn't an empty warehouse or a free plant but rather a field with structural supports that are intermittent with clustered gardens. This allows to combine a large range of activities (weddings, workout, meditation, concerts, fashion shows, etc), with the garden and its wildlife and weather.

Location / 地点:
Medellin, Colombia
Date of Completion / 竣工时间:
2006
Area / 占地面积:
4,200 m^2
Architecture / 建筑设计:
Plan:B Arquitectos (Felipe Mesa + Alejandro Bernal) + JPRCR Arquitectos (Camilo Restrepo + JPaul Restrepo)
Landscape / 景观设计:
Plan:B Arquitectos (Felipe Mesa + Alejandro Bernal) + JPRCR Arquitectos (Camilo Restrepo + JPaul Restrepo)
Photography / 摄影:
Sergio Gomez, Veronica Restrepo
Client / 客户:
Jardin botánico de Medellín (Medellin Botanical garden)

该项目采用了六角形的几何模块，从地块上拔地而起，为底部植物的生长营造了良好环境。此外，设计方案模拟树木的尺度及结构，将构造集中于主干上、采用混合结构和有机材料，扩展出半透明的枝叶状结构。

这些单体各装置了7个六边形的模块，形成具有特色的空间、结构和生物气候模式，适合打造重复性结构，控制植物生长，增强空间的灵活性，不但可以避免与现有树木接触，而且整个费用在有限的预算内。

这种样式或模块拥有一个中央六角形，起到中空树干的作用，里面集中容纳了配套设施（结构、电、水）、有机网络、热气出口以及内部花园的雨水和湿度调节系统。这些树干的形态被构想为向中心聚集的漩涡，为此，木质的内层采用了双曲抛物面结构，从而展现出一种集中的扭动力。该项目的设计力求将建筑结构和有机生命体有效地结合起来。

建筑的结构通过单体的重复延伸开来，但其空间和生物气候特征在每个单体中都有体现。可以说建筑的格局是由内向外发展而来的。

项目的场地不是一个空洞的结构，而是一处点缀着丛生植物的、拥有结构支撑的园地。这块场地既有花园和野生动植物，又有宜人的天气，拥有多种功能，可举办各种活动，如婚礼、运动、音乐会、时装表演等。

Lining: Harvested Wood Linings;
Road: Concrete Pavers;
Others: Steel and Polycarbonate Roof Tiling.

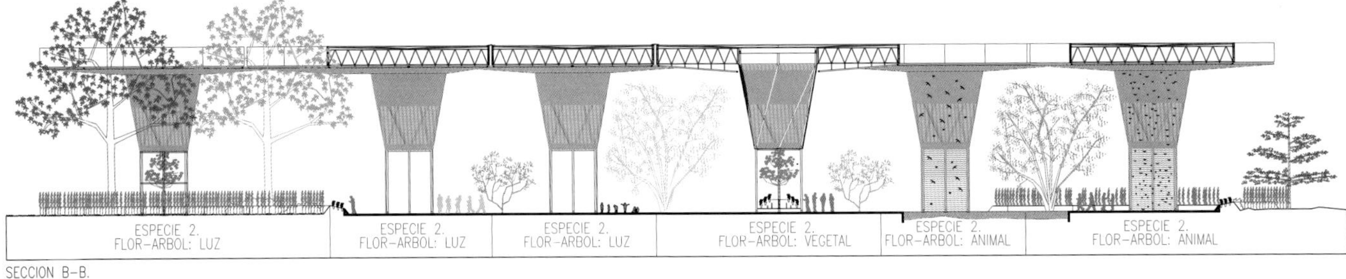
ESPECIE 2.
FLOR-ARBOL: LUZ
ESPECIE 2.
FLOR-ARBOL: LUZ
ESPECIE 2.
FLOR-ARBOL: LUZ
ESPECIE 2.
FLOR-ARBOL: VEGETAL
ESPECIE 2.
FLOR-ARBOL: ANIMAL
ESPECIE 2.
FLOR-ARBOL: ANIMAL
SECCION B-B.
0
5

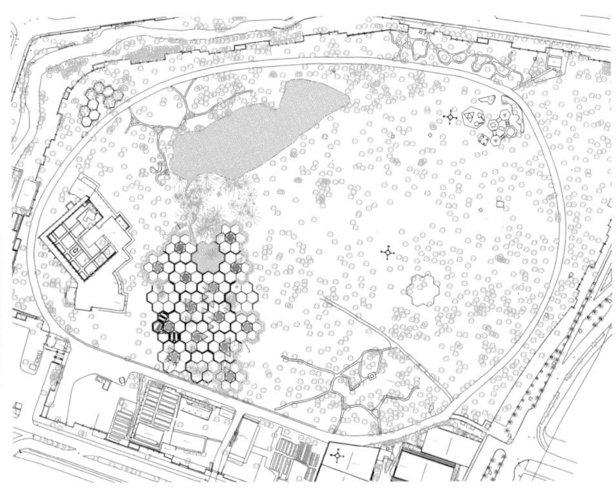

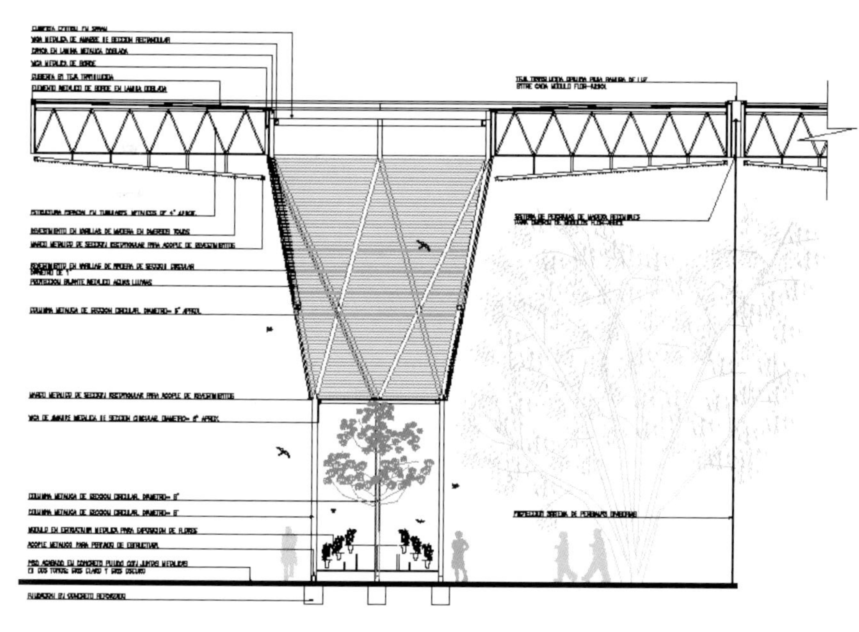

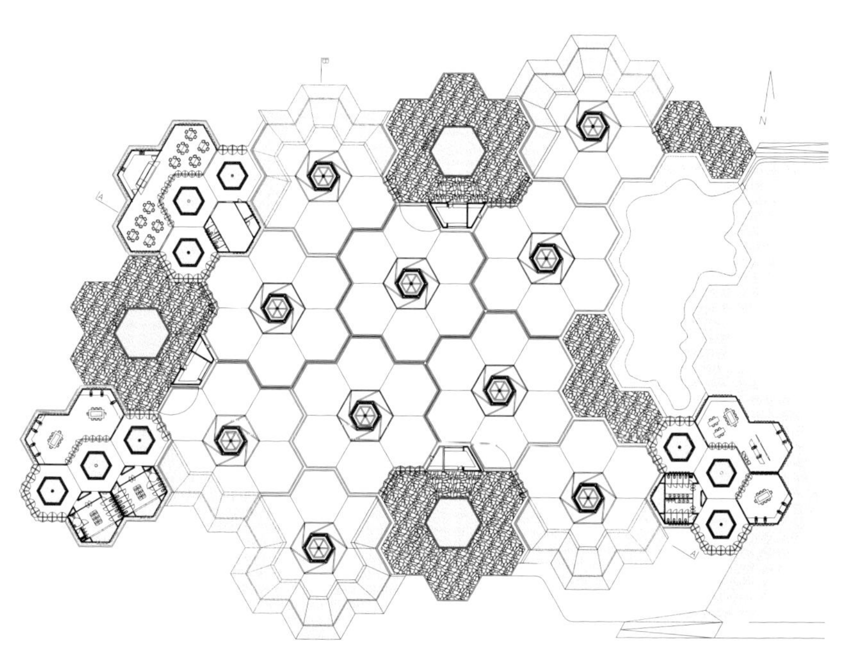

Red Rock Canyon Visitor Center

红岩峡谷游客中心

The Red Rock Canyon Visitor Center consists of a new Visitor Arrival building with 3576.8m^2 of outdoor exhibits, a remodel of the existing visitor center building to house administrative functions, and a new Visitor Contact/ Fee Station. Indoor and outdoor exhibits, administration, and the stunning landscape are integrated as one. The educational content of the exhibits and the building are inseparable as both introduce the up to 1 million visitors a year to water harvesting, greywater irrigation, climate specific architecture, and the use of shade, humidity, and air movement to temper outdoor space. The design significantly reduces the amount of indoor, conditioned space on this project by creating exterior, tempered microclimates for the majority of the topic-specific exhibit pavilions. In support of the Bureau of Land Management's mission to encourage stewardship for the land, the primarily outdoor experience instills in individuals a sense of personal responsibility for their environment's well-being.

Many resource-conserving ideas are incorporated into the LEED Gold certified facility. The Arrival Experience is sheltered by a "big hat" (a roof with ample overhangs) that creates an intermediate thermal transition zone and harvests rainwater used for interpretive exhibits and landscape irrigation. High-efficiency mechanical systems were specified, while daylighting, solar water heating, and a 55 kW photovoltaic array convert the region's intense sunshine into free energy. A transpired solar wall system provides heating for the rest rooms, allowing the mechanical system in these spaces to be eliminated.

The existing visitor center facility, constructed in 1982, was renovated to house administration for the National Conservation Area. In these new flexible office, meeting, and work spaces for staff, natural light is brought into the existing windowless space with the addition of overhead light monitors. High-efficiency lighting and a variable refrigerant volume (VRV) mechanical system replaced outdated existing equipments, offering flexible zoning and ease of installation in the existing structure.

红岩峡谷游客中心包括一个拥有3576.8m^{2}户外展区的新游客接待楼、一个由原有游客中心改造而来的用于容纳管理部门的建筑以及一个新的客服/收费站。室内外展览、管理部门和迷人的景观采用一体化集成的设计方案。展览和建筑是不可分割的两个部分，因两者共同向年客流量达10000人的游客展示着雨水收集、中水灌溉、特定气候建筑，以及通过运用光照、湿度和空气流动来驾驭户外空间。设计方案为大部分主题展馆的外围创造出局部微气候，大大减少了该项目中室内空调的数量。为响应土地管理局对于土地责任的号召，室外设计着重让人们体验到“保护环境，人人有责”的理念。

这座曾获得“能源与环境设计先锋奖”（LEED）金级认证的项目整合了多项资源节约型方案。“到访体验”处有一个巨大的帽状顶棚，不但成为温度过渡区域，还可以收集雨水，并将其用于解释性展品和景观灌溉。项目中使用了特定的高效机械系统，天然采光、太阳能热水以及55kW的太阳能光伏系统可以将该地区强烈的日照转化为可利用的能源。其他空间通过一个排气式集热墙系统供暖，因此无需安装其他机械系统。

该项目还对一座建于1982年的现有游客中心进行整修，将其用作国家保护区管理部门的办公区。这些新建的办公室、会议室和工作室采用灵活的设计，为原有的无窗室内空间的上方增加了光线监测器，将自然光线带进室内。高效的照明和VRV空调系统（变冷媒流量多联系统）替代了过时的老式设备，为当前的游客中心带来灵活的场地规划和方便的设备接入。

Location / 地点:
Nevada, USA
Date of Completion / 竣工时间:
2010
Area / 占地面积:
54,600 m^2
Architecture / 建筑设计:
Line and Space, LLC
Interior Design / 室内设计:
Line and Space, LLC
Landscape / 景观设计:
McGann & Associates
Photography / 摄影:
Henry Tom, Johnny Birkinbine, Robert Reck, Line and Space, LLC
Client / 客户:
United States Department of the Interior, Bureau of Land Management

Materials: Concrete Masonry, Steel Structural Frame, Storefront and Curtain Wall Systems with Insulated Low-E glazing.

FIRE

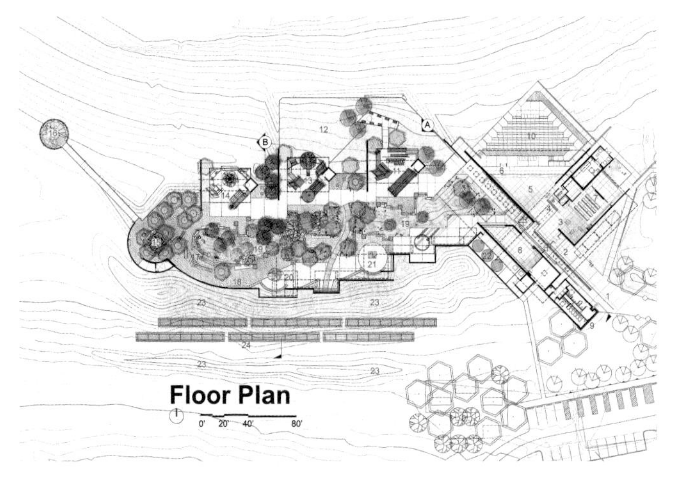

Key

1. Entry Plaza
2. Transition Space / Entry
3. Temporary Exhibits / Multipurpose
4. Information Desk
5. Arrival Experience
6. Panorama Window
7. Classroom with Outdoor Patio
8. Gift Shop
9. Transpired Solar Collector (NREL)
10. Outdoor Amphitheater
11. Earth Pavilion
12. Tortoise Habitat
13. Fire Pavilion
14. Air Pavilion
15. Four Elements Exhibit
16. 360° View Deck
17. Cliff Walls Exhibit
18. Water Walk
19. Natural Habitats
20. Desert Spring Exhibit
21. Desert Ecosystem Exhibit
22. Water Harvesting Storage
23. Earth Berms
24. 60kW Photovoltaic Array

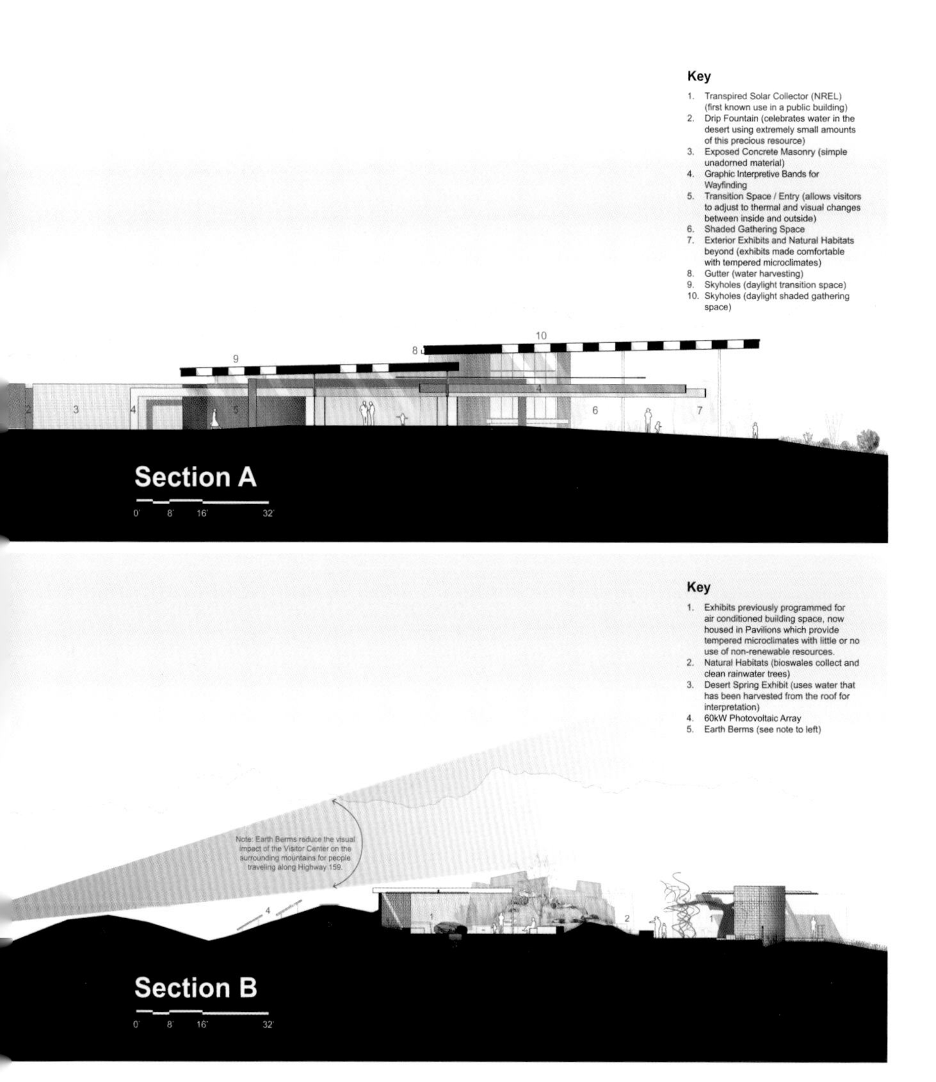
Key
1. Transpired Solar Collector (NREL) (first known use in a public building)
2. Drip Fountain (celebrates water in the desert using extremely small amounts of this precious resource)
3. Exposed Concrete Masonry (simple unadorned material)
4. Graphic Interpretive Bands for Wayfinding
5. Transition Space / Entry (allows visitors to adjust to thermal and visual changes between inside and outside)
6. Shaded Gathering Space
7. Exterior Exhibits and Natural Habitats beyond (exhibits made comfortable with tempered microclimates)
8. Gutter (water harvesting)
9. Skyholes (daylight transition space)
10. Skyholes (daylight shaded gathering space)
Section A
0' 8' 16' 32'
Key
1. Exhibits previously programmed for air conditioned building space, now housed in Pavilions which provide tempered microclimates with little or no use of non-renewable resources.
2. Natural Habitats (bioswales collect and clean rainwater trees)
3. Desert Spring Exhibit (uses water that has been harvested from the roof for interpretation)
4. 60kW Photovoltaic Array
5. Earth Berms (see note to left)
Note: Earth Berms reduce the visual impact of the Visitor Center on the surrounding mountains for people traveling along Highway 159.
Section B
0' 8' 16' 32'

DESERT AIR

Inspiria Science Centre

Inspiria 科学中心

Inspiria Science Centre is designed as one of the most advanced science centres in Northern Europe and is part of a long-term plan to make knowledge the most important asset of the Østfold Region in Norway. This ambitious plan is reflected in the architectural aspiration, as the trifold form is designed as a communication platform merging the environment, energy and health. The interior program includes 70 interactive exhibitions, workshops and the most advanced planetarium in Northern Europe and Inspiria. It is expected to attract more than 100,000 visitors a year, mainly school trips, families and tourists. Long-term plans to include the community would be addressed by annually offering over 40,000 students of the area a free trip to visit the venue.

Inspiria Science Centre combines communication and architecture into an inspiring and eventful whole, in order to provide new spaces for learning and support the idea of sustainability as a window of opportunity to increase the quality of life between humans and the environment. Inspiria Science Centre is thus designed as a passive house in close contact with both nature and the users, as glass enclosed wings extend from the focal circular atrium creating a dynamic heart to the building. By merging the architecture and the science centre's focus on the environment, energy and health, it is designed as a vibrant communication platform with a clear narrative. The narrative permeates the building design, as the trifold form symbolises nature's cyclical repetitions and spiral forms, which blend with the technology cycle expressed in the universal power of the circular basic form. The goal has thus been to create a striking building, which in itself constitutes an identity-laden branding of Inspiria Science Centre by uniting the activities of the science centre into a single concept.

Furthermore, Inspiria Science Centre is a unique example of how the public sector and the business community can come together and raise funds to enhance young people's interest in science. The architecture has thus been a significant icon in the branding and fundraising process of the €28.5 million science centre. For example, the science centre has been granted €5 million in subvention from the Norwegian Government and €7 million in subvention from the business community. Besides, the science centre has received subsidies from the government-owned corporation Enova that promotes environmentally friendly redistribution of energy consumption in the Norwegian construction sector.

Inspiria科学中心是北欧最先进的科学中心之一，也是挪威一项长期规划的一部分，旨在将科技逐步融入挪威东福尔郡地区。这项远大的计划在建筑的形式中得到了体现：三片翼状结构被设计为通信平台，与环境、能源和健康目标紧密关联。该中心内设70个互动展厅、工作室以及北欧最先进的天文馆，预计每年吸引10万人以上的游客，这里主要是进行学术参观、家庭参观和游客参观等。为提高社会知名度，该中心计划每年为当地超过4万名学生提供免费参观的机会。

Inspiria科学中心将交流和建筑融合成一个积极向上的、多元化的整体，提供新的学习空间，通过深化可持续发展的理念来提升人类与环境之间的和谐程度。因此，这个科学中心被设计成被动结构，与使用者及自然密切联系。一个玻璃包裹的翼状结构从中庭向外延伸，形成建筑充满活力的心脏。Inspiria科学中心集中融合环境、能源和健康目标，成为了一个具有叙述性风格、充满活力的通信平台。叙述性风格贯穿于建筑的设计之中，三片翼状结构象征着自然界的循环往复和螺旋形态，同时与旋回状的动力外观展示出的科技循环相融合。因此，该项目以建设一栋醒目的建筑为目标，通过将科技中心的活动进行一体化集成，为其带来醒目的个性特征。

此外，Inspiria科学中心的设计也是一项独特的方案，它向人们展示了公共部门和企业如何联合，如何筹集资金，以提升青少年对科学的兴趣。因此，该科技中心在其个性塑造和233.6万元的筹款过程中成为了一个标志性建筑。其中，该中心从挪威政府获得40.9万元的援助，从企业界获得57.4万元的的援助。此外，科学中心从国有机构Enova（挪威水资源和能源理事会）获得了资金补助，而Enova一直倡导挪威的建筑业进行环保型能源再分配。

Location / 地点:
Graalum, Norway
Date of Completion / 竣工时间:
2011
Area / 占地面积:
6,500 m^2
Architecture / 建筑设计:
AART Architects
Interior Design / 室内设计:
AART Architects
Landscape / 景观设计:
AART Architects
Photography / 摄影:
Adam Mørk
Client / 客户:
Borg Næring og Eiendom A/S

Ceiling: Metal;
Facades: Sea Water Proof Aluminium;
Glass facades: SAPA Glass Aluminium System;
Floors: Polished Concrete Floors.

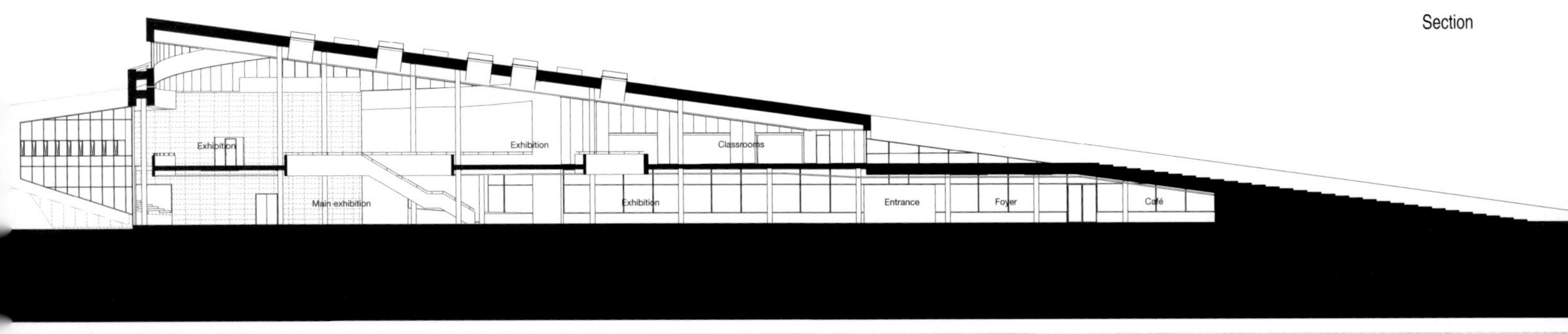
Section
Exhibition
Exhibition
Classrooms
Main exhibition
Exhibition
Entrance
Foyer
Café

inspiria

Elevation

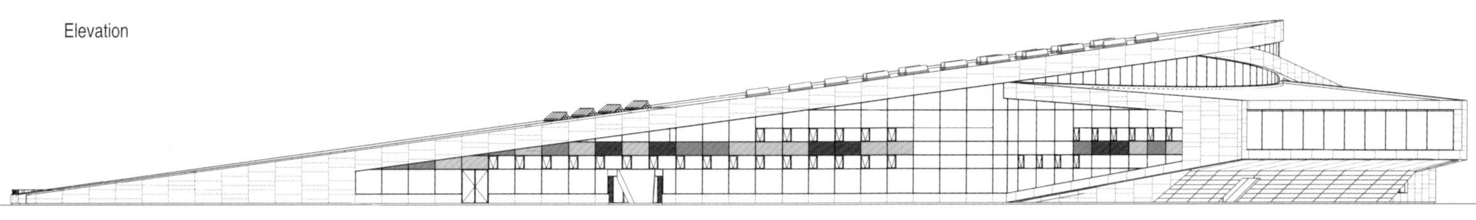

inspiria

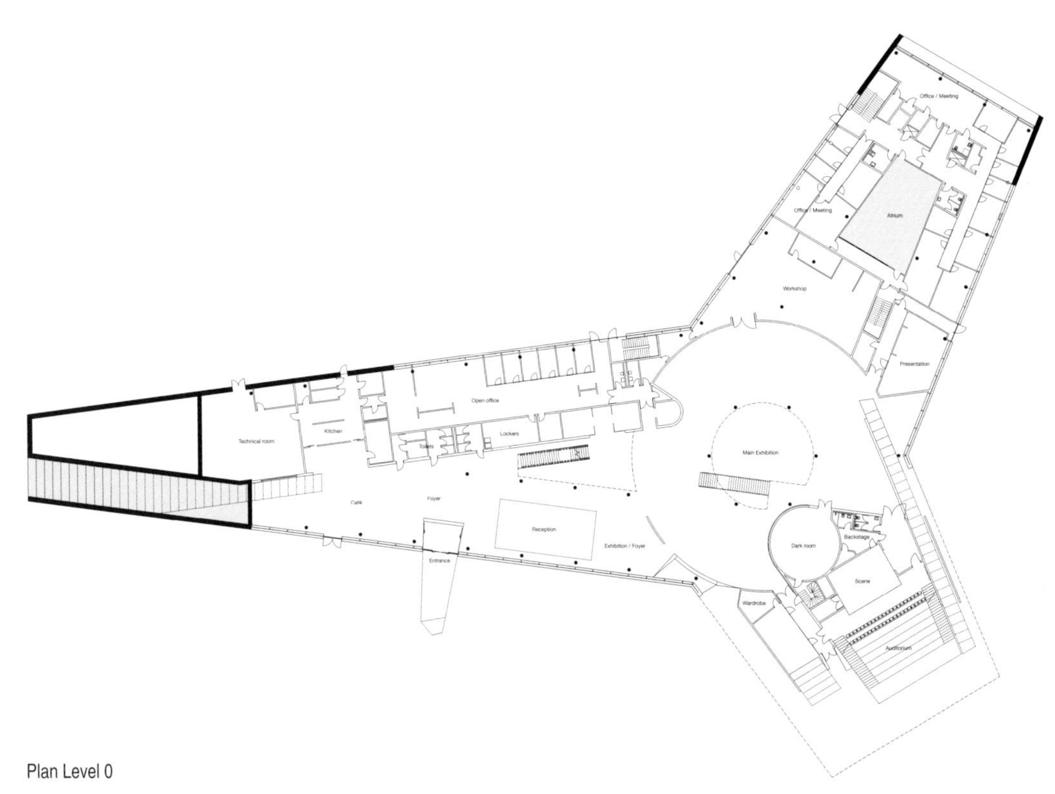
Office / Meeting
Office / Meeting
Atrium
Workshop
Presentation
Open office
Technical room
Kitchen
Lockers
Toilets
Main Exhibition
Cafe
Foyer
Reception
Entrance
Exhibition / Foyer
Dark room
Backstage
Scene
Wardrobe
Auditorium

Plan Level 0

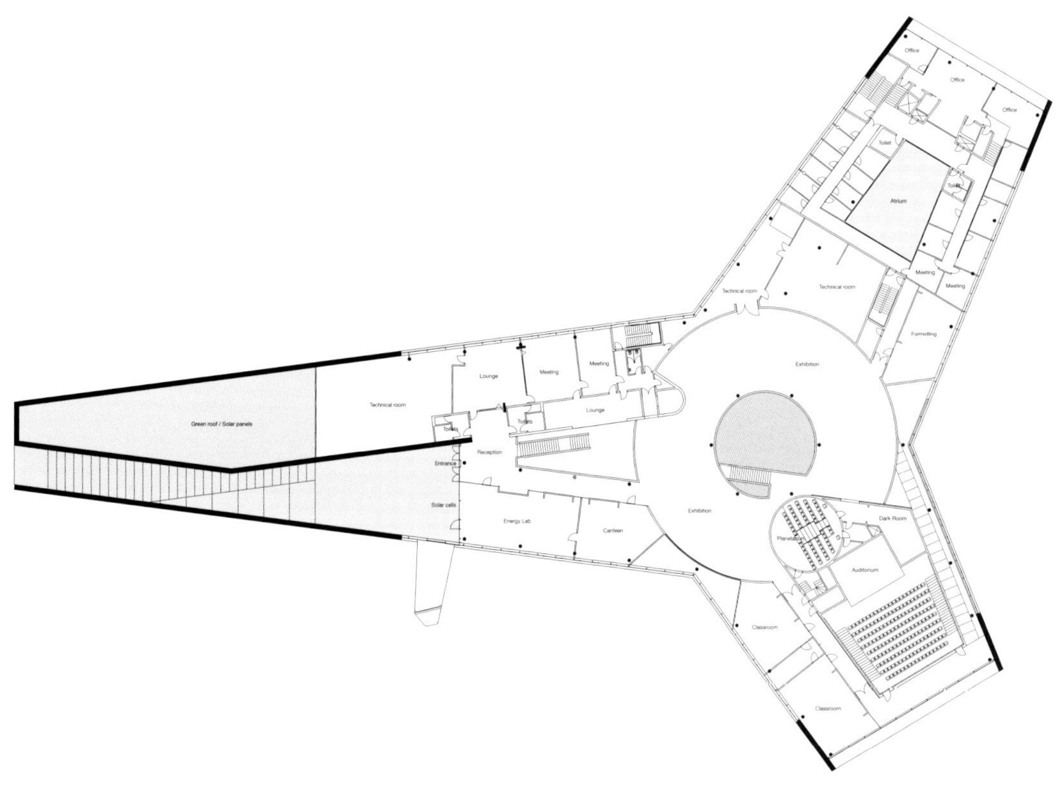
Office
Office
Office
Toilet
Toilet
Atrium
Meeting
Meeting
Technical room
Technical room
Formidling
Exhibition
Lounge
Meeting
Meeting
Technical room
Green roof / Solar panels
Toilets
Toilets
Lounge
Reception
Entrance
Solar cells
Energy Lab
Canteen
Exhibition
Dark Room
Auditorium
Classroom
Classroom

Plan Level 1

Museu

博物馆

NO.3 104-123

Carl-Henning Pedersen & Else Alfelt Museum

卡尔·亨宁·佩德森与埃尔斯·阿尔费尔德博物馆

Art museum exhibits the large collection of paintings by Carl-Henning Pedersen and his wife Else Alfelt. The small, but distinctive museum is situated within Birk Centerpark, a park area outside Herning containing various cultural, business and educational institutions. Birk Centerpark is a contemporary, international-class example of a functionally mixed urban area, in which the landscape, architecture and artistic elements are combined in exemplary fashion. The origins of Birk Centerpark lie in the establishment of C. Th. Sørensen's sculpture park and the construction of the Angli shirt factory for the businessman and art collector A. Damgaard in the 1960s. C. F. Møller Architects converted this spiral-shaped factory into a visual arts museum in 1976. Since then, several more landscaped areas have been added, and more developments have been built, each with its architectural basis in the white, stringent and geometric form language of the Angli factory.

The 1976 Carl-Henning Pedersen & Else Alfelt Museum is built up around a large collection of works by the Danish art couple Carl-Henning Pedersen and Else Alfelt, who were both active in the Cobra movement. The museum is characterised by the unity of art and architecture in the sculptural form. The facade is clad with ceramic tiles decorated with Carl-Henning Pedersen's colourful mythical beasts. The shape is entirely geometric, with a circular main building echoing the shape of the converted factory situated opposite. The geometry of the entire complex also interacts with the other buildings in the cultural and educational park, most of which were designed by C. F. Møller Architects.

In 1993 the museum was extended to provide more space for the arts collections. The geometrical vocabulary is continued in the extension, with a free standing structure in the surrounding park. A square extension underground, lit by a glass wall in the above-ground pyramidal skylight prism structure, and connected to the existing underground arts collections. The 1993-extension was designed in close cooperation between Carl-Henning Pedersen and C. F. Møller Architects.

该艺术博物馆展览着卡尔·亨宁·佩德森和他妻子埃尔斯·阿尔费尔德的作品集。这是个小型但极具特色浓厚的博物馆，它坐落在伯克中央公园内，这是海宁郊外一个包含各种文化、商业和教育机构的公园地带。伯克中央公园是一个混合功能的城市区域，作为现代化与国际社会的榜样，堪称融合景观、建筑和艺术元素的典范。伯克中央公园起源于C. Th. 索托森的雕塑公园，以及20世纪60年代为商人兼艺术收藏家A. 达姆高建造的安吉利衬衣厂。C. F. 摩勒建筑事务所于1976年将这座螺旋外观的工厂转换成了一座具有视觉艺术的博物馆。从那时起，这里又建立了更多的景观区和发展项目，并纷纷以安吉利衬衣厂的白色色彩、精细做工和几何状风格为设计灵感。

卡尔·亨宁·佩德森与埃尔斯·阿尔费尔德博物馆建于1976年，陈列着艺术家卡尔·亨宁·佩德森和埃尔斯·阿尔费尔德夫妇大量的作品，二人都曾在眼镜蛇艺术运动中扮演重要角色。该博物馆通过雕塑式设计取得了艺术和建筑的统一。建筑的外立面覆盖着瓷砖，上面装饰了卡尔·亨宁·佩德森丰富多彩的玄幻野兽图案。建筑带有很强的几何风格，楼的主体结构呈圆形，同对面改造工厂的建筑形态相呼应。整个综合设施的几何形态还与公园内其他建筑产生互动，这些建筑大部分都是出自C. F. 摩勒建筑事务所的设计。

1993年，博物馆进行了扩建，从而可以容纳更多的艺术藏品。扩展部分延续了原有的几何形态，以独立的结构屹立在主楼附近。方形的扩展结构位于地下，和现有的地下艺术品展厅相连，而地上部分采用玻璃幕墙并设有天窗，可以为地下扩展部分提供采光，其棱柱结构呈现出金字塔形的特色外观。1993年的扩展建筑由卡尔·亨宁·佩德森和C. F. 摩勒建筑事务所密切合作完成。

Location / 地点:
Herning, Denmark
Date of Completion / 竣工时间:
2007
Area / 占地面积:
2,950 m²
Architecture / 建筑设计:
C. Th. Sørensen; Sven-Ingvar Andersson; C. F. Møller Architects
Landscape / 景观设计:
C. Th. Sørensen; Sven-Ingvar Andersson; C. F. Møller Architects
Photography / 摄影:
C. F. Møller Architects
Client / 客户:
Herning Municipality and The Carl-Henning Pedersen & Else Alfelt Museum

Floors: Exposed Concrete;
Furniture: Cast Concrete Benches/Tables;
Illumination: Downlights, Wallwashers for Artwork Illumination, Exterior Façade Wallwashers; Plants: Grass Roof;
Others: Cobblestones, Water Trench, Concrete Ramp.

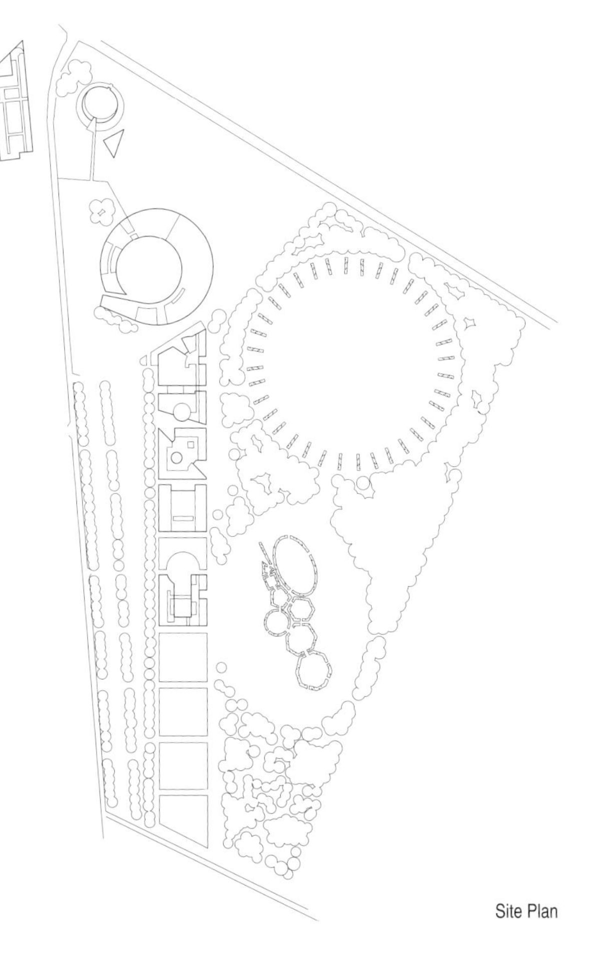

Site Plan

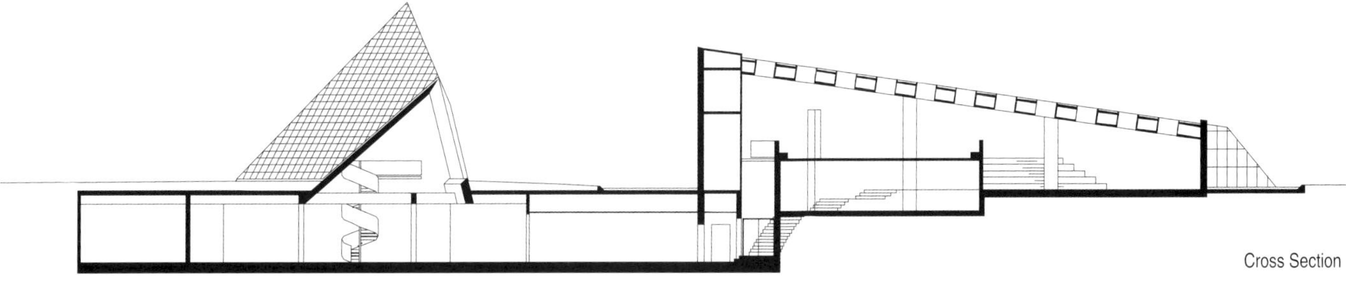

Cross Section

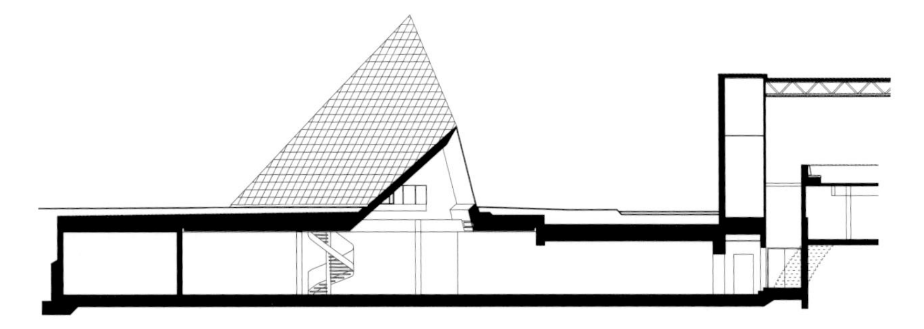

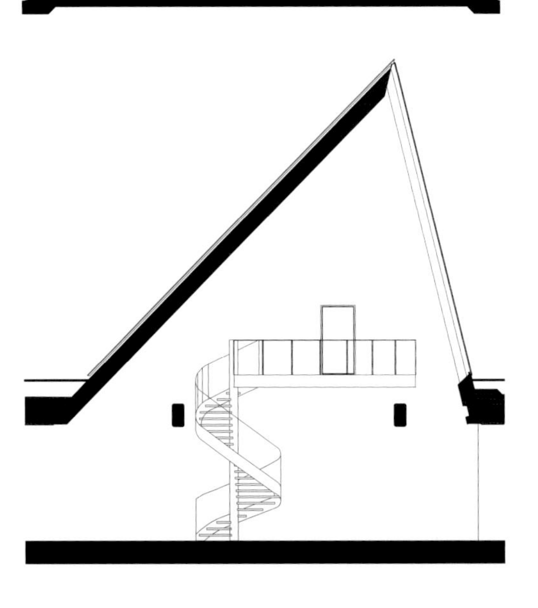

Sections of Pyramid

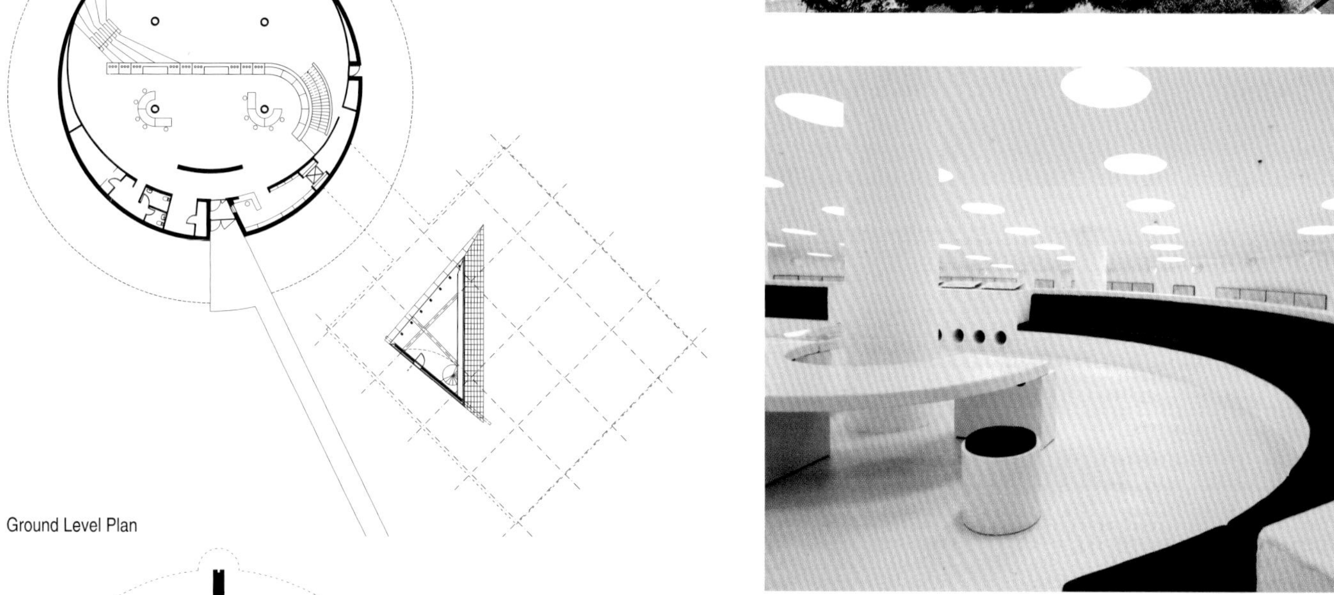

Ground Level Plan

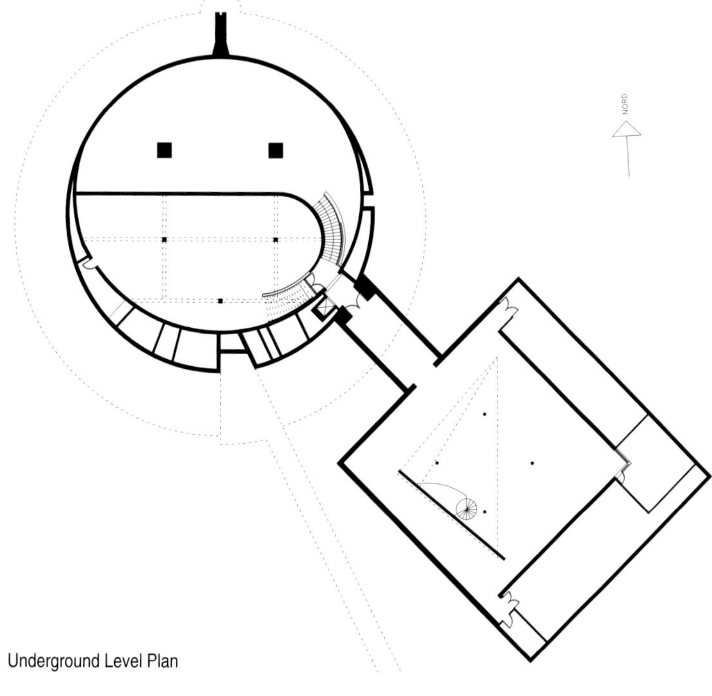

Underground Level Plan

National Maritime Museum

国家海事博物馆

Location / 地点:
London, UK
Date of Completion / 竣工时间:
2011
Area / 占地面积:
7,300 m^2
Architecture / 建筑设计:
C. F. Møller Architects
Landscape / 景观设计:
Churchman Landscape
Photography / 摄影:
C. F. Møller Architects
Client / 客户:
National Maritime Museum, London

The National Maritime Museum in London contains the world's largest maritime collection, housed in the historic buildings part of the Maritime Greenwich World Heritage Site. The project encompasses the creation of a new wing for the National Maritime Museum, named the Sammy Ofer Wing after the international shipping magnate and philanthropist Sammy Ofer, who has funded most of the project. The aim has been to open up and reveal for everyone the fascinating stories of people and the sea.

The goal of the expansion has been to open up the museum and allow the display of more collections than ever before. The museum's collections range from e.g. a toy pig that survived the sinking of RMS Titanic to Lord Nelson's last letter to his daughter.

The main idea of the extension has been to ensure minimal interventions in this sensitive historic site and yet give the museum a new, distinctive main entrance and the necessary additional exhibition space, as well as a new café, restaurant, library and archives that meet the particular demands for storage of historical documents.

The design solution creates a new main entrance emerging from the terrain. Most parts of the new building, however, are located underground. The roof of the new wing is a green, public landscaped terrace overlooking the Park, accessed at all levels by gentle ramps, even more so causing the building to blend with the park landscape. The extension has a contemporary aesthetic, but is inspired by the Baroque buildings' rhythmic sequence of windows, and the profile of the new extension has been kept low to allow the Grade I listed Victorian facade of the existing south west wing of the museum to be appreciated as a backdrop to the striking new building.

位于伦敦的国家海事博物馆拥有世界上数量最多的航海收藏品，是格林威治世界海事遗产的一部分。该项目包括在国家海事博物馆以及在其一侧新建的相连的建筑。国际航运巨头和慈善家萨米・奥弗是项目的主要资助者，因此，新建筑被命名为萨米・奥弗之翼。项目的宗旨是为众人揭示人与海背后引人入胜的故事。

扩建建筑的目标是让博物馆能够展出比以往最多的收藏品。该博物馆的藏品范围甚广，有从泰坦尼克号幸存下来的小玩具猪，也有纳尔逊勋爵给女儿的最后一封信。

其主要设计思想是确保对这个敏感的历史遗迹形成最小化的介入，同时为博物馆带来全新而又与众不同的主入口和必要的更大的展览空间，以及新的咖啡馆、餐厅、图书馆，还有能满足存储历史文件特殊需求的档案馆。

设计方案创造了一个随地形而浮现的全新主入口，然而，新建筑的大部分结构都位于地下。这侧新建筑的屋顶为一个绿色、服务公众的景观露台，人们在此可以俯瞰公园的全貌。从各个楼层均可通过缓坡到达露台，这种设计更使建筑融入公园景观之中。该扩展项目虽然具有当代美学特征，却是以巴洛克式建筑富有韵律的窗口布置为灵感而设计的。新建筑的外形被控制在一定高度，这样，博物馆南翼原有的一级保护建筑——维多利亚外墙可以被大众所欣赏，并成为新建筑醒目的背景。

Floors: Exposed Concrete;
Ceilings: Concrete, Plasterboard, Stainless Steel 3D Shaped Panels;
Illumination: Fluorescent Strip Lights, Downlights, Concealed Perimeter Lighting of Steel Ceilings;
Plants: Clipped Hornbeam, Grass;
Others: Aquifer Thermal Storage System.

HIGH ARCTIC
HIGH ARCTIC

Sammy Ofer Wing
NATIONAL MARITIME MUSEUM

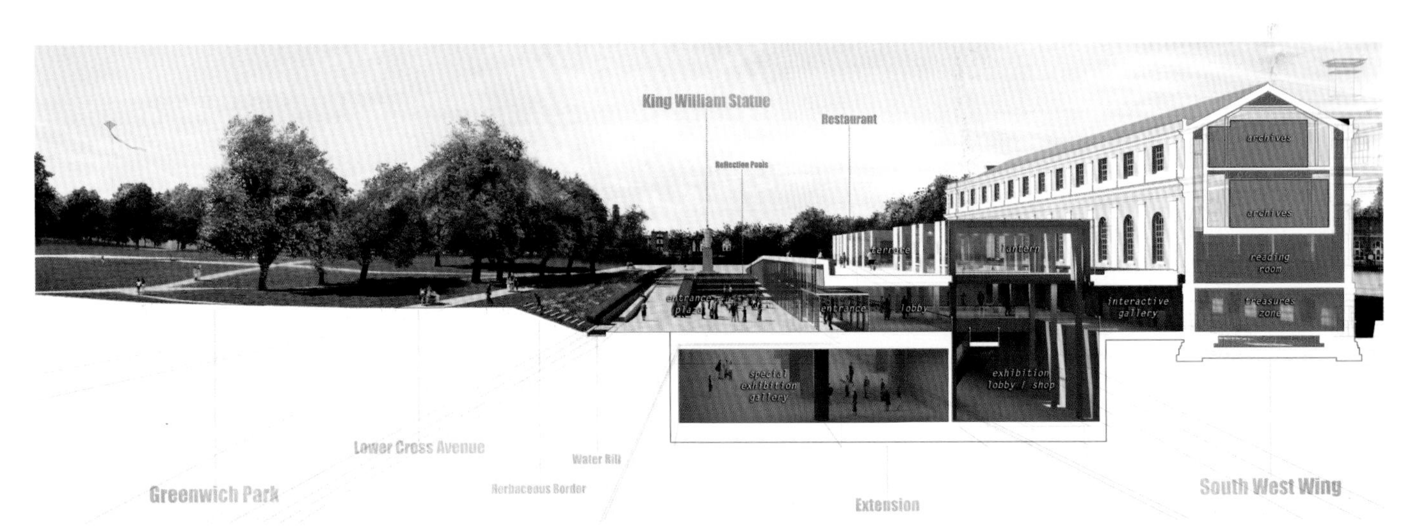
King William Statue
Restaurant
Reflection Pools
archives
archives
terrace
lantern
reading room
entrance plaza
entrance
lobby
interactive gallery
treasures zone
special exhibition gallery
exhibition lobby / shop
Lower Cross Avenue
Water Rill
Greenwich Park
Herbaceous Border
Extension
South West Wing

Sammy Ofer Wing
NATIONAL MARITIME MUS

Sammy Ofer Wing
NATIONAL MARITIME MUSEUM

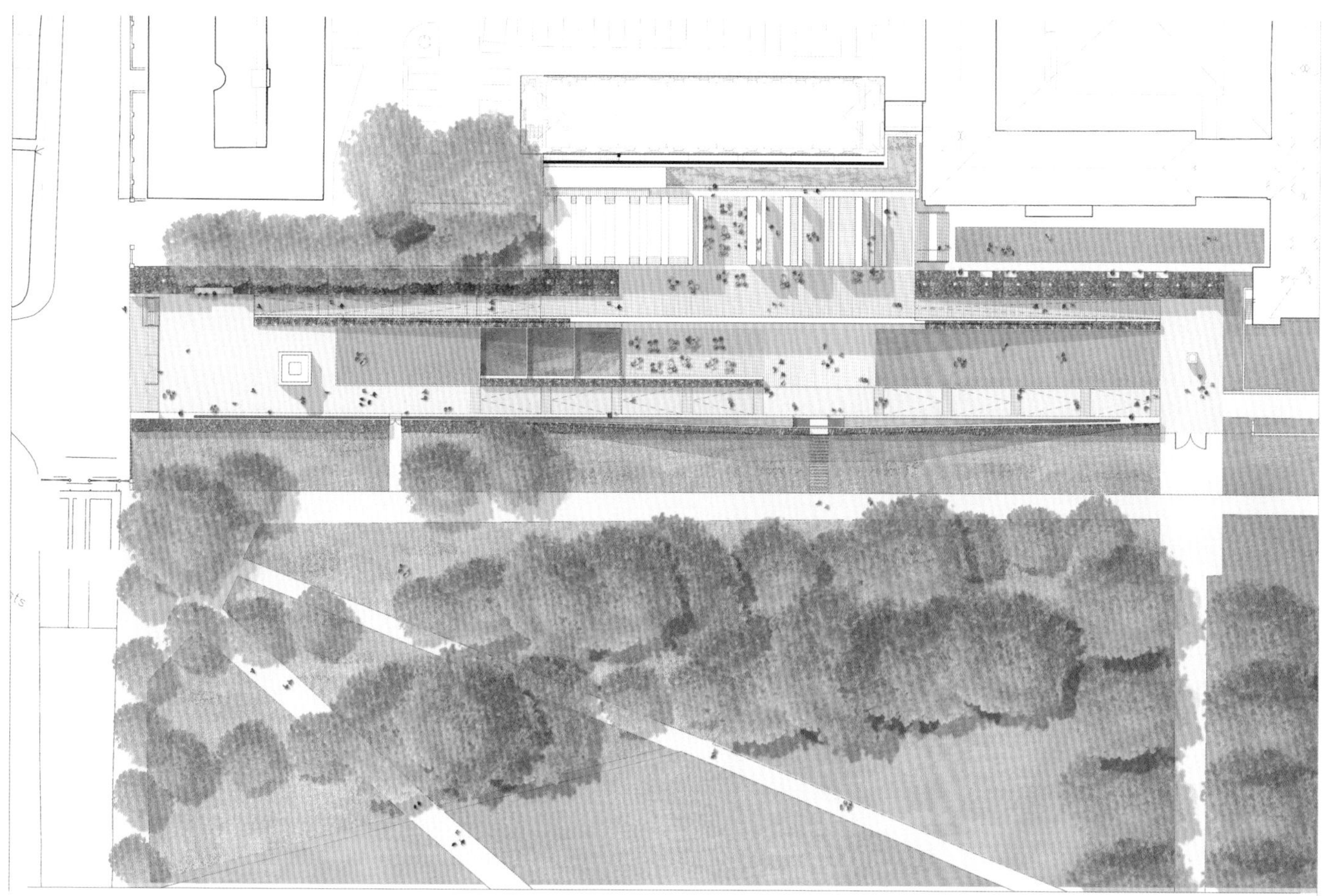

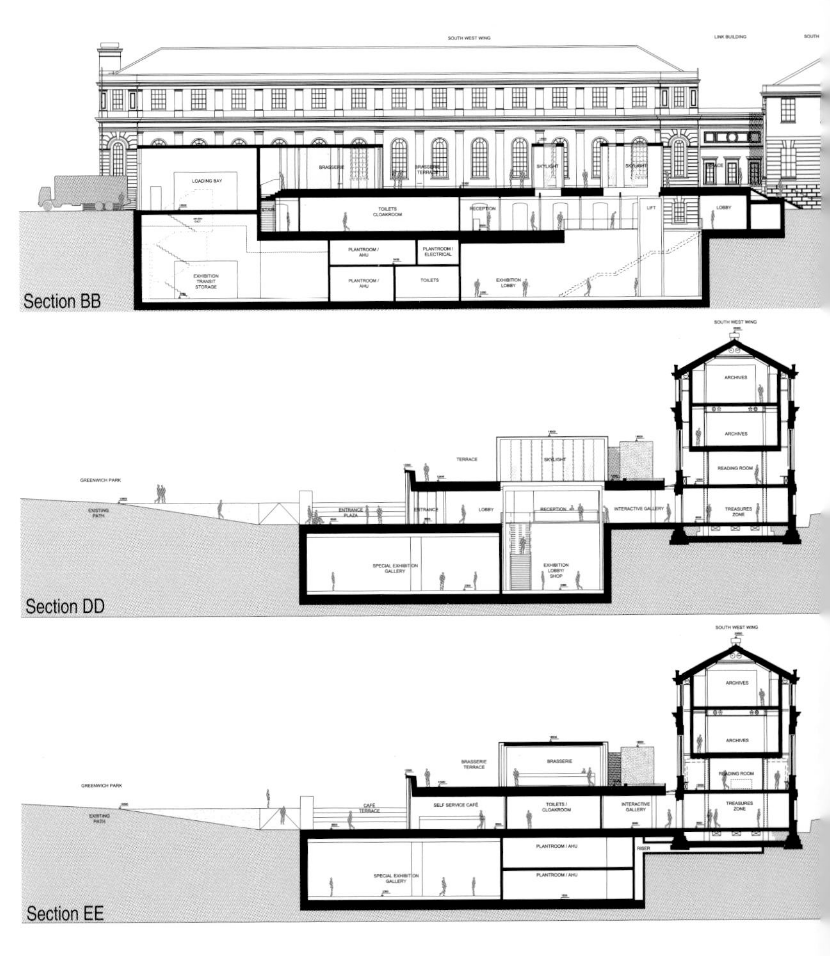
Section BB
Section DD
Section EE

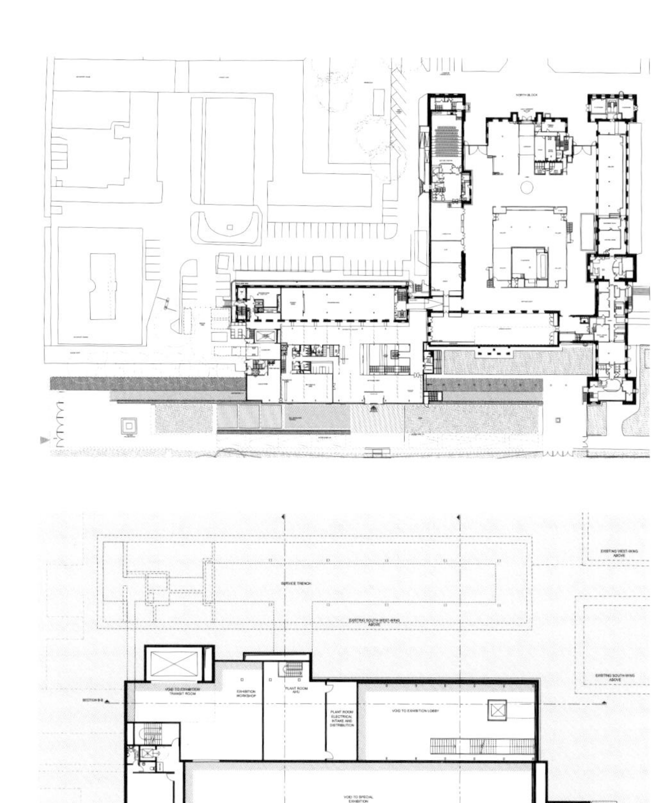

Plan Level 1

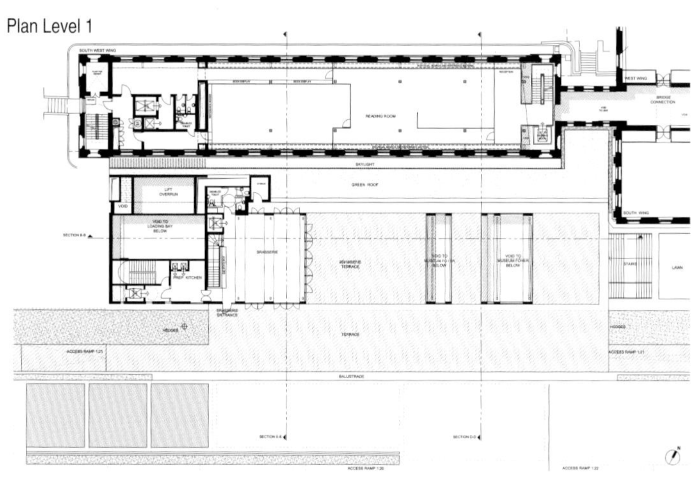

Plan Level 2

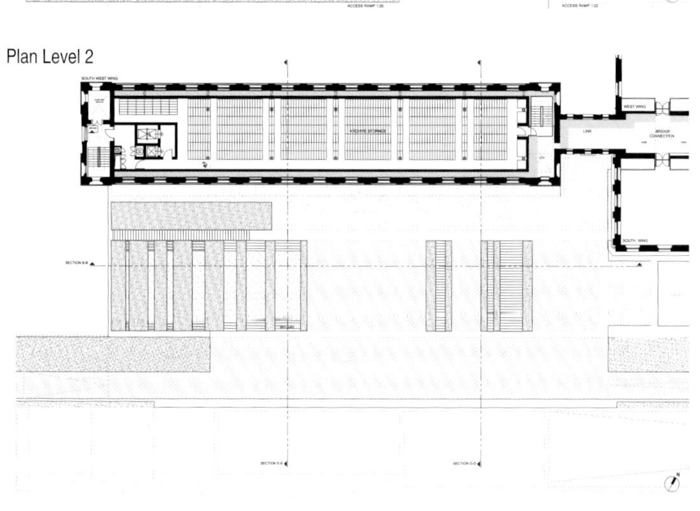

Open-air Exhibition Grounds of the Estonian Road Museum

爱沙尼亚道路博物馆的露天展台

Location / 地点:
Varbuse, Estonia
Date of Completion / 竣工时间:
2010
Area / 占地面积:
13,000 m^2
Architecture / 建筑设计:
Salto AB
Interior Design / 室内设计:
Salto AB
Landscape / 景观设计:
Salto AB
Photography / 摄影:
Nathan Willock, Karli Luik, Pelle-Sten Viiburg
Client / 客户:
Estonian Road Museum

The Exhibition Grounds of the Estonian Road Museum are remarkable for its location, surroundings, typology, scale and architectural solution. It's an object where the informative level – i.e. the long history of roads –, as well as leisure activities and landscape design have all been given equal treatment.

The exhibition grounds are located on an open field, built across the road from the historical Varbuse postal station. Dating back to 1830s, the old complex is now used as the museum's main venue.

The concept of the additional outdoor exhibition area is based on a road – while strolling through, your route will be surrounded by different landscapes. The chosen solution forms a long 8-shaped path, where functions with different characters and scales are placed in succession like a comic strip. The exhibition begins with an overview of traffic signs through history, continuing with segments of different types of historical roads, lined with objects related to travelling by road, as well as all kinds of machines used for maintaining or repairing them. A late-19th century steel bridge has been removed from its original location and given a new use linking the two parts of the exhibition above the entrance area.

Nearly all space necessary for the museum is scooped into the hilly South-Estonian landscape, leaving the rest of the environment as untouched as possible: natural and artificial landscapes are clearly separated. A hollow ranging from 10 cm to 4 m deep forms more than 13000 m^2 of open-air exhibition space which is barely visible from the remote surrounding areas. For the most part, the structure is built of reinforced concrete, with wood-panelled 'nests' (ticket and souvenir booth, picnic area) and concrete walls with printed graphic images depicting roadside landscapes and scenes softening the object that is in itself a piece of infrastructure as well as architecture.

Phase 2 of the museum extension foresees the reconstruction of an existing exposition hall. Currently an inapt eyesore, the hall is to become fully integrated with the rest of the complex. In addition to exhibition space, it will also include a blackbox theatre and an office.

爱沙尼亚道路博物馆的露天展台因其位置、环境、类型、规模和建筑设计等诸多因素而卓而不凡。本案的设计目标是将教育功能——如讲解道路的悠久历史——与娱乐功能和景观设计予以同等重视。

展示区坐落在一片开阔的场地内，距离历史悠久的Varbuse邮局不远。这处古老的历史遗迹可追溯到19世纪30年代，如今被用来作为博物馆的主要场地。

闲步其中，所到之处被不同风景环绕，这便是加设户外展示区的创意初衷所在。最终选定的方案包括具有八种不同功能且各具特色的小路，如同一幅风景连环画一般分布在场地内。展示以贯穿各个历史展区的总体交通导视图开始，依次连接不同的参观路线。路线周围设有相关路牌和多种道路维护设备。一座19世纪末期的钢构桥被从博物馆的原址移至此项目的入口，将两部分展区连接起来。

博物馆露天展台的场地几乎全部是在南爱沙尼亚丘陵地上进行挖掘而成的，设计时尽量不破坏周边环境，同时使自然和人工景观清晰地分开。凹陷区深度为4~10cm，所形成的1.3hm^{2}户外展区从远处看去隐约可见。展区结构主要由钢筋混凝土和木制嵌板"温床"（售票口、纪念品货摊和野餐区）构成。展示有各种图像的混凝土墙和路边的风景以及场地内的景观有机地结合起来，将项目本身的建筑功能和基础设施融于其中。

改造原有博物馆大厅是本次扩建工程的重点。现在看来，该大厅将与其他部分融为一体。除了展示部分，它也包括一个剧场和一个办公室。

Materials: Concrete, Wood, Graphic Concrete.

GUESS

Concept Plan

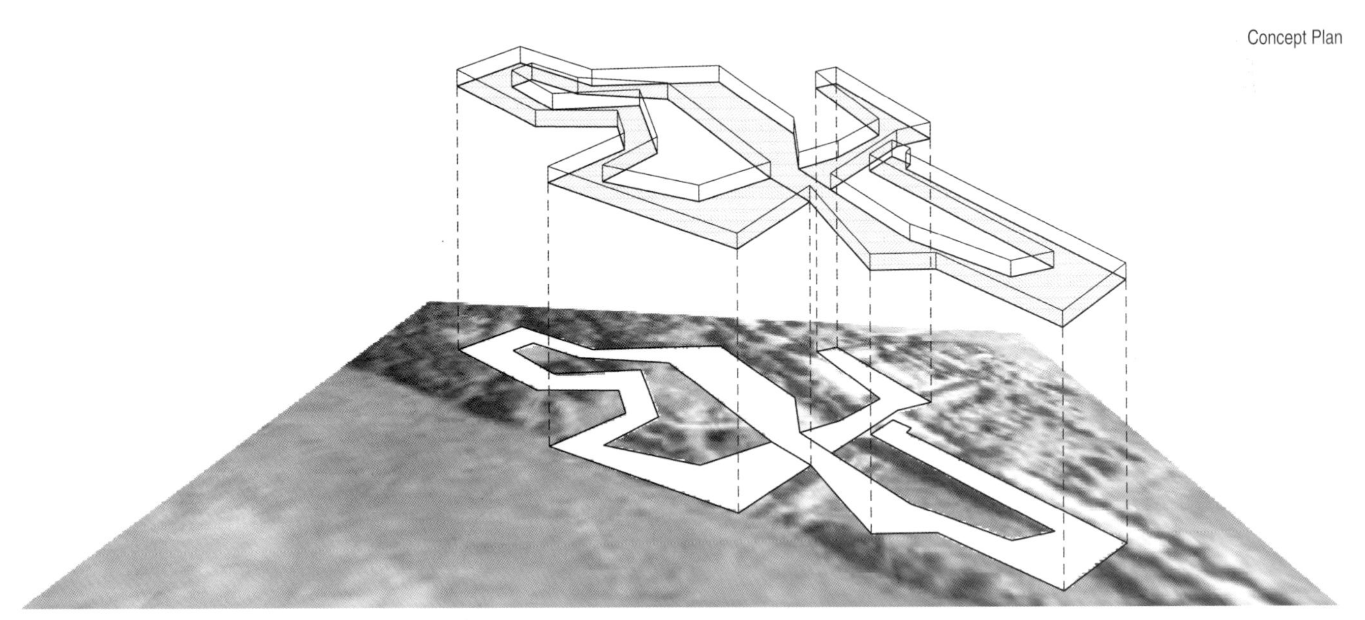

Section Traffic City

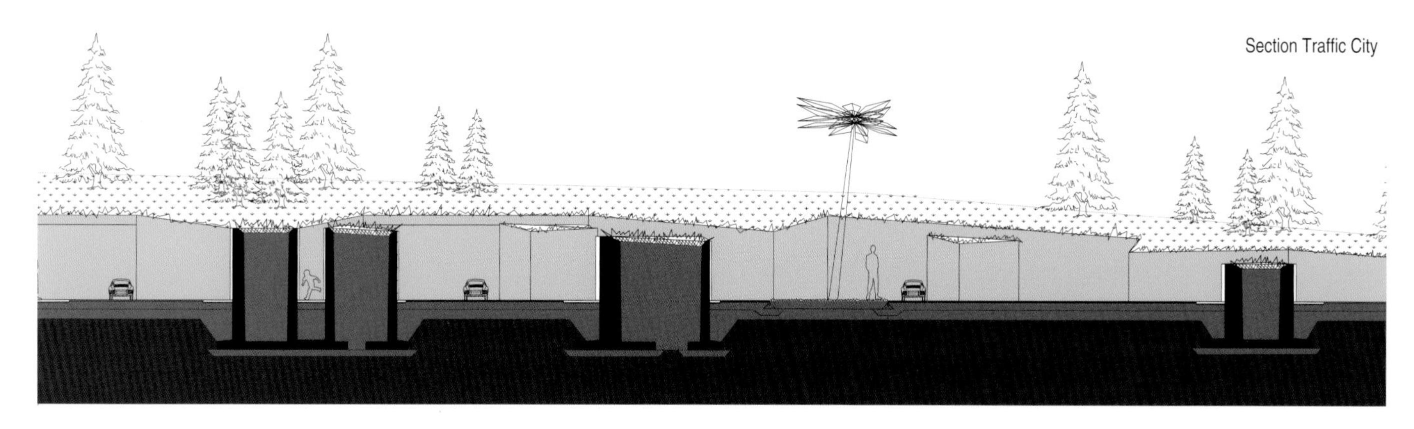

Section Picnic Area

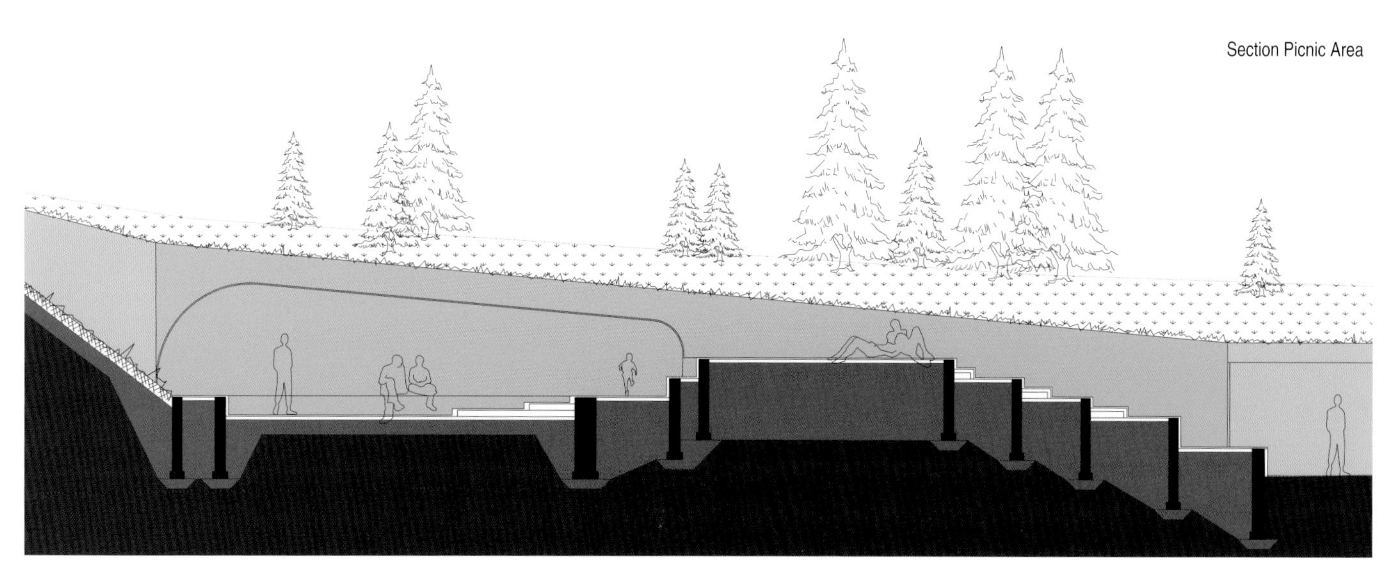

Section Hollow Road

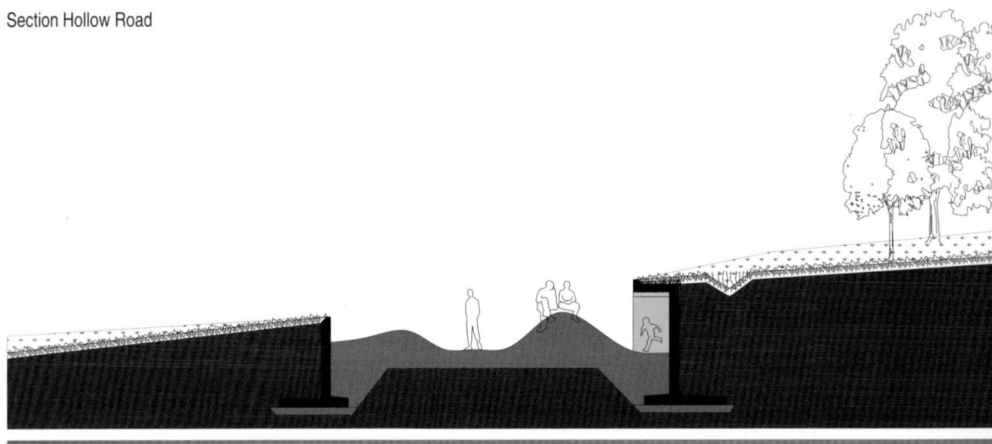

Piirsalu 5.5 km. 11 kl.
Reval 198 km
Narva 192 km
Weissenstein

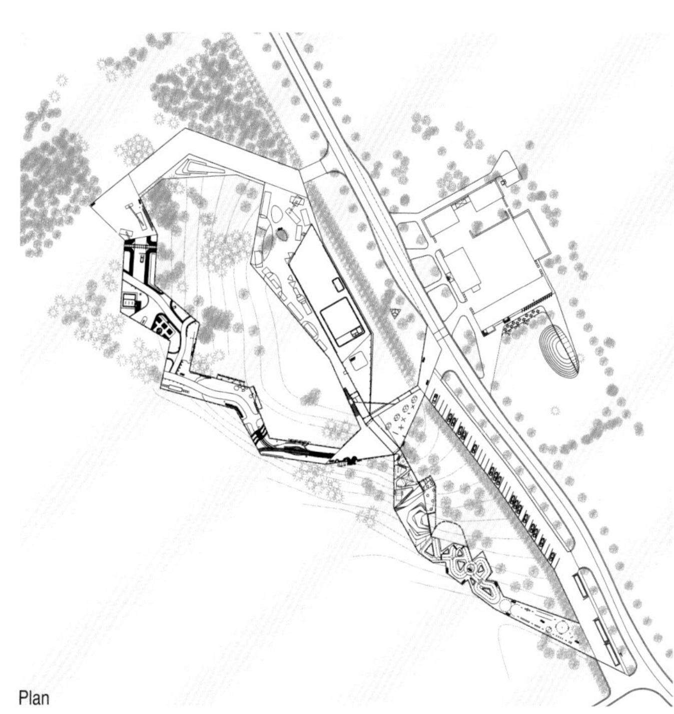

Plan

Reval 198 km
Narva 192 km
PIHKVA
PÕLVA

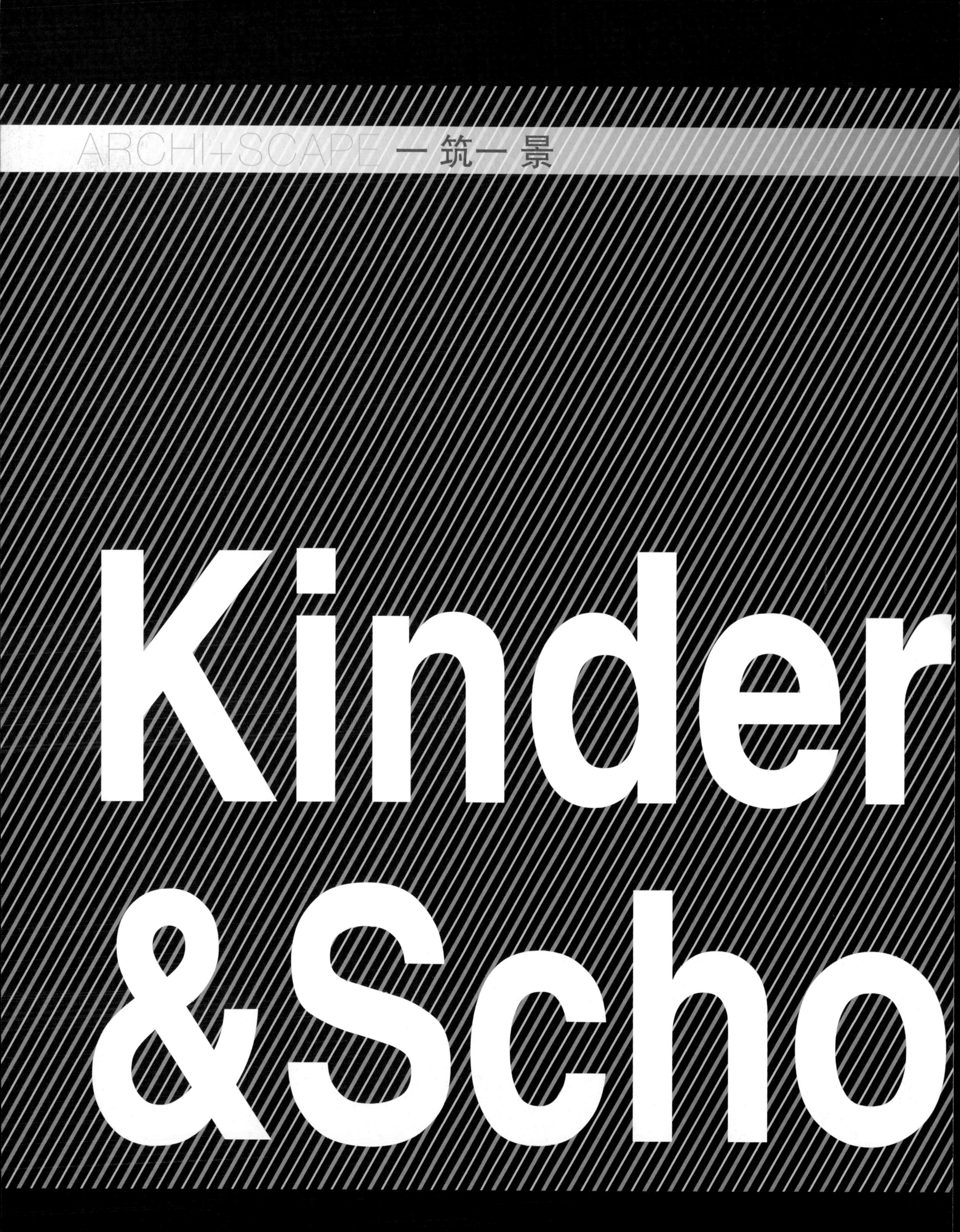
ARCHI+SCAPE 一筑一景
Kinder
&Scho

garten

ol

幼儿园与学校

NO.4 124-203

High School Levi Strauss in France

法国利维·斯特劳斯中学

The college Levi Strauss is settled in the heart of an urban growth district, between its ancient housing, warehouses and the port district of Lille, North of France. The main building is settled on the urban boulevard, the main hall, highly transparent, is opened on the front square, this gives an institutional feature to the high school playing a major role within the district. The main entrance is through a porch at the intersection of Boulevard de la Lorraine and Rue Lestiboudois.

Very sunny and sheltered from the winds, the playground's mainly mineral and generously planted. Opened on the playground, the entrance of the dining hall and club. Dedicated to the pupils facilities, those spaces have been thought like spaces in the bricks oriented towards the trees of the playground. On top of the covered playground situated on the southern side, the scientific classrooms offer a large view on the nearby urban environment. Connecting to these specialized classrooms, the library occupies a central position on the first floor with direct access to the school hall. The asymmetrical alignment of the variously sized square windows bring light into the classrooms and offer pupils large views of the city. On the southern part of the site outdoor sporting facilities and a gymnasium operate independently.

As in many regions of northern Europe, the brick is the only material used for the facades. The architects wanted rounded corners, so that the high school looks soft. There's no sharp angle. The bricks are rendered in 3 stratums corresponding to the 3 shifted levels of the building which create open spaces and identify the entrance of the pupils.

利维·斯特劳斯中学坐落在法国北部城市里尔的市区发展核心地带，附近是古老的住房、仓库及港口区。这所中学在当地“扮演”重要的角色，主楼位于城市大道上，正厅的通透感极强，它面对着前方的广场。进入主入口的通道是一条门廊，主入口位于洛林大道（Boulevard de la Lorraine）与莱蒂布杜瓦街（Rue Lestiboudois）的交叉口处。

操场光线充足，拥有良好的小环境，主要由石材铺就而成，周围栽种了丰富的植被。餐厅和俱乐部的入口正对着操场开放。科学教室位于被围合的操场南侧，从那里可以欣赏到广阔的周边城市景观。与这些专用教室相连的是图书馆，它位于一楼的中心位置，直接通往学校大厅。正方形的窗户大小不一，不对称地排列着，一方面为教室带来了自然光线，另一方面让学生们能一览城市的景色。在场地的南部，还设置室外运动设施和一座体育馆。

如同欧洲北部的许多地区一样，该建筑将砖作为外墙的唯一材料。这里的建筑没有尖锐的棱角，因为建筑师采用了圆滑的墙角设计，从而使这所中学的教学楼看起来更柔和。砖块抹有三层底灰，以此与建筑的三个楼层呼应，这些楼层创造了一个开放的空间，并凸显了学生入口的位置。

Location / 地点:
Lille, France
Date of Completion / 竣工时间:
2010
Area / 占地面积:
16,560 m²
Architecture / 建筑设计:
Tank Architectes
Interior Design / 室内设计:
Tank Architectes
Landscape / 景观设计:
Paysages
Outside Design / 室外设计:
Atelier Télescopique
Photography / 摄影:
Julien Lanoo
Client / 客户:
Conseil General du Nord

Buildings: Bricks;
Floor: Quartz Concrete;
Stairs: Welded Steel.

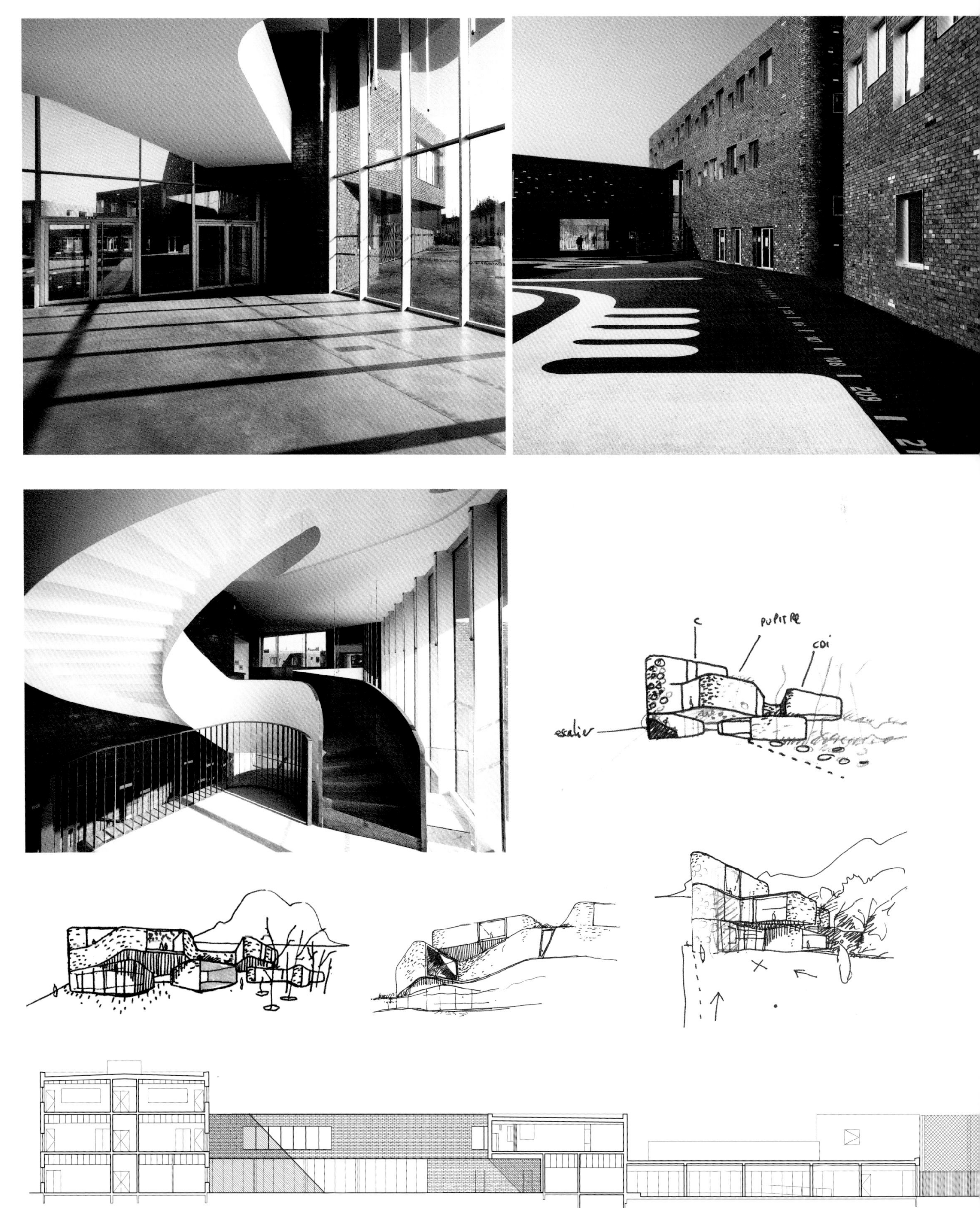
C
PUPITRE
CDI
escalier

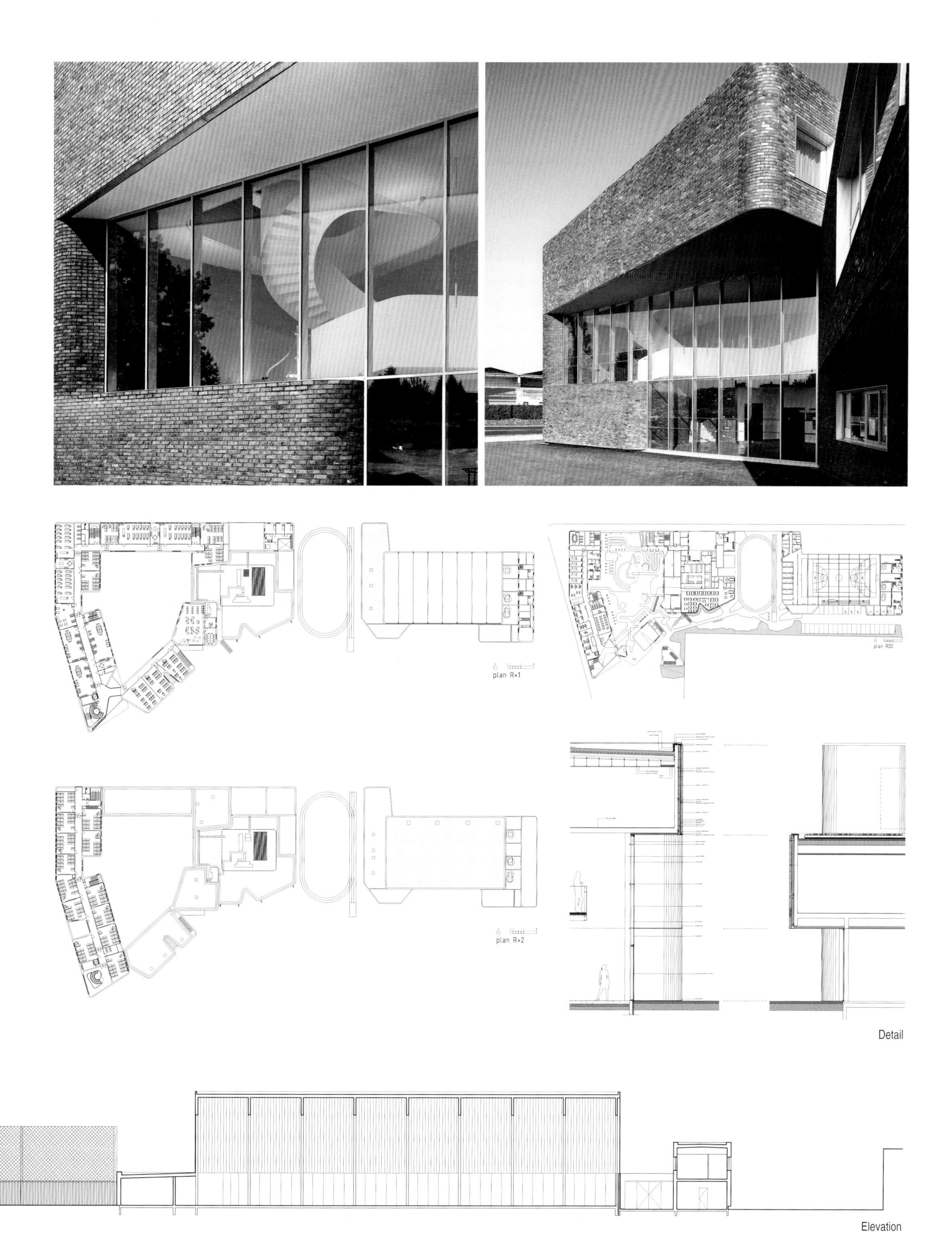

Detail

Elevation

Kindergarten Pajarito la Aurora

"小鸟的极光"幼儿园

The designers consider the kindergarten as a garden. People can grow flowers, vegetables and trees here. where you can give the soil and the necessary conditions for the plants growth and a garden where children can grow and be cultivated. The designers think of a garden at the scale of children, where they can play, learn, grow and run among their classrooms, courtyards and gardens.

They are faced to building a fragment on the western slope of the valley of Aburrá. For this they have a plot drawn with geometric rigor in the middle of a mountain. It seems, will host a partial plan. How to face an important natural landscape knowing that it will disappear quickly? They like working with the natural landscape and try to understand it in its complexity. When an architect has these interests, it ends up looking like a gardener. Therefore, they have considered this request, close to that would give a gardener.

They understand the kindergarten as a structure that can have multiple and diverse gardens inside. For this they study a mathematical structure that demonstrates these conditions: Voronoi diagrams. The decomposition of a plane into multiple convex polygons is given by the fact that each polygon has a center generator and all the points that are in it are closer to the center point than to any other polygon.

The inner garden (classroom) is the smallest unit and the previous landscape unit in the first garden. Its geometry is close to the petal of a flower, is a six-sided polygon in which opposing lines are parallel. This allows people to be repeated, rotated or multiplied many times and manner as required programmatically and geographically, in open or closed configurations.

There are many outdoor gardens, each identified with a topographic and motivated by a child's play: Garden valley and sand games, rubber mountain and small hill jumps, plain wooden and wheel-wheel games, mountain cave and hideouts, canyon maze and persecutions, tall grass mountain and walks, slide rocky and climbing, wood swamp and water games, flower garden and vegetable and crops.

设计师的设想是将这个幼儿园建设成一个花园。这里可以种植花卉、蔬菜和树木，既为植物的生长提供土壤环境，又为儿童提供一个娱乐的场所。设计师从儿童的角度出发，力求为孩子建立一个可以玩耍、学习、成长以及在教室、庭院和花园间跑闹嬉戏的公园。

项目的建设场地位于Aburrá山谷西部的坡地上。设计师在山地当中以几何法精确地划分出一块用地。设计师综合考虑项目所处位置的自然景观，从近似于园艺家的角度来考虑这个项目。

设计师将幼儿园构想成一个内含多个不同花园的结构。于是，他们研究了一项能够展现这种构架的数学模型：Voronoi图形——将平面分解成若干个凸多边形。这些多边形都遵循以下规则：每个多边形都有自己的中心点，而在对应的多边形内的任意一点到该中心点的距离小于任意一点到其他多边形中心点的距离。

内部教室空间是最小结构的单元。其几何形态近似于花瓣状，是对边平行的六边形。这种或开放或闭合的空间结构能够让人们根据不同的活动需求选择不同的场地。

外部空间包括许多花园，每个花园地形各异，这里有不同的游戏设施，包括花园山谷和沙坑、橡皮山和跳跃小丘、纯木轮盘花卉、山洞和藏匿空间、峡谷迷宫和曲径、草坡和走道、岩石坡和攀爬处、沼泽和戏水空间，以及花卉、蔬菜和谷物种植空间。

Location / 地点:
Medellin, Colombia
Date of Completion / 竣工时间:
2011
Area / 占地面积:
1,400 m^2
Architecture / 建筑设计:
Viviana Peña, Catalina Patiño, Eliana Beltrán (Ctrl G arquitectos) + Federico Mesa (Plan:b arquitectos)
Interior Design / 室内设计:
Viviana Peña, Catalina Patiño, Eliana Beltrán (Ctrl G arquitectos) + Federico Mesa (Plan:b arquitectos)
Landscape / 景观设计:
Michael Gilchrist
Photography / 摄影:
Ctrl G architects
Client / 客户:
Mayoralty of Medellin 2008-2011 operated by the Urban Development Corporation-EDU.

Walls: Concrete;
Classroom floors: Purple, Blue and Green Vinyl Flooring;
Outside Floors: Concrete Paving;
Roof: Artificial Grass;
Glasses: Colored Red, Yellow Glasses Red;
Columns: Green Painted Metal Column;
Flower Plants: Besitos, Margaritas, Hortensias, Orquídeas, Lirios;

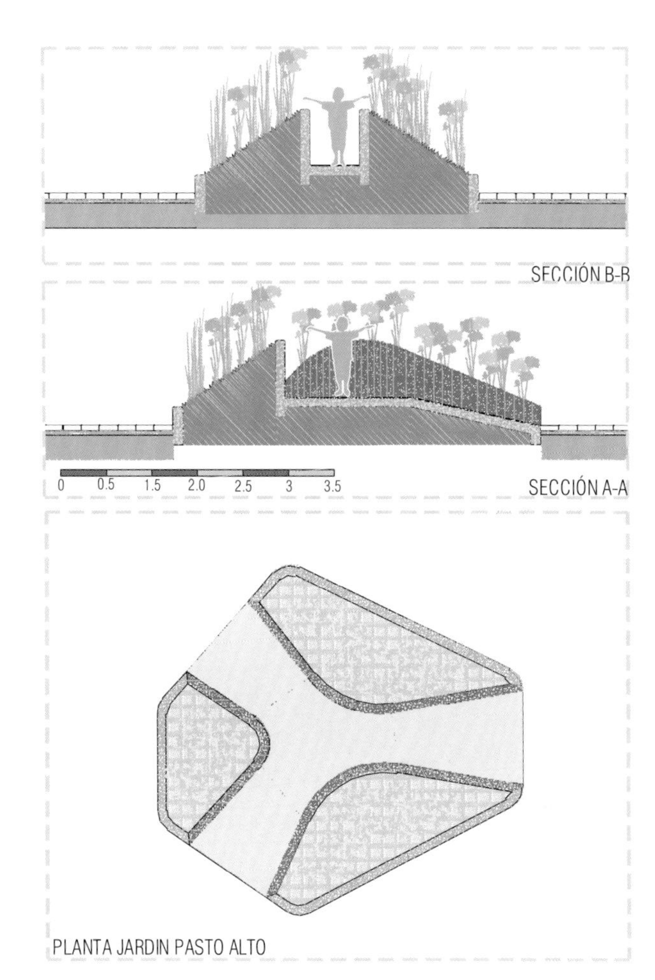
SECCIÓN B-B
0 0.5 1.5 2.0 2.5 3 3.5
SECCIÓN A-A
PLANTA JARDIN PASTO ALTO

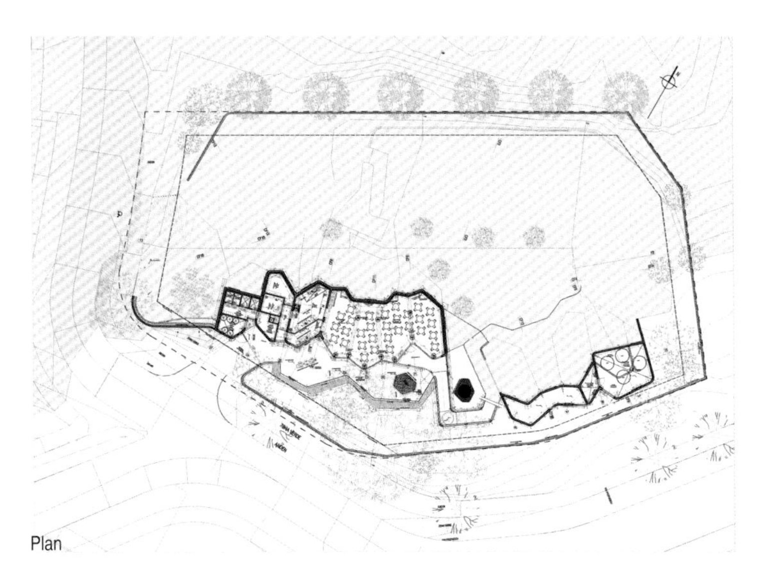
Plan

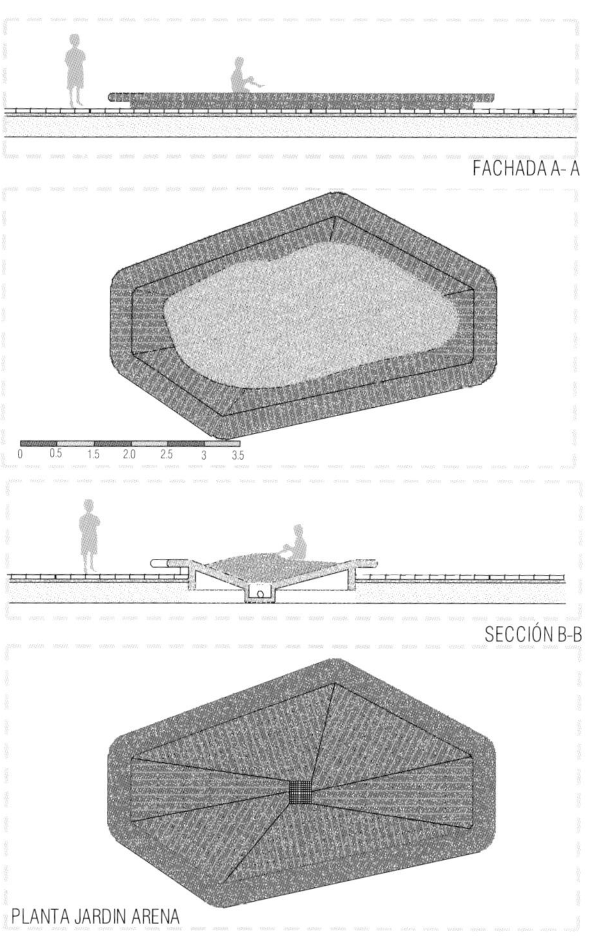
FACHADA A- A
0 0.5 1.5 2.0 2.5 3 3.5
SECCIÓN B-B
PLANTA JARDIN ARENA

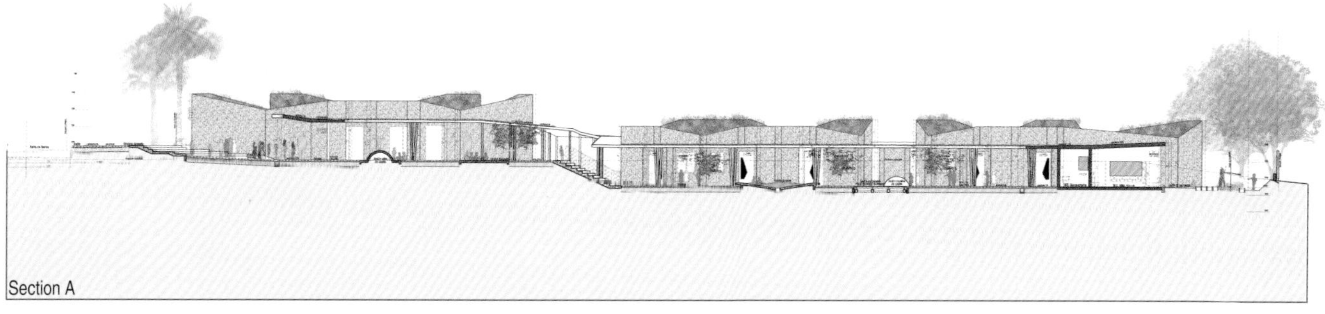
Section A

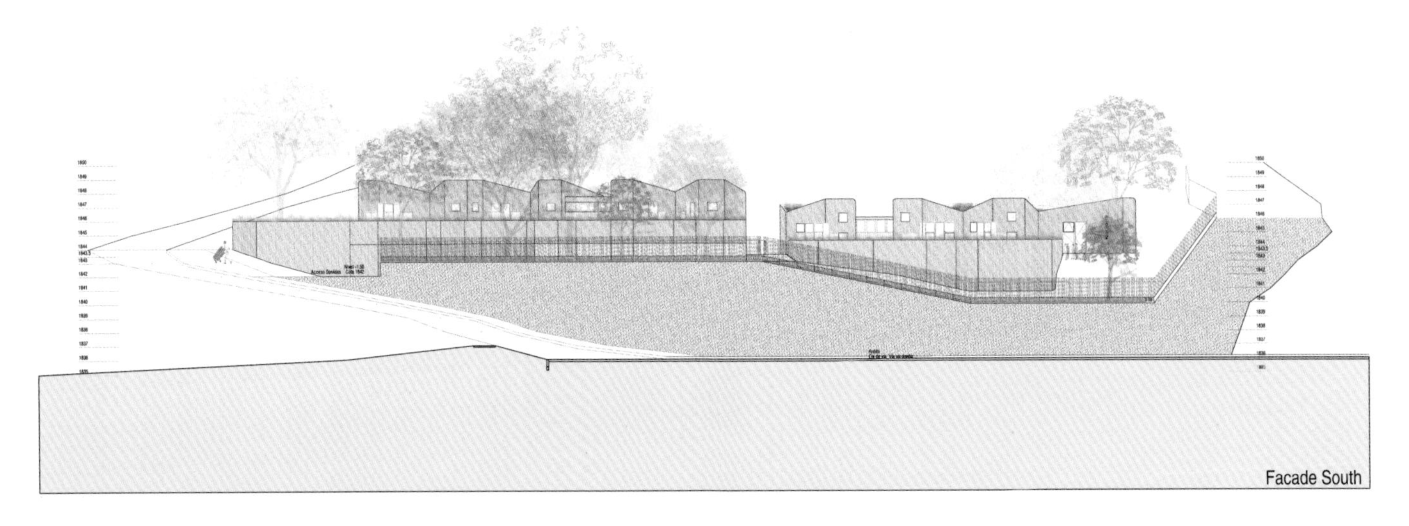
Facade South

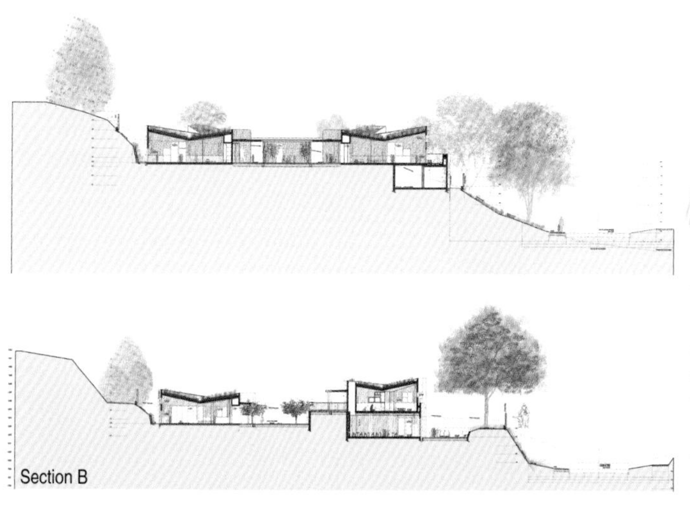

Section B

Marcel Sembat High School

马塞尔·桑巴中学

The Marcel Sembat high school in Sotteville-lès-Rouen has the particularity to be crossed by a street and so is separated to two parts reconnected by a bridge building. A public park is located straight on the south side of the site. The entire school is divided in six buildings of different periods, from 30's to 90's.The project is a rehabilitation and extension of the high school.

They are presenting here the biggest part of this project, the one that gives his visual identity: the new workshops building. This part of the Marcel Sembat high school is dedicated to technical teaching about motors and mechanic vehicles. It demands big spaces with high heights and important surfaces. They opted for simple and easy understandable project including workshops at the park with smooth lines which are connected to the park floor with their declivity. They proposed the demolition of two buildings, the reorganization of the workshops and the teaching spaces of industrial technology. They gather in a building unit all functions related to the workshops. The building starts at the boundary of the park and fits naturally to the site by the wavy design of its vegetated roof to varying curves and low height. The slight oscillations reveal the fringes of a park in motion in the trees' shades. The main ideas of their proposal are, to find a unity and identity of the school in the whole site, to integrate and connect the high school and the park with the particular shape of the "blades".

They also wanted to create a public space around the crossing street by making a great plaza in front of the new workshops building. And finally their purpose was restoring the views and transparency to the outdoors, and perspectives on public space and landscapes. For that they used glass and translucent polycarbonate on the frontages. They also created some little patios between the “blades” to make skylights. The unity between the two sites around the crossing street is ensured by the creation of a new public space which will be surrounded by a new library on the east side. This future "plaza" will be in the same time a meeting space, a walking or restoring area and the main access way to the high school.

Location / 地点:
Sotteville lès Rouen, France
Date of Completion / 竣工时间:
2011
Area / 占地面积:
12,764 m^2
Architecture / 建筑设计:
archi5 & B. Huidobro
Interior Design / 室内设计:
archi5 & B. Huidobro
Landscape / 景观设计:
archi5 & B. Huidobro
Photography / 摄影:
Thomas Jorion
Client / 客户:
Haute Normandie district

马塞尔·桑巴中学位于法国市镇Sotteville-lès-Rouen，一条街道横贯其中，将校园分为两个部分，并通过一座桥梁连接。场地南部紧挨着一个公园。整个校园分布着建于不同时期的六栋大楼，时间为20世纪30－90年代。该项目是对这所中学校园的一次修复和扩建。

图中展示的是该项目最大的组成部分，也是学校的标志性建筑，即新落成的办公楼。这是一处有关汽车和交通工具技术培训的场所，因此它需要较高的层高和严谨的墙体结构。设计师设计手法简单明了，采用线条流畅的斜坡将工作室与公园地面相连，从而使它们融为一个整体。他们计划拆除原有两座建筑，重组办公楼和工业技术教学场所。所有与办公相关的功能设施都被整合在大楼内。这栋大楼以公园边界为起点。波浪形的绿色屋顶有着不同的高度，同场地空间自然地结合在一起。项目的主要设计理念是在整片场地中表现出学校的统一性和特征性，通过绿色屋顶特别的形态将校园与公园完美融合。

设计师还希望在新落成的工作室楼房前设立一个大广场，从而建立一处公共场所。最后，该项目力求恢复户外空间的景色，使建筑具备通透性。为此，设计师在建筑的正立面设计中使用了玻璃和半透明聚碳酸酯材料。他们还在绿色屋顶之间建立起若干个小庭院，并在其中建造天窗。街道两边的场地中间将开辟出一个新的公共广场，并绕其东部新建一个图书馆，借以增强两边的统一性。这个未来的“广场”同时将作为集会空间、步行或整修区以及学校的主要通道。

Structure : Galvanizes Steel;
Roof : Steel Tray, Vegetated Roof;
Interiors : Concrete, Apparent Steel Structure.

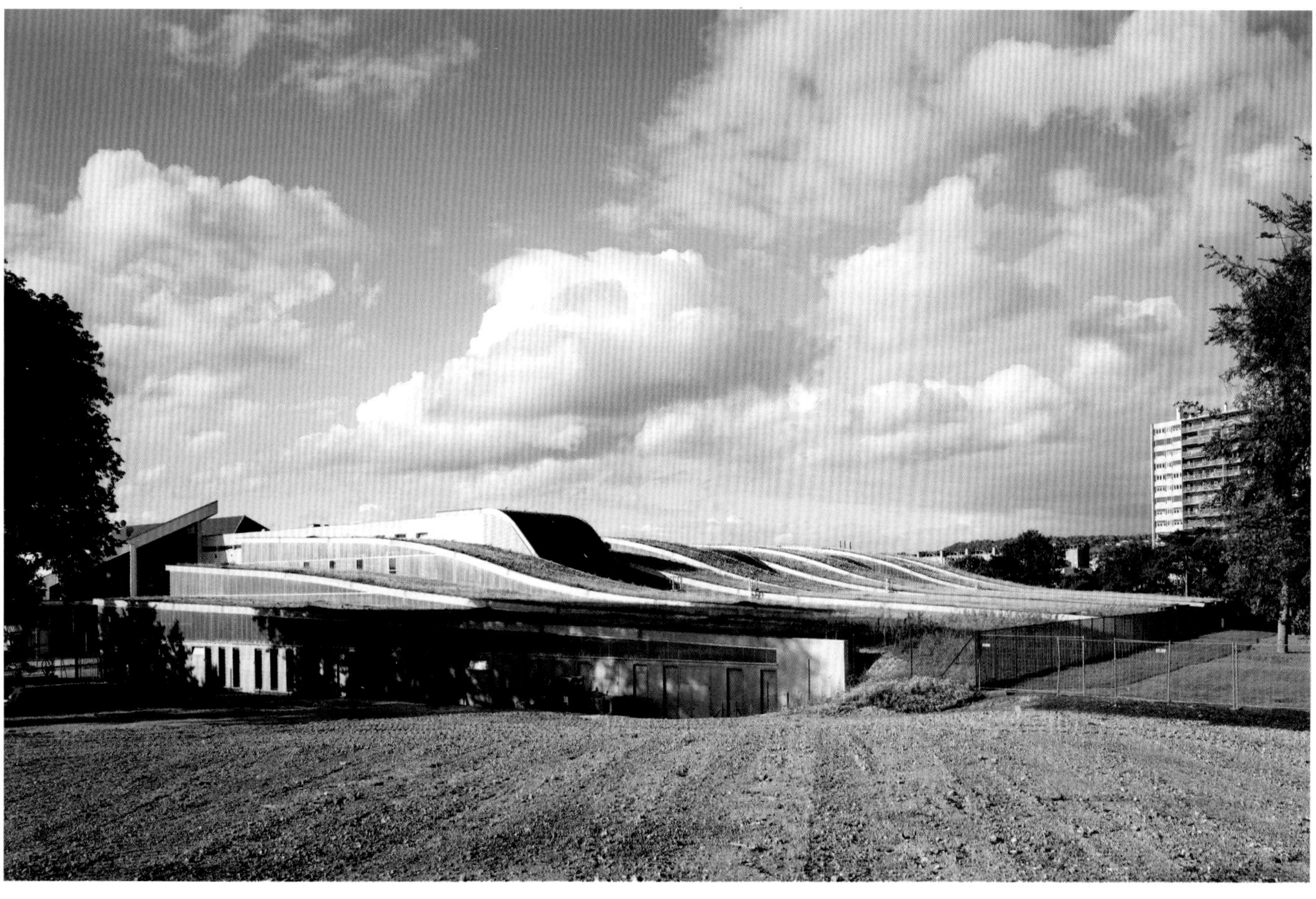

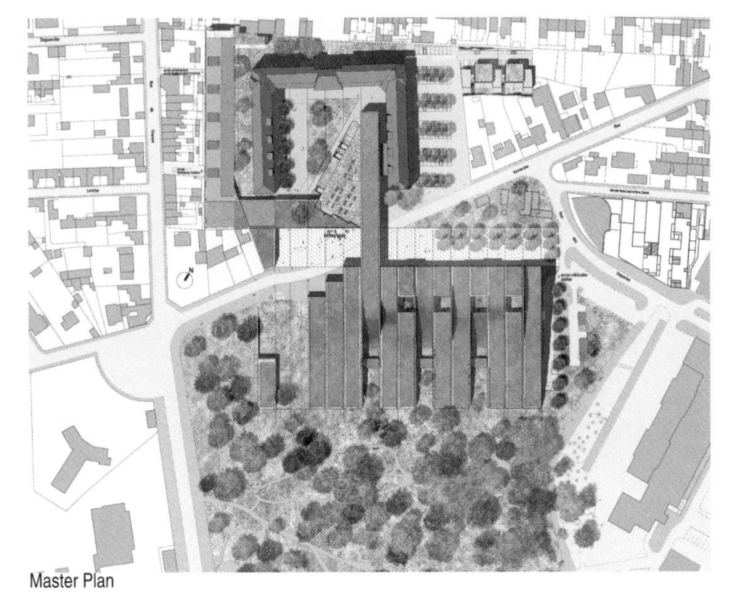
Master Plan

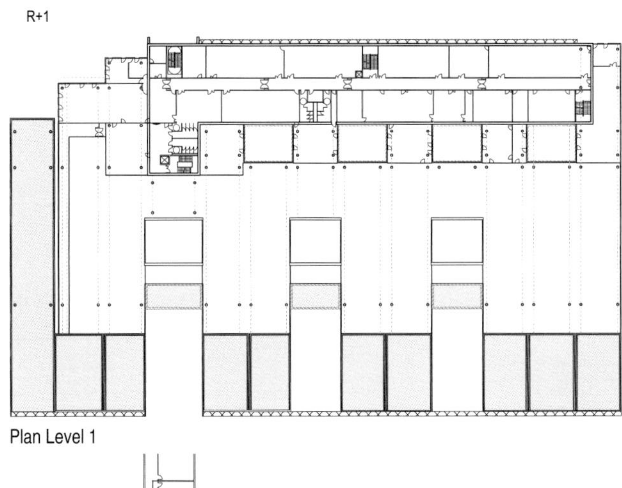

Plan Level 1

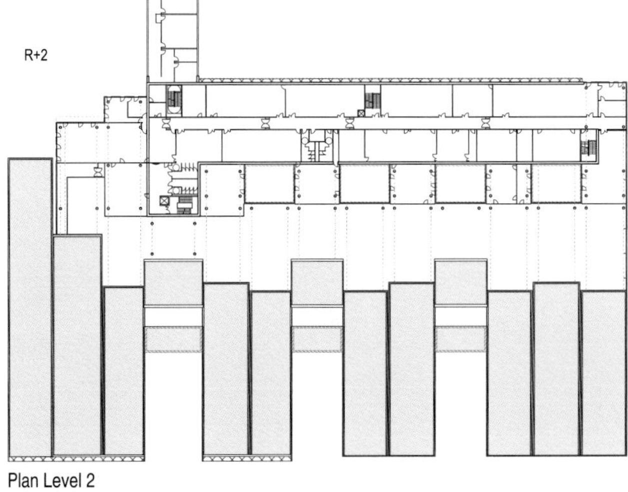

Plan Level 2

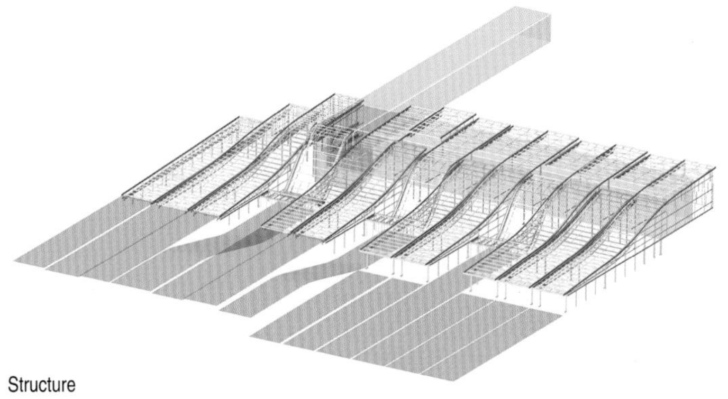
Structure

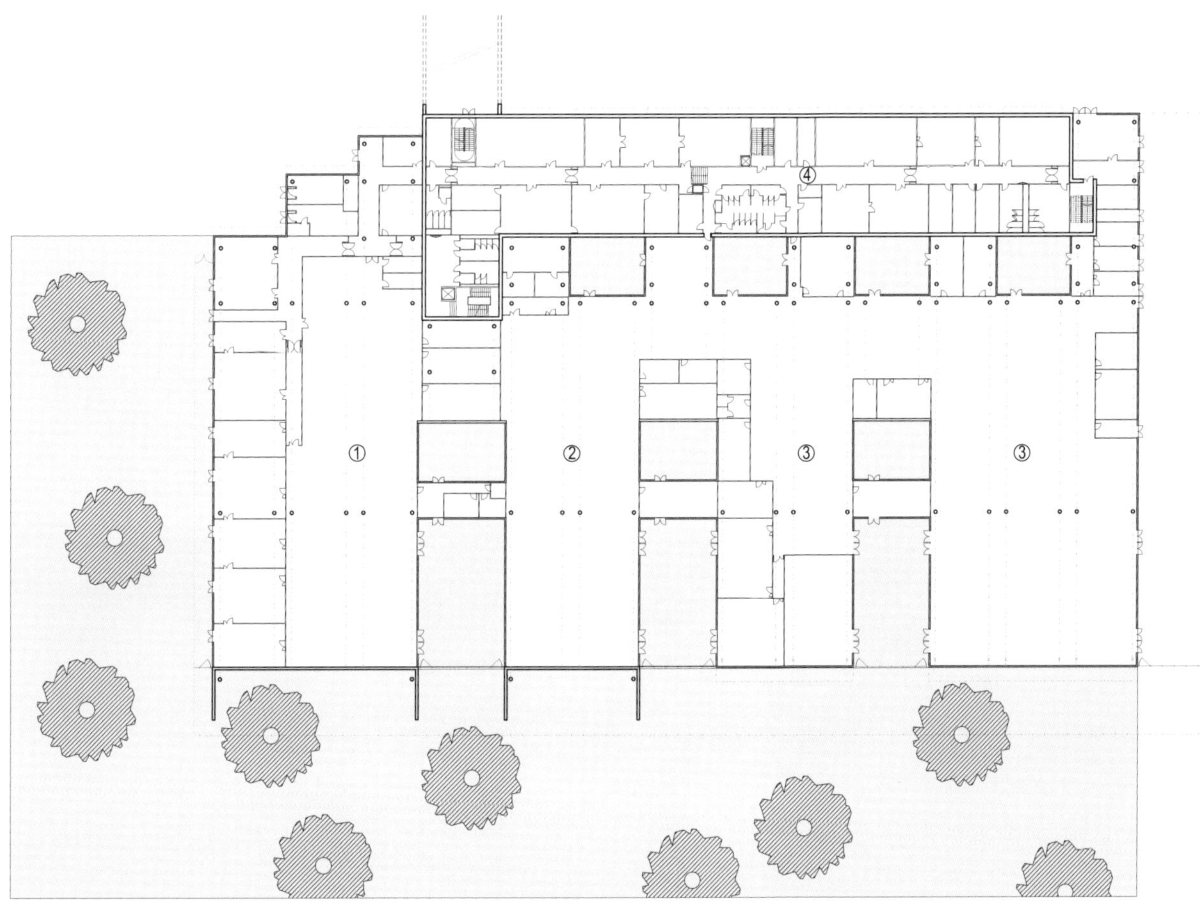

PLAN RDC

LEGENDE

1 ATELIER ELECTROTECHN

2 ATELIER CARROSSERIE

3 ATELIER MECANIQUE

4 SALLES DE COURS
(bâtiment existant)

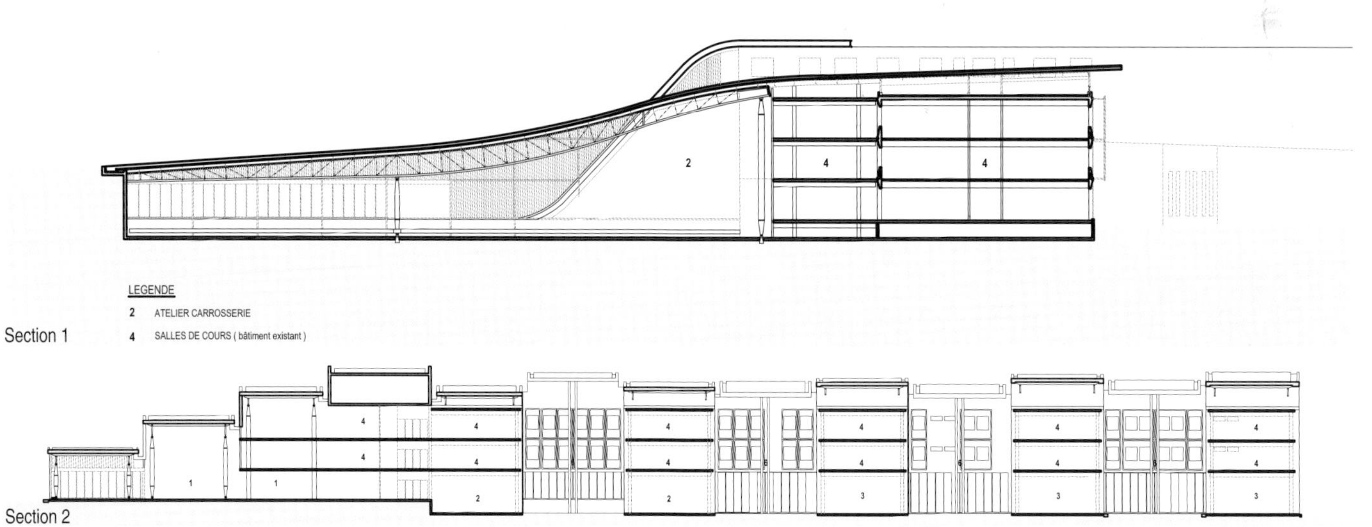
LEGENDE
2 ATELIER CARROSSERIE
4 SALLES DE COURS (bâtiment existant)
Section 1
Section 2

Paichai University Howard Center

培材大学 霍华德中心

The Howard Center of Paichai University is positioned at the long and narrow site located in the eastern end of the campus and creates a boundary with surrounding buildings: 21th Century Hall, Woo Nam Hall, Appenzeller Memorial Hall and International Hall. The linear mass is placed on the street side going along with the site morphology while the outdoor space for the children located at the rear side of the building has the minimum effects from the campus.

The Howard Center is basically an education center from the Department of Early Childhood Education, and four institutions which are kindergarten, nursery, nurturing school and Institution of Childcare Teacher occupy this building. Using the existing topography, four different programs have their own entry levels. The main entrance creates an outdoor gathering space using the different levels which make smooth connection towards outdoor space. The section reflecting the tight/loose relationship between the programs and the inner circulation is projected through the transparent façade skin while it is maximized on the West side of façade with vertical louvers.

The main entrance between the main building mass and the separate mass at the back side of the site leads people to the courtyard which was created by restoring the small hill on the South side. The in-between space of the two buildings has the light and transparent spatial condition with the construction of polycarbonate and glass panels. It allows the kindergarten, playground and nursery to have optical connection and mutual relationships. It is also intended to make the space function as the urban square in the middle of the six classrooms so that the children could create their own space by gathering, scattering, looking, associating and communicating with each other.

培材大学的霍华德中心位于校园东端一片狭长的场地上，与周围的21世纪厅、Woo Nam厅、阿彭策纪念馆和国际大厅形成一道边界。该中心细长的外形与街道的方向平行，顺应了场地的形态，而儿童户外活动空间位于建筑的后方，使得它对校园的影响降至最低。

霍华德中心主要是幼儿早期教育系的教学中心，建筑中共容纳了四个机构：幼儿园、托儿所、培育学校以及儿童护理教师研究所。这四个机构的入口相对独立。设计师在主入口处创造了室外聚会空间，使其与户外环境和谐相融。透明的建筑立面反映了各个机构与内部流通线之间或紧密、或松散的关系，这种关系在安装了垂直百叶窗的西侧立面最为直观。

主楼与副楼之间的主入口位于建筑背面，可将人们引向庭院。这个庭院是通过修复南侧的小土丘而建成的。两座建筑之间的空间安装了聚碳酸酯和玻璃板，使建筑间的空间通透。建筑师还特意在六间教室的中央设计了一个类似城市广场的空间，使其成为孩子们集散、观望和沟通的场所，使他们拥有属于自己的空间。

Location / 地点:
Daejon, South Korea
Date of Completion / 竣工时间:
2010
Area / 占地面积:
4,998 m^2
Architecture / 建筑设计:
iArc Architects
Interior Design / 室内设计:
iArc Architects
Landscape / 景观设计:
iArc Architects
Photography / 摄影:
Youngchae Park
Client / 客户:
Paichai University

Materials: Reinforced Concrete, Steel.

114
114

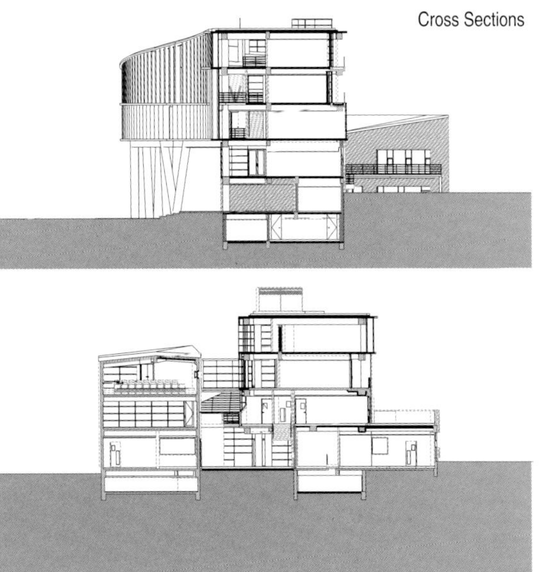

Cross Sections

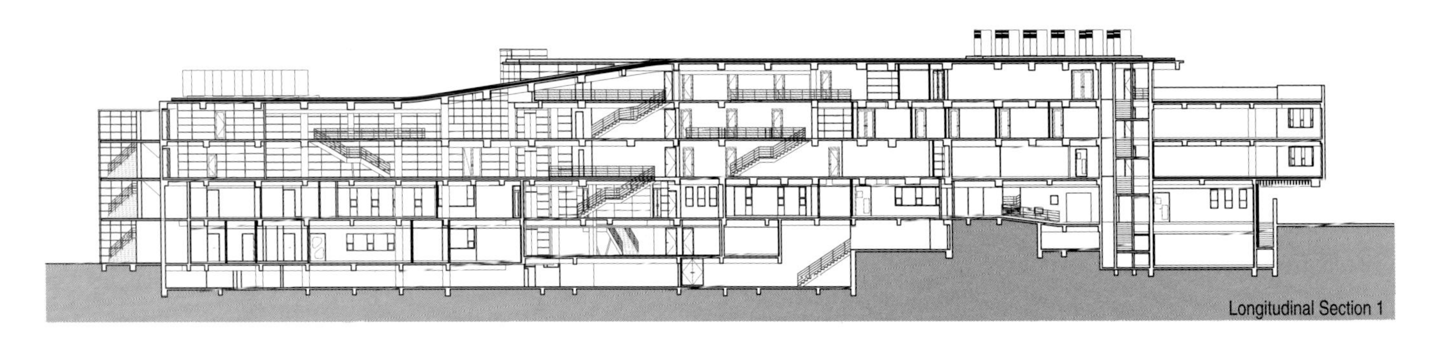

Longitudinal Section 1

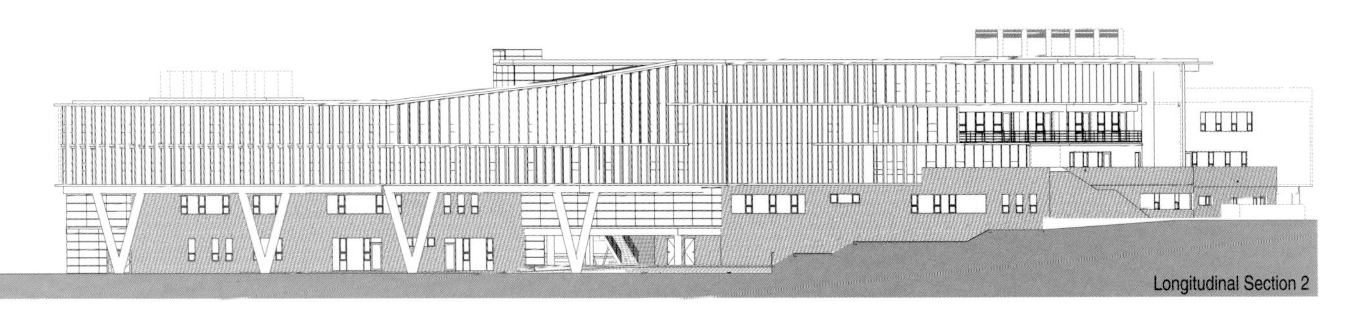

Longitudinal Section 2

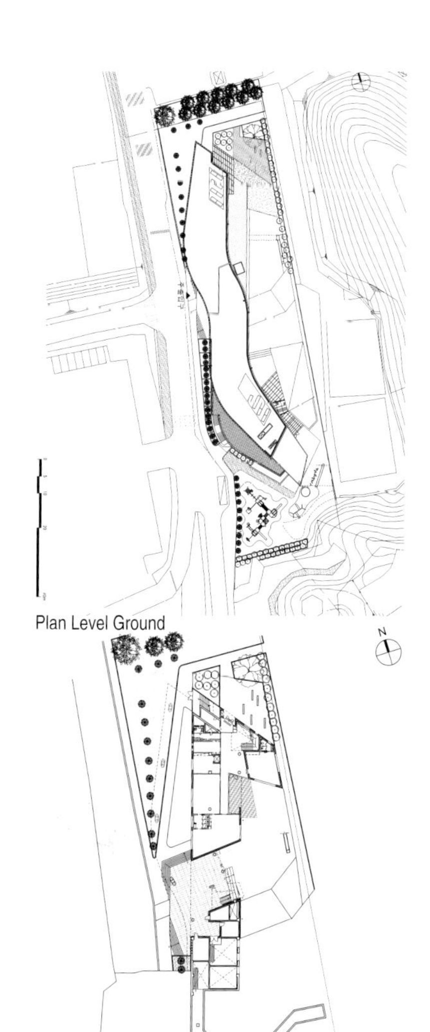
Plan Level Ground

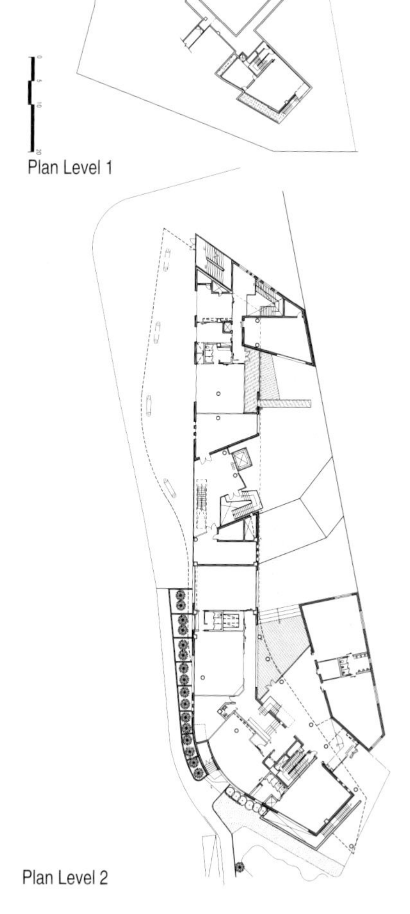
Plan Level 1

Plan Level 2

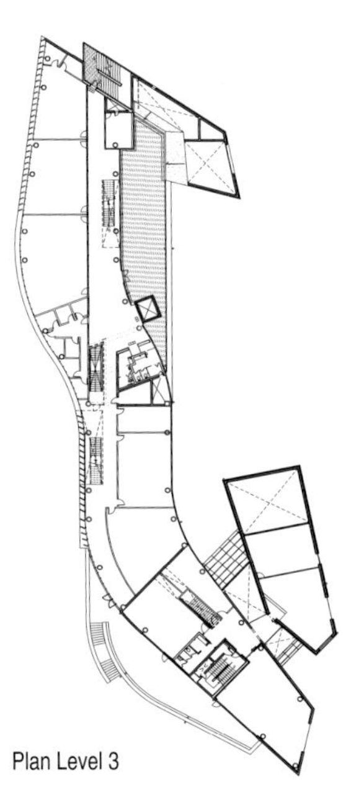

Plan Level 3

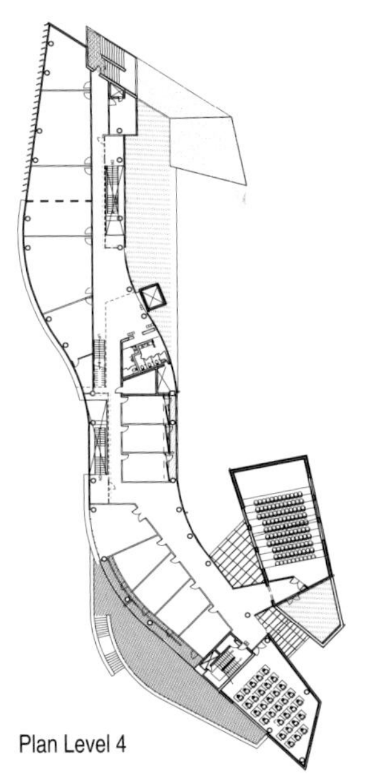

Plan Level 4

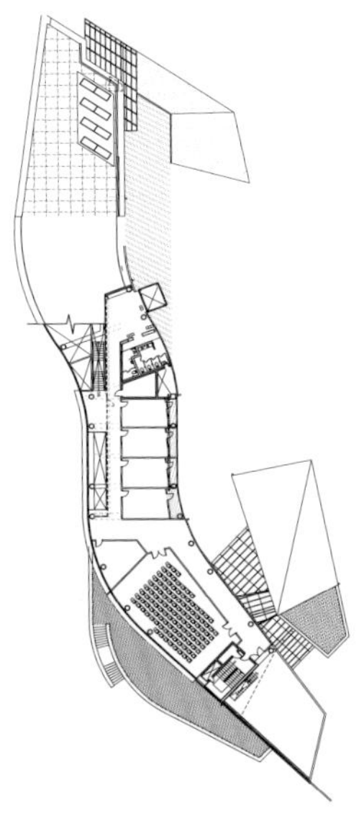

Plan Level 5

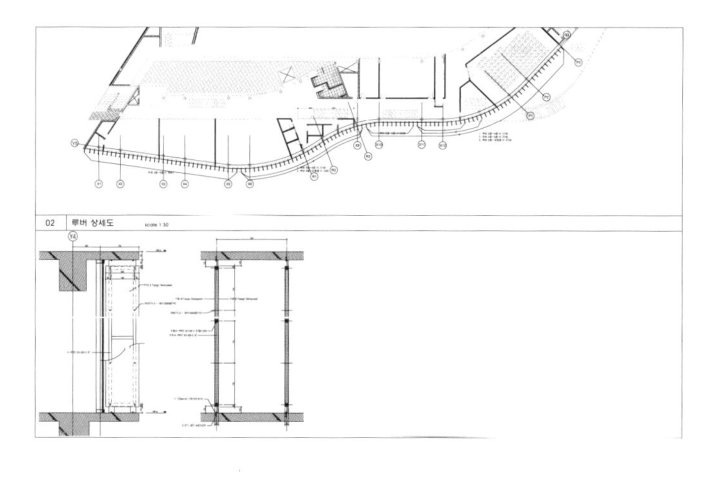
02
루버 상세도

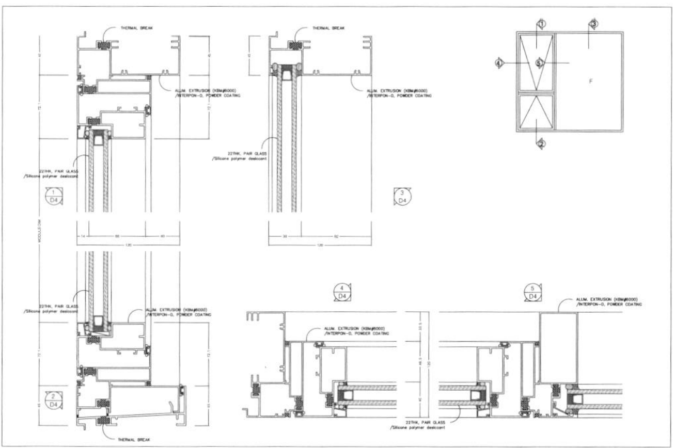
THERMAL BREAK

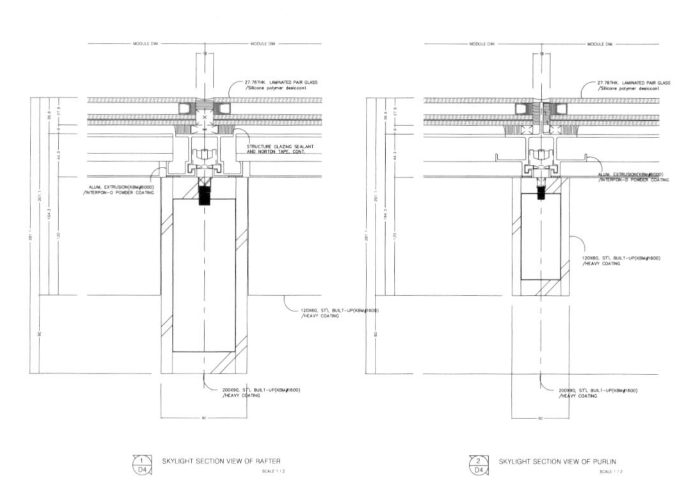
SKYLIGHT SECTION VIEW OF RAFTER
SKYLIGHT SECTION VIEW OF PURLIN

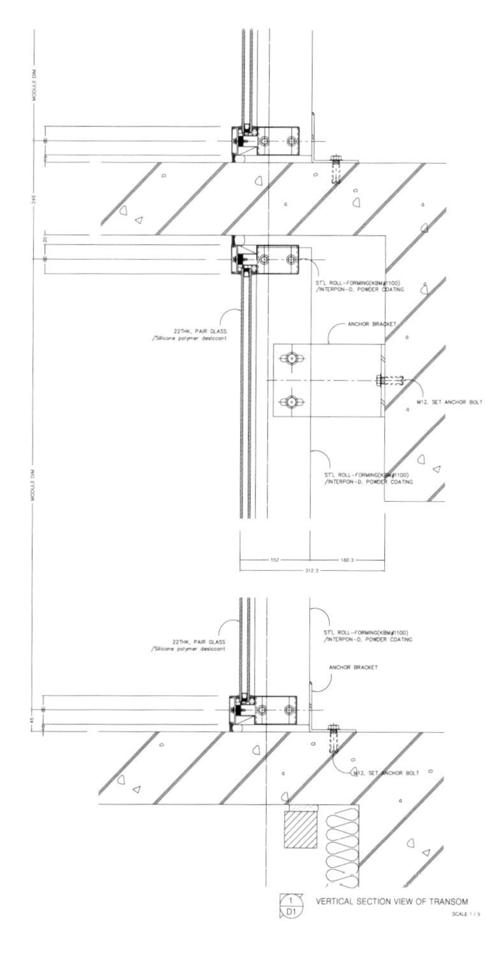
22THK. PAIR GLASS
/Silicone polymer desiccant
ANCHOR BRACKET
M12. SET ANCHOR BOLT
VERTICAL SECTION VIEW OF TRANSOM

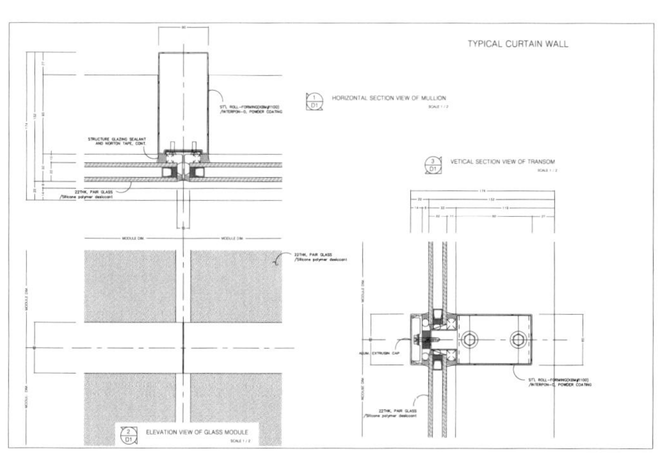
TYPICAL CURTAIN WALL
HORIZONTAL SECTION VIEW OF MULLION
VETICAL SECTION VIEW OF TRANSOM
ELEVATION VIEW OF GLASS MODULE

Tartu Kesklinna School

塔尔图市中心学校

The extension to Tartu Kesklinna School provides additional functions and opportunities to the historical building, doing this without overshadowing the old volume but rather as an intelligent accessory to it. The new volume avoids pretentiousness, interacting with the user as an equal partner, creating a diverse, friendly and light-hearted environment. The building tries to activate the surroundings, both in terms of making active use of the courtyard with the outdoor auditorium as well as domesticating the previously underused park-like back yard. In a playful way, it is possible to move either over the building or beneath it, where the main entrance lies. Blueberry pattern is enlivening the ascetic facade shell. The building is an attempt at being child-friendly without being childish.

To make the new part in consistent with existing building, the extension building should fit the existed architecture style while at the same time provide the folk with a brand new feeling. Considering the surrounding environment, the designer created an amazing facade which blended into the view perfectly and at the same time highlighted the exist building.

As a primary school building, the extension part could give children enough safe fieled for creative and exploratory playing. Different with some childish primary school buildings, the Tartu Kesklinna school extension based on existing historical building integrated contemporary design concept, technology and materials.

位于塔尔图的市中心学校通过该扩展项目为这个历史建筑带来了新的功能和机会，扩展部分并未遮挡原有的建筑外立面，而是成为老建筑的附加体。新建筑摒弃了华而不实的风格，创造了一种舒适、多样化的环境。该建筑力求带动周围环境，为此，庭院空间被地用作户外礼堂，之前未充分发挥作用的后院花园也被利用起来。正门处的设计饶有趣味，人们可以选择从上方或下方通过建筑。蓝莓图案使朴素的外墙显得生动活泼。该建筑的设计目标是贴近儿童但不显稚气。

为了和原有建筑保持一致，扩建部分须符合现有的风格，同时要给大众带来全新的感觉。设计师衡量了周围环境后，建造出令人称奇的外墙，不但与场地景观完美融合，还很好地突显了原有建筑。

作为一所小学教学楼，其扩展部分提供了足够安全的空间，这些空间可以供孩子们进行创造性和探索性的游戏。塔尔图市中心学校不同于一些带有稚气的小学建筑，其扩展楼房以原有建筑为基础，整合了当代的设计理念、技术和材料。

Location / 地点:
Tartu, Estonia
Date of Completion / 竣工时间:
2007
Area / 占地面积:
2,500 m^2
Architecture / 建筑设计:
Salto AB
Interior Design / 室内设计:
Salto AB
Landscape / 景观设计:
Salto AB
Photography / 摄影:
Martin Siplane, Reio Avaste, Karli Luik
Client / 客户:
Tartu City Government

Materials: Concrete, Wood, Steel.

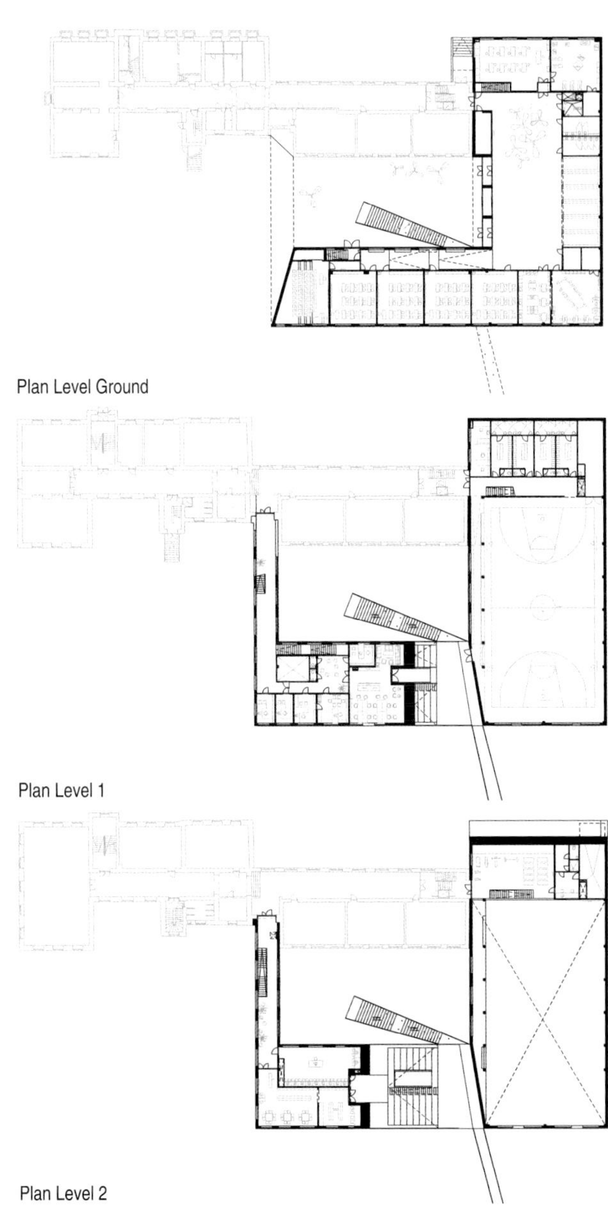

Plan Level Ground

Plan Level 1

Plan Level 2

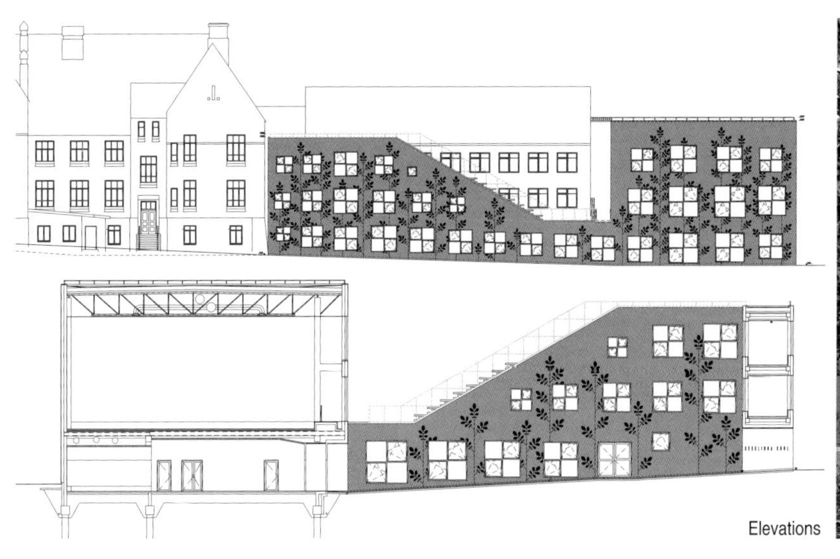

Elevations

USJ Campus Innovation

圣·约瑟夫大学校园改造

This new campus takes a contextual approach, integrating physically, culturally, and historically with Beirut's urban tissue. Conceptually an urban block with sculpted voids, the building's hollow spaces define six autonomous blocks and construct multiple viewpoints across Beirut, connecting students to their dynamic setting. The voids also generate a street-level meeting space, which flows fluidly to the top floor in the form of a massive staircase. It concludes at a landscaped terrace overlooking the city. Light is a vital element in oriental architecture and one that shapes its style and identity; the campus exposes alternate light qualities through Moucharabieh-inspired perforations and a polycarbonate volume. Such manipulation presents a striking contrast in filtered light and luminescence. A stylized random-opening treatment is a snapshot of the Lebanese War, lending a poetic glimpse into the reality of destruction and violence.

In order to accommodate with local environment and scenery, the exterior of the USJ Campus de L'Innovation et du Sport is built in a particular form which at the same time adapts to the local climate. The exterior stairs provide the users an alternative way to appreciate the view of the whole city besides connecting the two buildings. The beautiful facades of the building are the design highlights.

该项目从背景因素入手，将建筑从物质、文化和历史角度融入贝鲁特的城市结构之中。大楼是具有雕塑风格的孔洞结构，创造了具有概念城市街区的景观。整个建筑被中空的空间划分为6个独立的区块，创造出可以观赏贝鲁特景观的多个视角。孔洞结构还形成一个临街的集会空间，并通过一个大型的楼梯与顶层相连。楼梯的尽头是一个景观露台，在这里学生们可以俯瞰全城的景色。光是东方建筑中不可或缺的元素，能够塑造建筑的风格和特征。这座建筑采用了以阿拉伯式窗孔为原形的孔洞结构和聚碳酸酯材料，从而使光线可以穿过立面。这种手法使直射光和漫射光形成了丰富的光影效果。这些特色孔洞无规则地排列，向人们展示了战争带来的残酷现实。

该项目的外观采用了特定的建筑形态，使得建筑与当地环境、风景相协调，同时建筑能够适应当地的气候条件。外部的楼梯一方面作为连接两栋大楼的通道，另一方面为人们提供一种欣赏城市景观的独特方式。精美的建筑外墙是项目设计的亮点所在。

Location / 地点:
Beirut, Lebanon
Date of Completion / 竣工时间:
2011
Area / 占地面积:
1,500 m^2
Architecture / 建筑设计:
109 Architects in collaboration with Youssef Tohme
Interior Design / 室内设计:
109 Architectes
Landscape / 景观设计:
109 Architectes
Photography / 摄影:
Albert Saikaly
Client / 客户:
Université Saint-Joseph (USJ)

Floor: Concrete, Ceramic, Hardwood;
Walking Paths: Ceramic;
Furniture: Concrete Benches, Classroom Furniture;
Illumination: Floor Washlights, Fluorescent Light, Spotlights, Wall Fixtures;
Elevation: Concrete, Polycarbonate, Glazing.

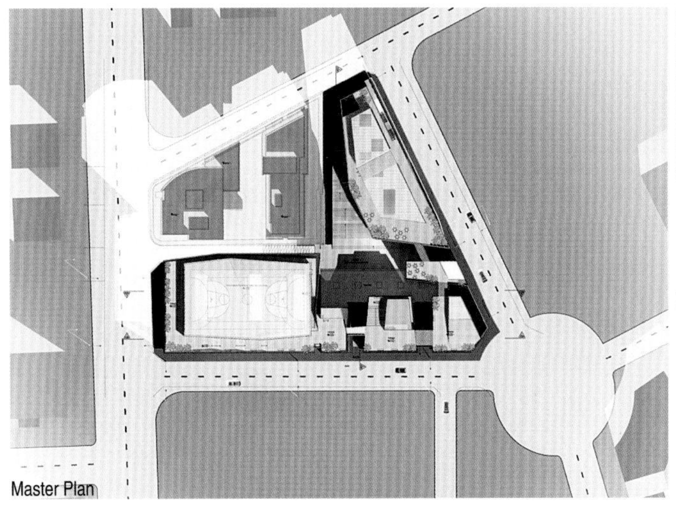

Master Plan

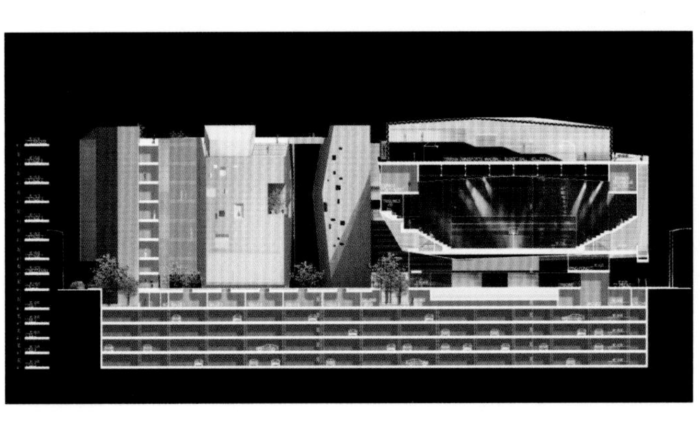

Facade DD

Plan Level 1

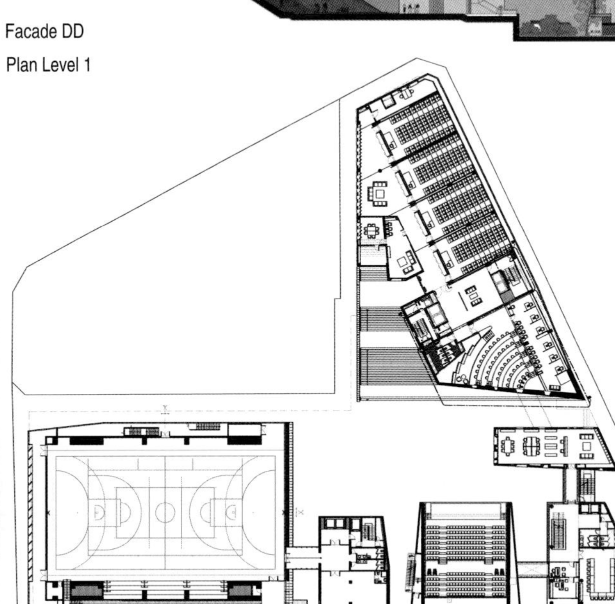

Section BB

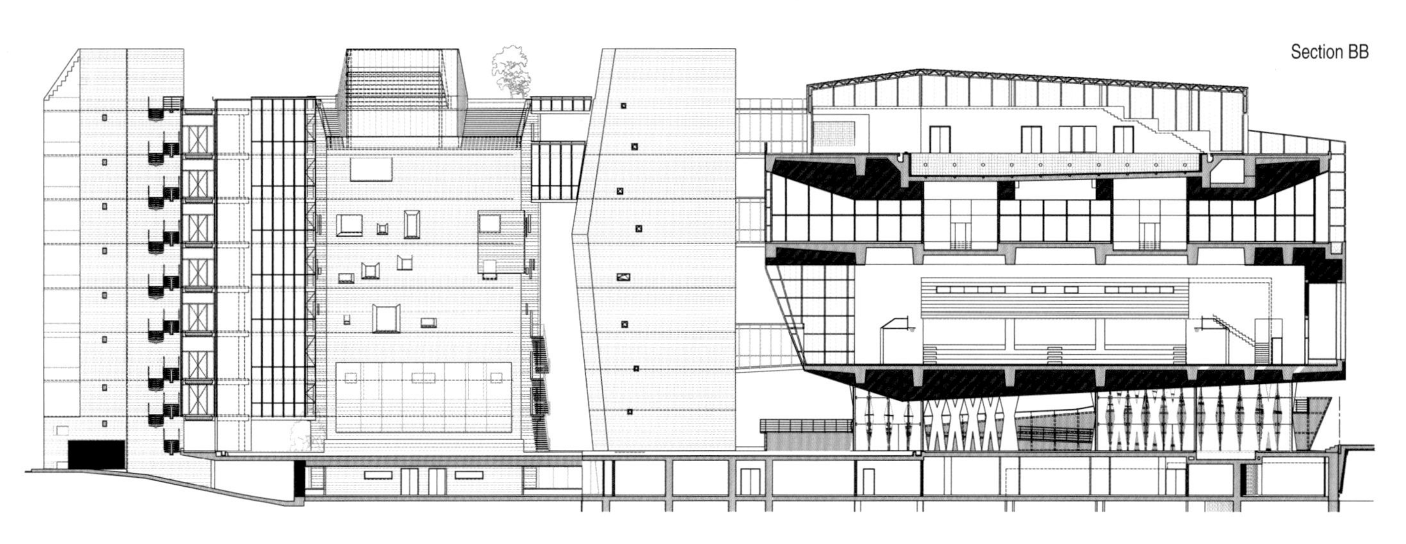

Plan Level 0

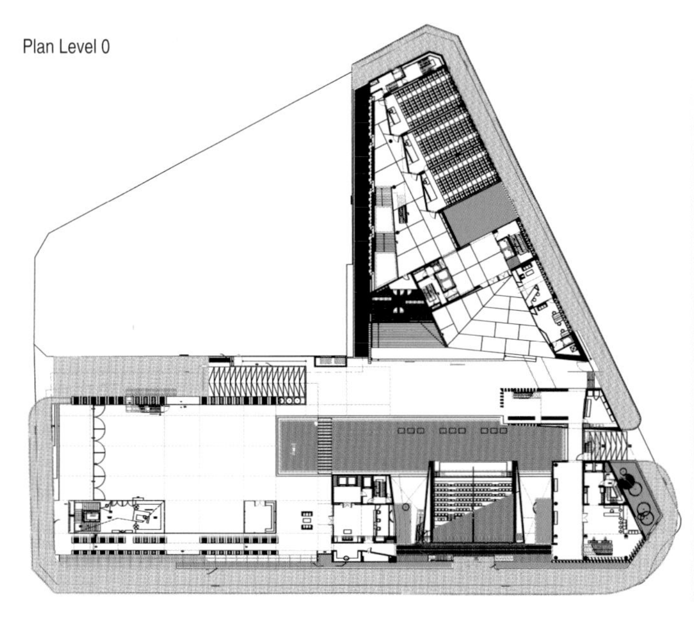

Facade C3

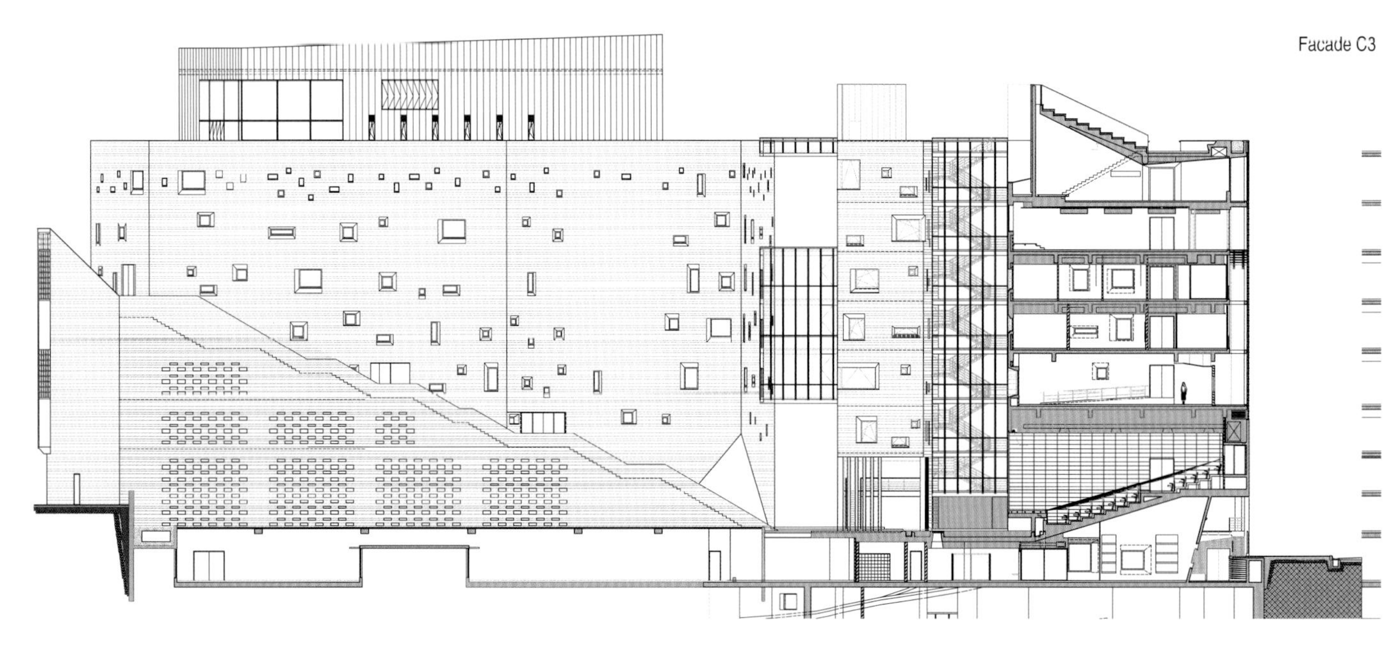

Master Plan

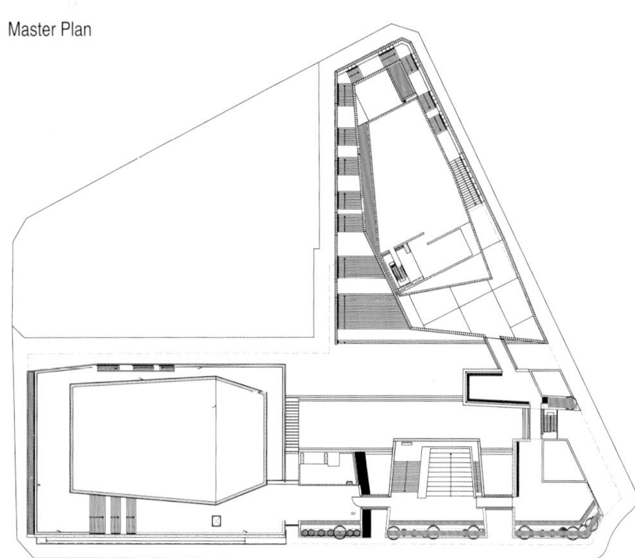

Facade B1

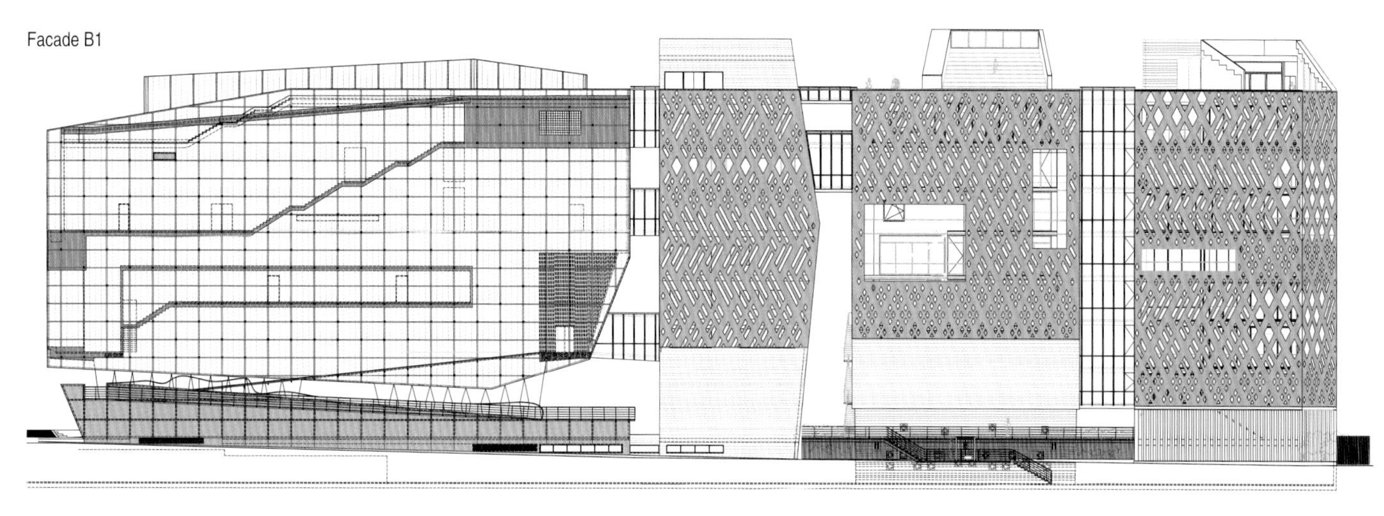

The 2nd Phase of Dormitory for ITRI Southern Taiwan Campus

台湾工业技术研究院台湾南部校区二期工程

The site for the second phase is part of the 470,000 m^2 campus of ITRI in Southern Taiwan. The mid-to-low density development plan of the campus started in 1995, and proved in 2002. The first-phase construction launched in 2003 was carried out under the guidelines of green building technology and ecological construction. The whole campus is programmed to be a research environment including building hardware and landscaping software for 1500 people. The overall layout comprises research buildings, cafeteria, dormitories, ecological ponds, bamboo forest, organic green house, and an art district for bamboo kiln.

The site is surrounded by hills in three directions, and fronted by lakes to the west. The layout is aimed not only to have the building cluster fit in the environment but also to make the place for the habitat of the existing eco-system. Further, the built site, along with its networked landscape within the whole campus, is expected to be a great place for ecological observation. The building group is formed as a part of the landscape vista dialoguing with the topography, and its roofs serve for viewing Chiayi Plain.

The ecological pond in the center of the courtyard may adjust the micro-climate and co-work with a nearby retention pool. With future efforts to cultivate the lake area in the vicinity of the construction site, the experience of strolling around the site and its neighborhoods will be characterized by the scenic water body in stepped elevation and in varied scales from manmade to nature.

Bamboo forest is a prosper ous scene in the campus area, and utilizing this local material helps to reduce carbon footprint of the new construction. The application of bamboo ranges from planting to architectural elements like exterior screening of the staircase, soft partition in the entrance area, and in the courtyard to define the outdoor corridor. Further walking path is planned to circulate to the bamboo forest around the site and to the Bamboo Art Section in the southern campus. Locally produced brick is adopted for the pavement of the semi-outdoor walkway.

这项二期工程的场地是台湾工业技术研究院（ITRI）台湾南部47hm^{2}校区中的一部分。校区低密度的发展计划起草于1995年，于2002年被通过。一期建设工程于2003年动工，以绿色建筑和生态建设的原则为导向。整个校区通过硬件建设和环境美化为1500人提供了研究场所。整体规划包括科研楼、食堂、宿舍、生态池塘、竹林、有机绿色住房和竹木工艺区。

该场地三面环山，西面临湖。布局的目的在于不仅使建筑融入到环境中，还要使其成为现有生态系统的一部分。此外，建筑场地和校园内网络化的自然景观将是生态观察的理想场所。建筑群构成景观远景的一部分，与地形产生互动，其屋顶可供观赏嘉义的平原。

庭院中心的生态池可以调节局部微环境，并与附近的储水池一起工作。场地附近的湖泊区域今后将被开垦，这样，场地的周围和街区将以阶梯式的景观以及人工与自然相结合的景观模式呈现风景秀丽的水边景区。

竹林是校园充满活力的一角，而且这种采用本地材料的设计方法有助于减少新建筑的排碳量。对于竹子的运用包括种植竹林和在建筑中采用竹材，其中，竹材被广泛地用于楼梯的外饰面、入口区的软隔断和庭院的室外走廊。校园的道路也得到疏通，以方便师生到达场地周围的竹林和南校区的竹木工艺区。本地生产的砖料被用于铺设半户外走道。

Location / 地点:
Taiwan, China
Date of Completion / 竣工时间:
2010
Area / 占地面积:
52,792 m^2
Architecture / 建筑设计:
Bio-architecture formosana
Interior Design / 室内设计:
Bio-architecture formosana
Landscape / 景观设计:
Ecologue Landscape Architecture, Environmental Planning & Design; Bio-architecture formosana
Photography / 摄影:
Bio-architecture formosana
Client / 客户:
Industrial Technology Research Institute

Floor: PVC Pavement, Permeable Brick;
Walking Paths: Local Brick;
Furniture: Bamboo Screen;
Illumination: Wall Light, Bollard Light;
Plants: Banana, Long'an, Makino Bamboo, Water Lily, Ginger Lily.

Elevation

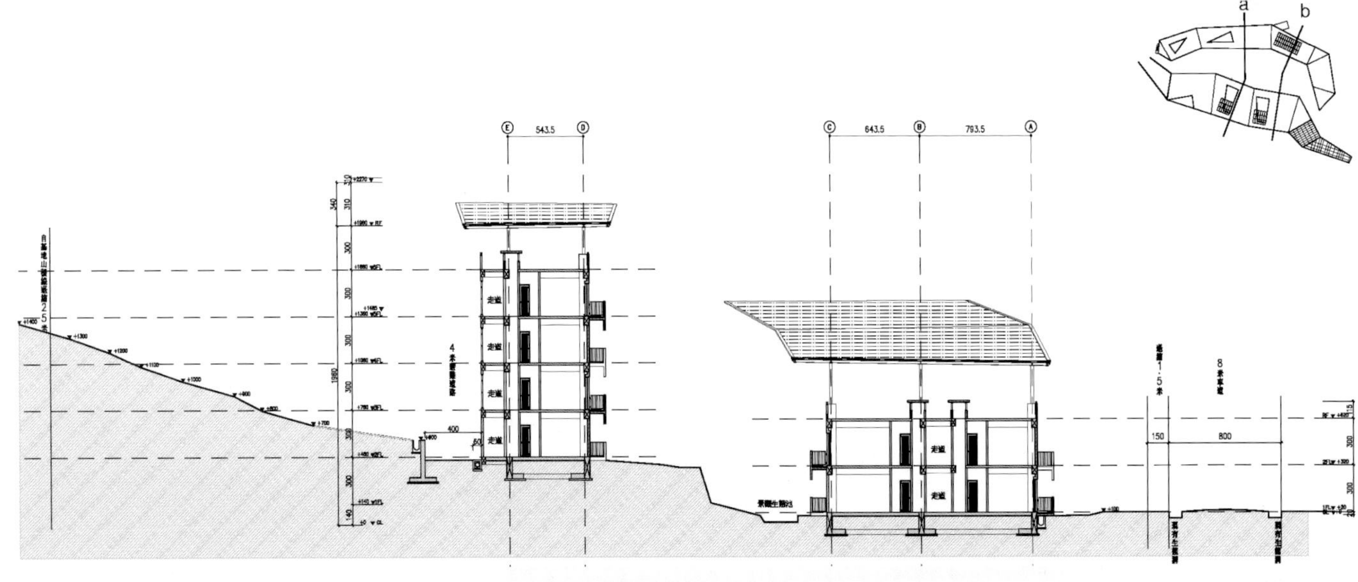

section a

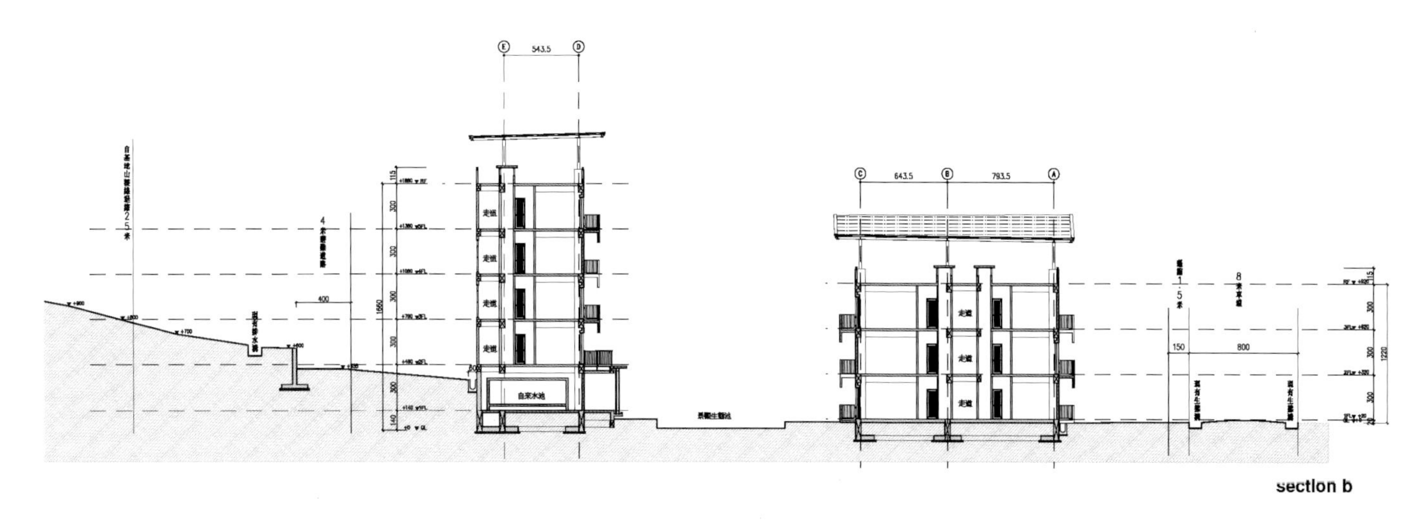

section b

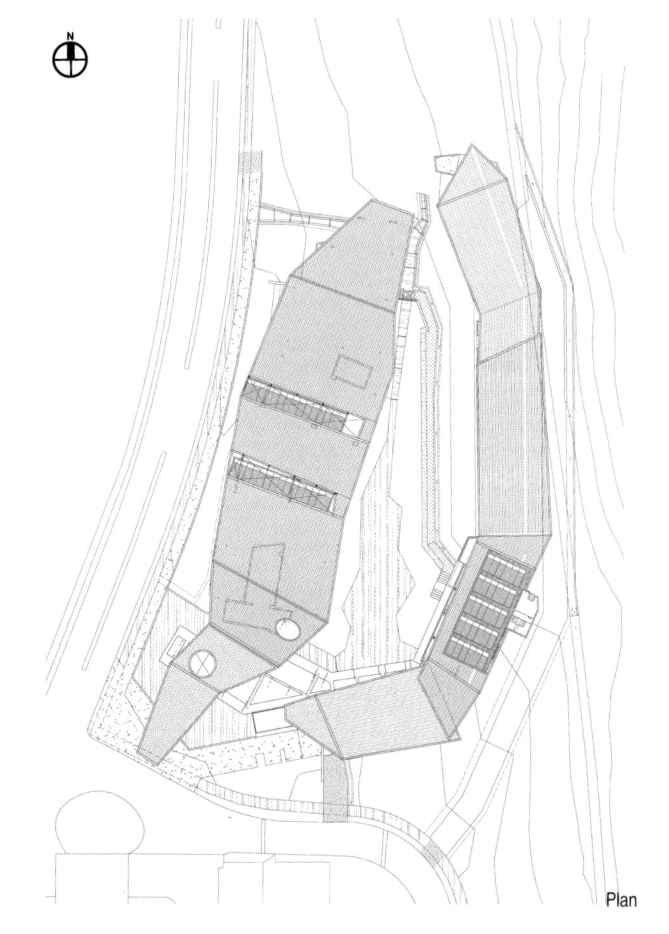

Plan

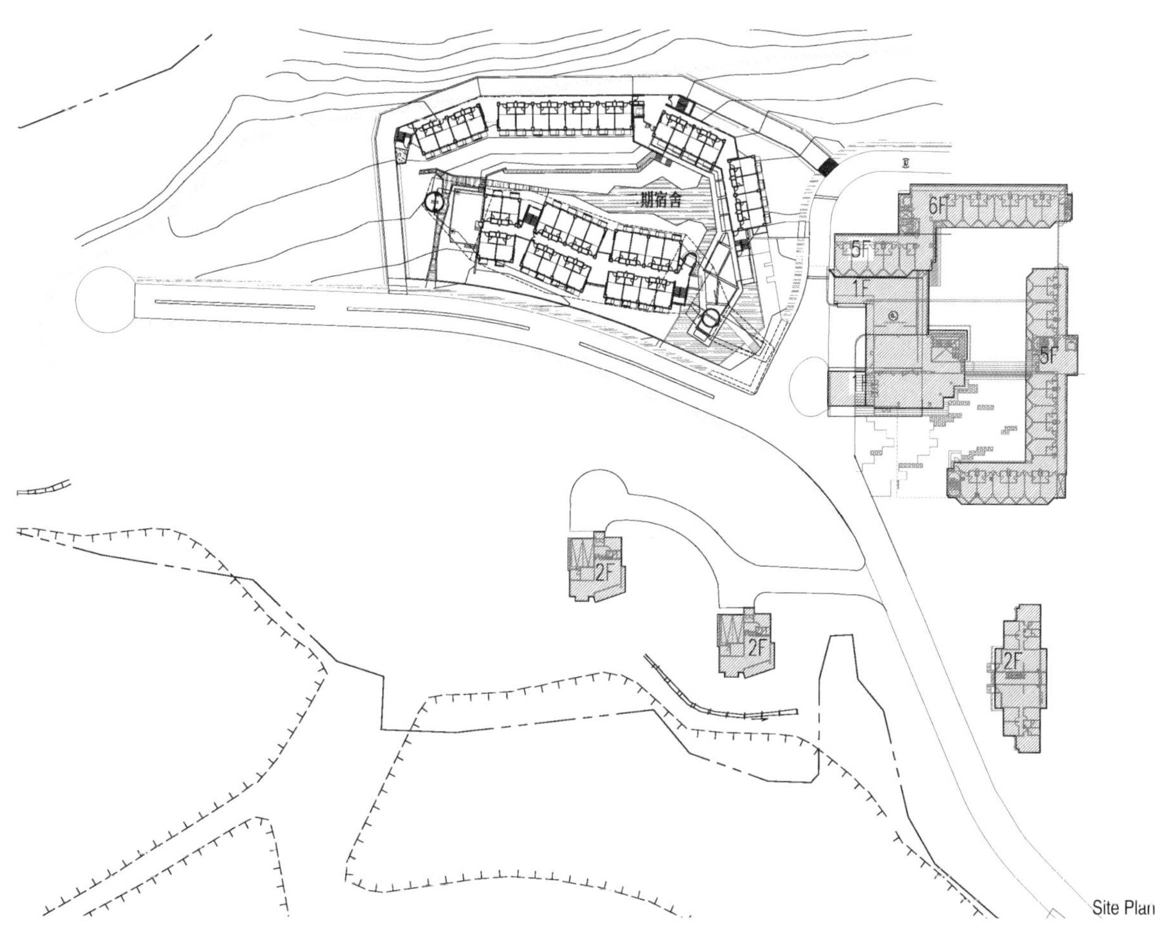

Site Plan

Gates Center and Hillman Center

盖茨中心与希尔曼中心

Location / 地点:
Pittsburgh, USA
Date of Completion / 竣工时间:
2009
Area / 占地面积:
19323.83 m^2
Architecture / 建筑设计:
Mack Scogin Merrill Elam Architects with Gensler and EDGE Studio
Interior Design / 室内设计:
Mack Scogin Merrill Elam Architects
Landscape / 景观设计:
Michael Van Valkenburgh Associates
Photography / 摄影:
Timothy Hursley
Client / 客户:
Carnegie Mellon University

The Gates Center for Computer Science will complete a computer science complex on Carnegie Mellon University's west campus. The building will house the three departments of the School of Computer Science providing offices, conference rooms, open collaborative spaces, closed project rooms and a reading room for more than one hundred and twenty faculty, three hundred and fifty graduate students, one hundred researchers or postdoctoral fellows and fifty administrative staff members along with a more public component of ten University classrooms, a two hundred and fifty seat auditorium, a café and two University computer clusters.

The following design principles have been used to guide the project conceptually. The empowerment of the individual is fundamental to the mission of an academic institution. Intelligent, creative people possess an innate desire for the freedom of choice. They privilege order over the systematic and demand the maintenance of individuality within a respected collective. Architecture and Landscape Architecture have the capacity to sponsor the uncompromised coexistence of seemingly irreconcilable differences. At Carnegie Mellon, interdisciplinary and collaborative work between both its academic units and its faculty has been and continues to be one of its defining strengths. As the School of Computer Science grows in size and diversity, its physical facilities must sustain and encourage a successful environment of collective difference.

The site chosen for the Gates Center has an unusually complex set of conflicting technical, functional and aesthetic challenges that, with the addition of the Gates Center, can serve to transform the West Campus area into a visually and physically integrated campus precinct. The Carnegie Mellon Campus plan and many of its buildings have distinctive, enduring characteristics that embody the highest ideals of the institution's founders. They were exceptional works of their time and served as exemplary benchmarks for an architecture and landscape architecture of our own time. An architecture that represents Carnegie Mellon University's and the School of Computer Science's exceptional status among the world's leading academic institutions will be best informed from within the project's own situation. An inclusive process that reinforces reciprocity within the university, strengthens relations with its neighboring communities and deepens its commitment to the stewardship of our natural resources will add richness and credibility to this project.

盖茨计算机科学研究中心将在卡耐基梅隆大学西校区建立一座计算机科学综合大楼。该大楼将容纳计算机学院的三个部门，内设办公室、会议室、开放式合作空间、封闭式项目研究室和阅读室，供超过120名教师、350名研究生、100名研究员或博士后以及50名行政人员所使用，同时配有更加公共化的构成空间，包括10间大学教室、1间可容纳250人的礼堂、1家咖啡馆和2个大学计算机机房。

该项目的设计原则是为单体注入活力，这是学术机构建设的重中之重。建筑设计和景观设计有能力让看似不可调和的差异和谐共存。跨学科学术交流和教师间的协同合作从来都是以后更会是卡耐基梅隆大学独特的优势之一。随着计算机学院规模的不断壮大，其硬件设施需要更多的设备。

盖茨中心所选用的场地在技术、功能和美观上具有冲突性。而盖茨中心的建立将使西校区转变成视觉上、物质上互相协调的校园区域。卡耐基梅隆大学的校园规划和其中的许多建筑物都具有独特、持久的特征，体现了机构创始人的终极理念。它们既是昔日的杰作，也是如今建筑和景观设计的典范。该建筑代表着卡耐基梅隆大学及其计算机学院在世界领先学术机构中的杰出地位。加强校园内设施的互益发展，强化学校与周边地区的关系，深化其自然资源管理等进程，将增加该项目的丰富性和可靠性。

Floor: Polished Concrete, Wood, Slate, Broadloom Carpet, Modular Carpet;
Interior Walls: Wood, Fabric Wrapped Panels, Gypsum Board, Acoustic Plaster;
Exterior Walls: Zinc, Anodized Aluminum Curtain Wall;
Window Assembly: Black Zinc Shingles, Folded Natural Zinc Panels, Zinc Slip Sheet; Exterior Cement Board with Laminated Aluminum Backing, Fixed and Operable Window with Shared Mullion, Unfaced Rigid Insulation, Exterior Cement Board Sheathing, Light Gauge Metal Framing and Interior Gypsum Board;
Green Roof: Tray Green Roof System, Intensive Green Roof;
Pausch Bridge: LED Lighting, Custom CNC Routed Aluminum Panels.

Site Plan

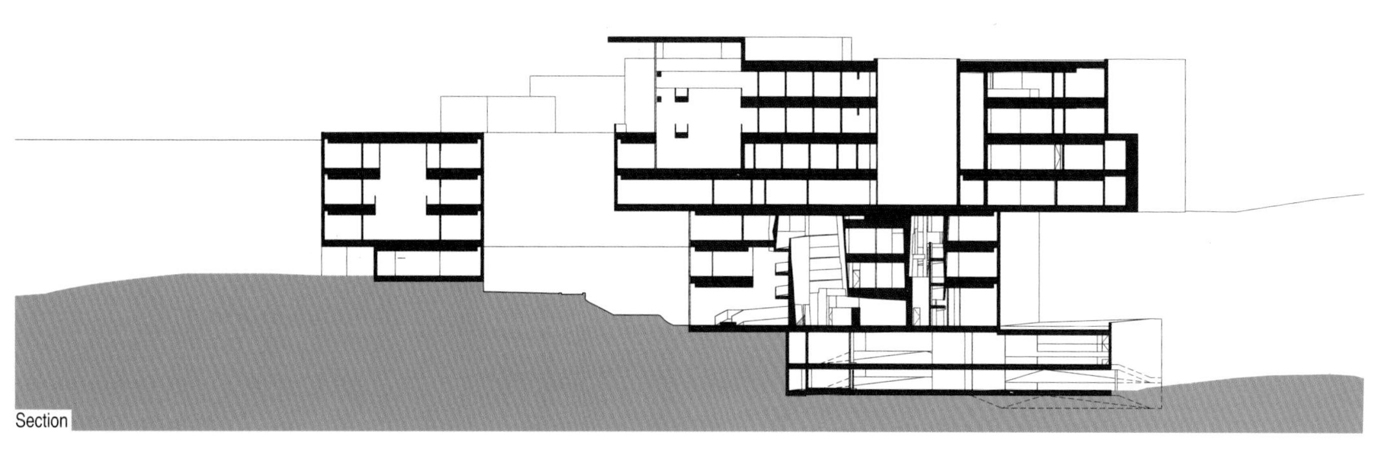

Section

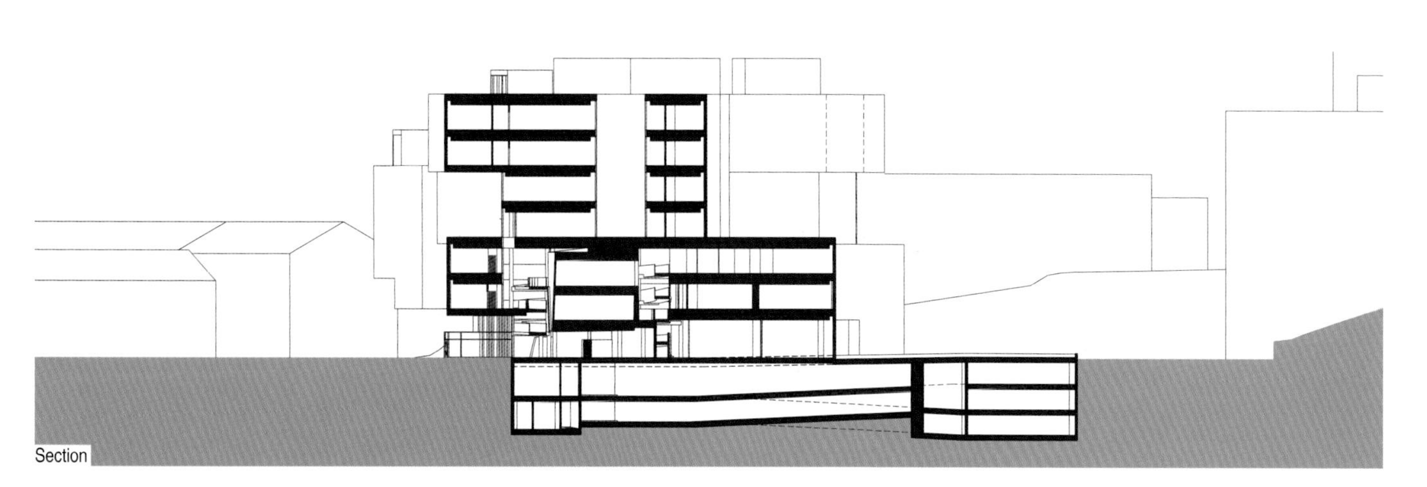

Section

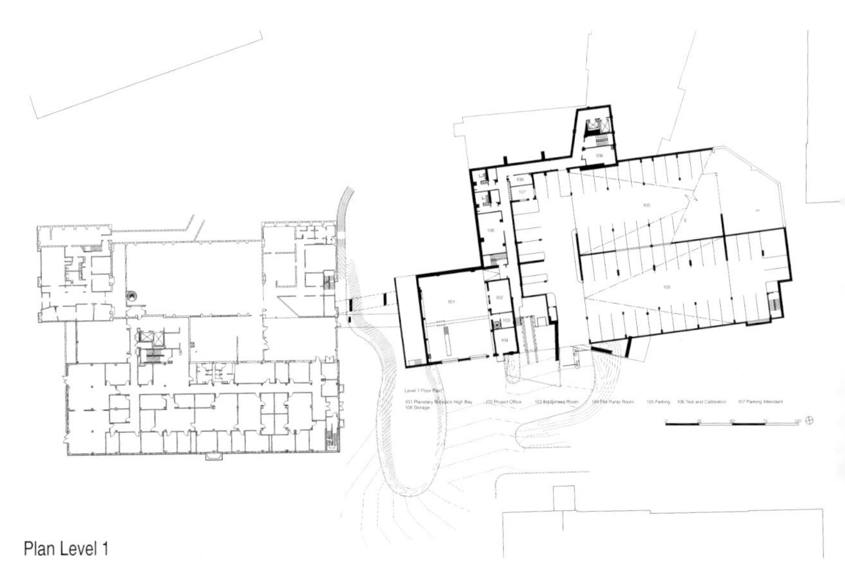

Plan Level 1

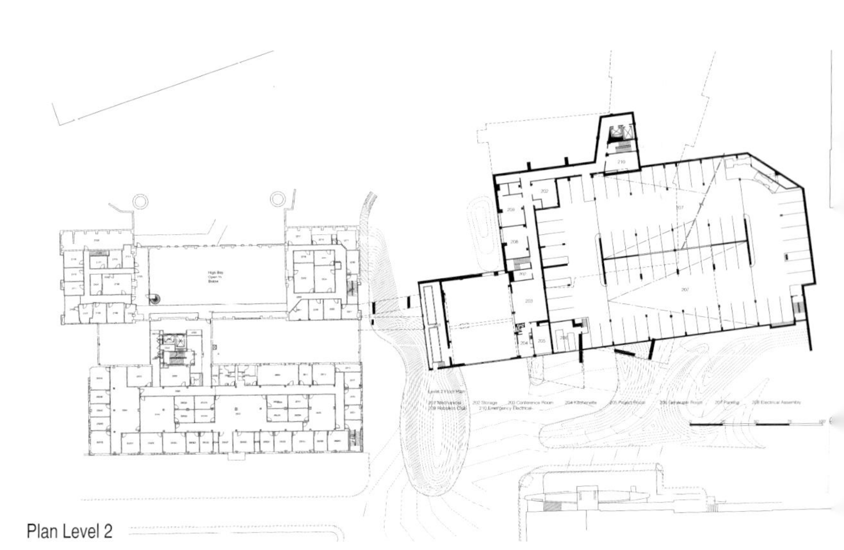

Plan Level 2

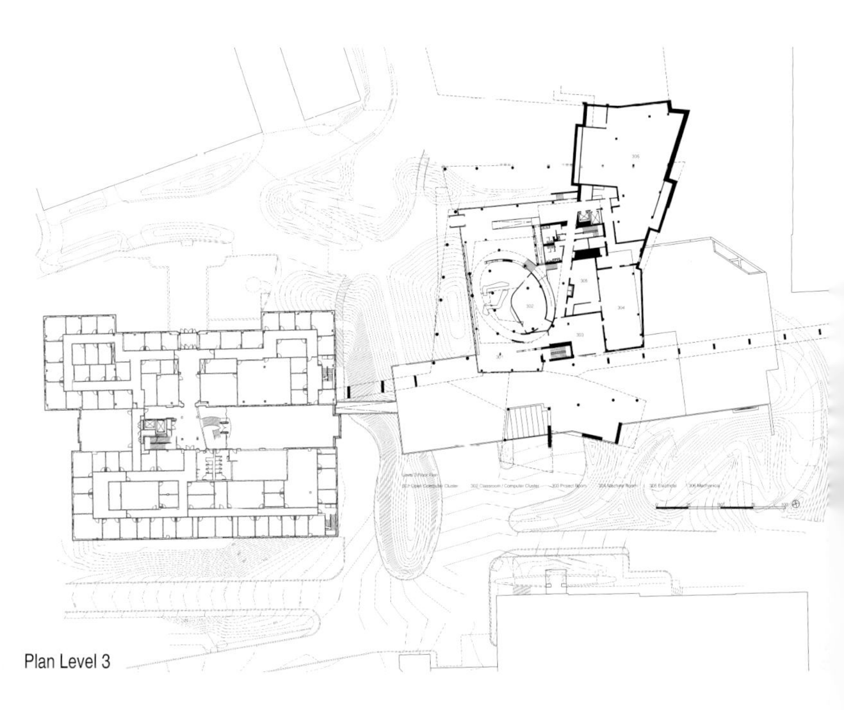

Plan Level 3

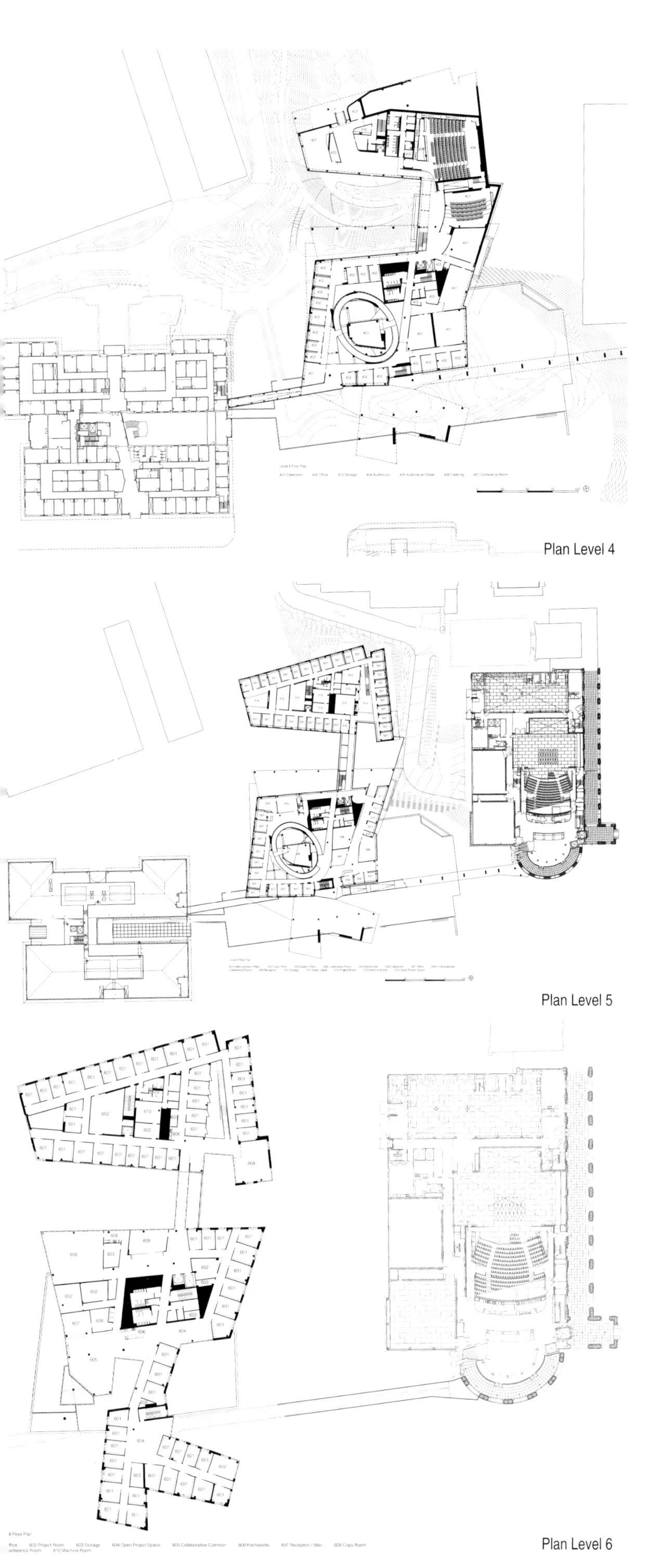

Plan Level 4

Plan Level 5

Plan Level 6

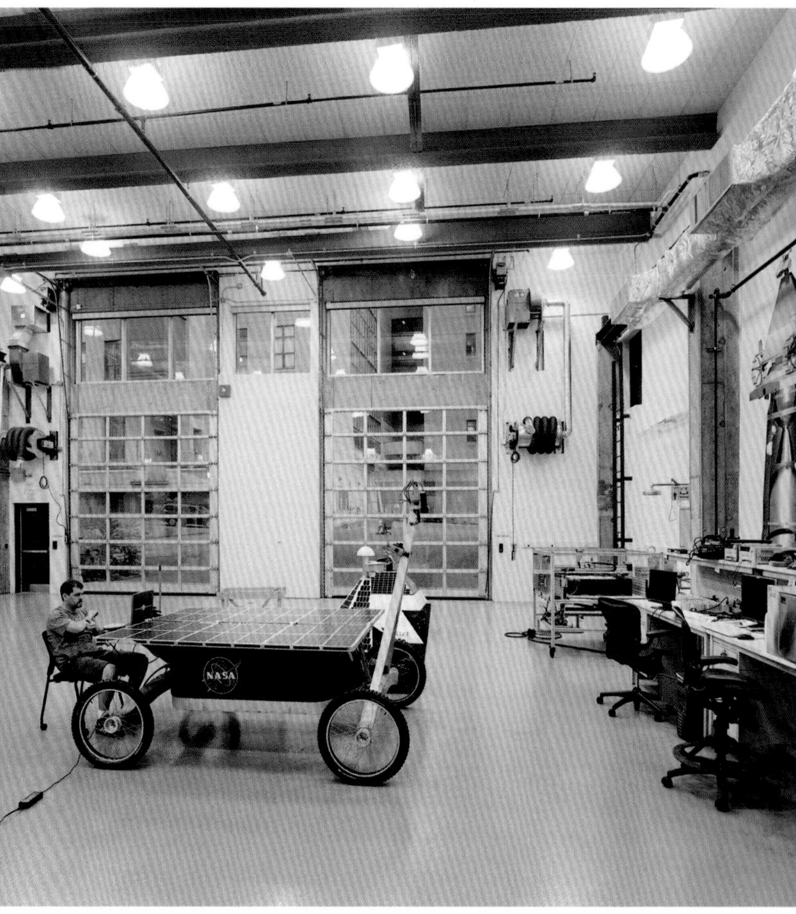

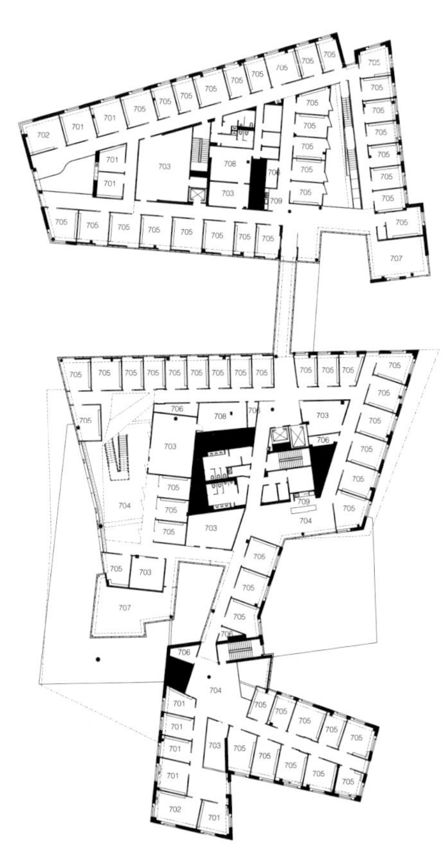

Level 7 Floor Plan

701 Administrative Office 702 Department Head Office 703 Project Room 704 Open Project Space 705 Office 706 Storage 707 Conference Room
708 Machine Room 709 Kitchenette

Plan Level 7

Level 8 Floor Plan

801 Office 802 Open Project Space 803 Kitchenette 804 Project Room 805 Conference Room 806 Storage 807 Department Head Office
808 Administrative Office 809 Copy Room 810 Machine Room 811 Nursing Room

Plan Level 8

Level 9 Floor Plan
901 Reading Room
902 Office
903 Open Project Space
904 Storage
905 Project Room
906 Copy Room
907 Conference Room
908 Kitchenette
0'
25'
50'
75'
100'

Plan Level 9

Aspen Middle School

阿斯彭中学

The new Aspen Middle School building enhances the Aspen School District emphasis on creative classroom learning, outdoor education, and environmental responsibility. The comfortable and stimulating environment has been achieved by integrating ample day-lighting, high indoor air quality, high-efficiency building & technology systems, better access and security. Additionally, views and spaces connect the building's users to the dynamic surrounding mountain landscape.

The architectural vocabulary of the new building reinforces the image established by other buildings on the school district's three school campuses. Regionally manufactured brick, metal panels, translucent panels, and aluminum-framed high performance glazing are modern, durable and low to no-maintenance materials that sheath the building's exterior surfaces. The extended canopy and adjacent arcade identifies the main east-facing entry, while at the same time, provides a safe, visible, protected shelter for student drop-off and pick-up. Light Louvers, solar shades and "tuned" high-performance glazing control the effects of the sun around the facility.

Natural light and views as well as operable windows are provided to all educational spaces. Window openings and sunshade devices respond to the solar orientation, and create a distinctive aesthetic feel. Wherever possible, sustainable, green materials such as bamboo, recycled content flooring and ceiling tiles are incorporated into the design. All new casework was formaldehyde free. Interior materials use low VOC paints, finishes and adhesives. High efficiency mechanical, electrical and plumbing systems integrate innovative products and techniques such as solar air heating, waterless urinals, occupancy sensors and solar tubes to make this the most energy efficient classroom building on the school campus. These combined strategies result in a high performance building that reduces almost 1 million pounds of CO_2 per year, reduces water usage by 40% and reduces storm water runoff by 25%. The school was awarded Gold LEED certification by the U.S. Green Building Council.

新落成的阿斯彭中学教学楼加强了该校区注重创造性课堂学习、户外教育和环境保护的特点。室内采光充足，空气质量较高，环境舒适，设计整合了高效的建设和技术系统。此外，建筑设计巧妙地将人们同周围动态的山地景观联系在一起。

新大楼的建筑形式强化了校区其他建筑固有的外观形象。建筑的外表面采用现代、耐用、低维护或零维护的材料，包括生产于当地的砖、金属板、透光板和铝框高性能玻璃窗。顶棚和拱廊凸显了朝东的入口，同时为学生提供了一个安全、醒目、受保护的场所。百叶窗、遮阳棚和可调控的玻璃窗调节了设施周围的光影效果。

所有教学场所都拥有自然光线、景观视野以及活动窗。窗户和遮阳结构会随日照方向的变化而变化。绿色可持续的环保材料，如竹、回收地板和天花板瓷砖被尽可能多地应用在设计中。所有的新建筑空间都是无甲醛环境。内饰材料使用的是低挥发性的涂料、饰面和黏合剂。高效的机械、电气和管道系统集成了创新产品和技术，如太阳能空气加热装置、无水便池、占位传感器和太阳能管，这使得这栋教学楼的节能程度位居校园所有建筑之首。这些综合战略措施形成了一个高性能的节能建筑，每年可减少近454吨的二氧化碳排放量、降低40％的用水，并减少25％的雨水径流。该学校已经得到美国绿色建筑委员会授予的“能源与环境设计先锋奖”（LEED）金级认证。

Location / 地点:
Aspen, USA
Date of Completion / 竣工时间:
2007
Area / 占地面积:
12,555 m^2
Architecture / 建筑设计:
Studio B Architects: Scott Lindenau, FAIA Design Principal, Gilbert Sanchez, AIA Principal, Mike Piche, AIA, Kevin Heath, AIA, Hutton Ford Architects
Interior Design / 室内设计:
Studio B Architects
Landscape / 景观设计:
Design Concepts
Photography / 摄影:
Aspen Architectural Photographers+Time Frame Images
Client / 客户:
Aspen School District

Exterior Materials: Over-sized Textured Brick, Galvalume Inter-locking Metal Panels, Aluminum Clad Windows and Doors and Tranluscent Walls;
Interior Materials: Bamboo Ceilings and Walls, Textured Brick, Galvalume Metal inter-locking Panels, Gypsum Wall Board, Concrete Floors, Acoustical Ceilings, Rubber Floor Tiles, Suspended Light Fixtures, Chrome Plumbing Fittings and Reclaimed Carpet Squares.

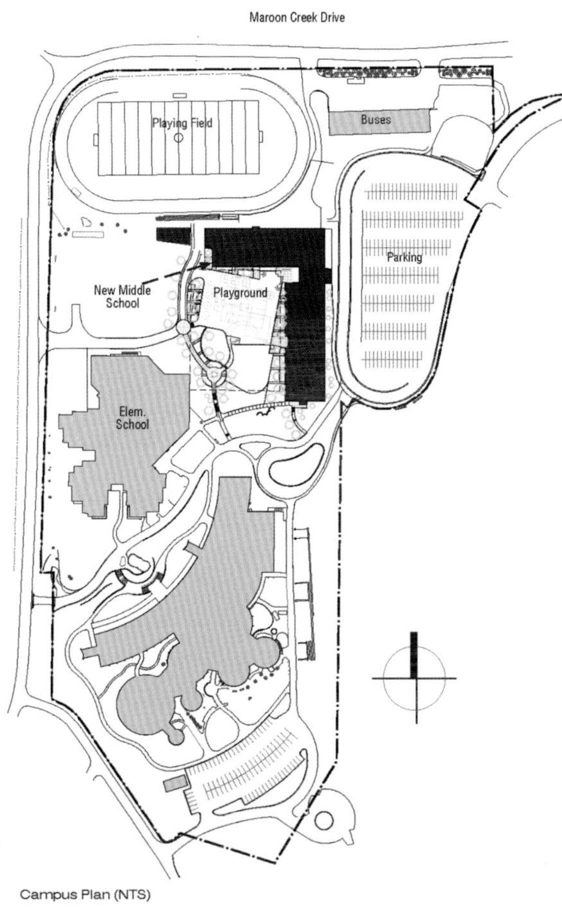

Campus Plan (NTS)

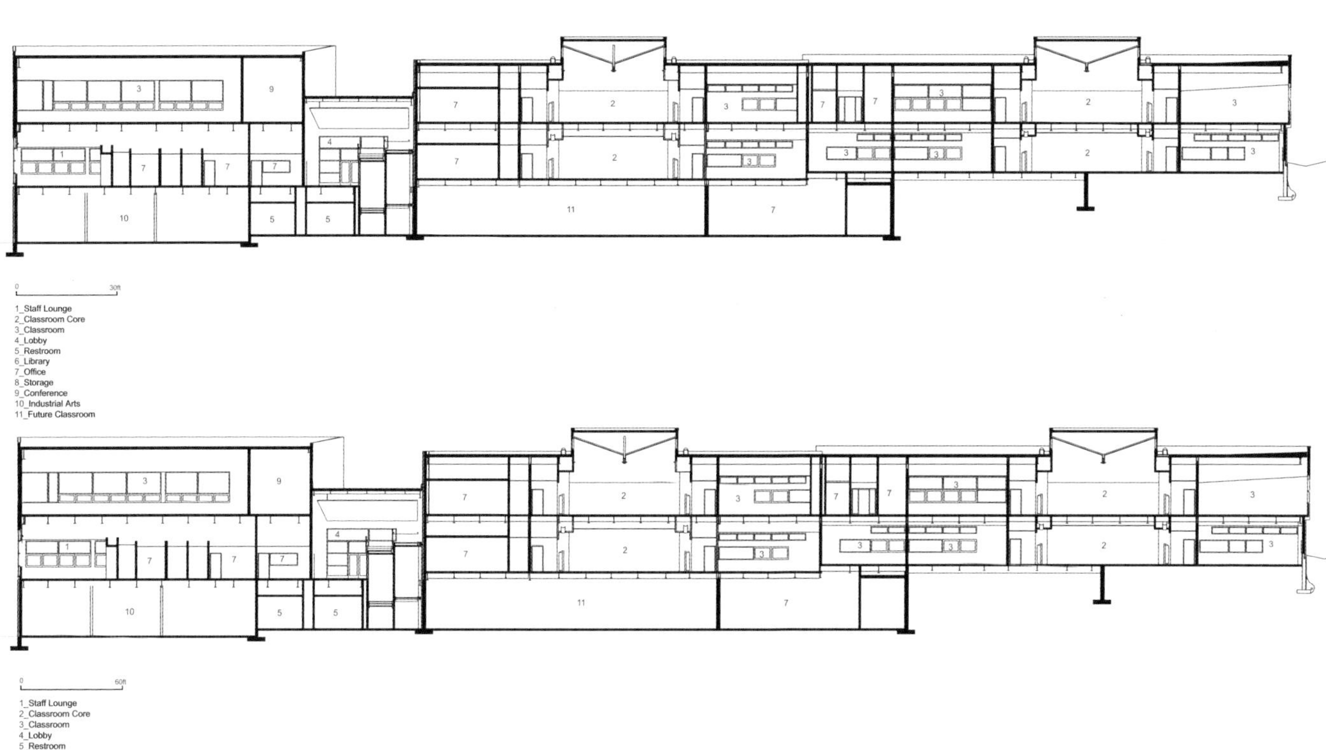

0 30ft

1_Staff Lounge
2_Classroom Core
3_Classroom
4_Lobby
5_Restroom
6_Library
7_Office
8_Storage
9_Conference
10_Industrial Arts
11_Future Classroom

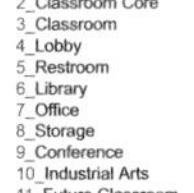

0 60ft

1_Staff Lounge
2_Classroom Core
3_Classroom
4_Lobby
5_Restroom
6_Library
7_Office
8_Storage
9_Conference
10_Industrial Arts
11_Future Classroom

Third Floor Plan

1_Gym
2_Classroom Core
3_Classroom
4_Lobby
5_Restroom
6_Library
7_Office

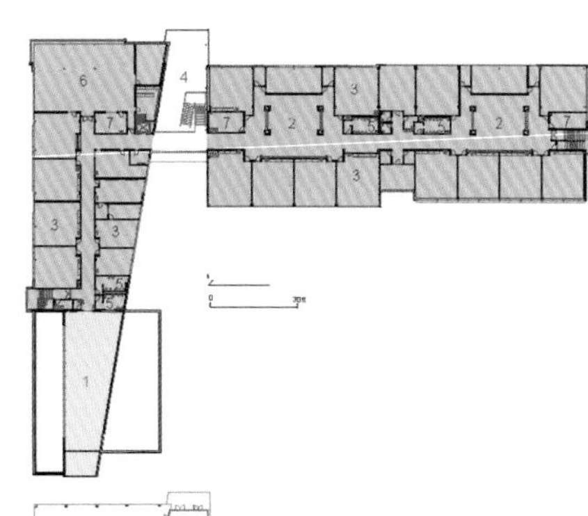

Second Floor Plan

1_Gym
2_Cafeteria
3_Administration
4_Classroom Core
5_Classroom
6_Lobby
7_Restroom
8_Terrace

First Floor Plan

1_Gym
2_Outdoor Ed.
3_Arts
4_Locker Room
5_Music
6_Future Classroom
7_Mechanical
8_Playground

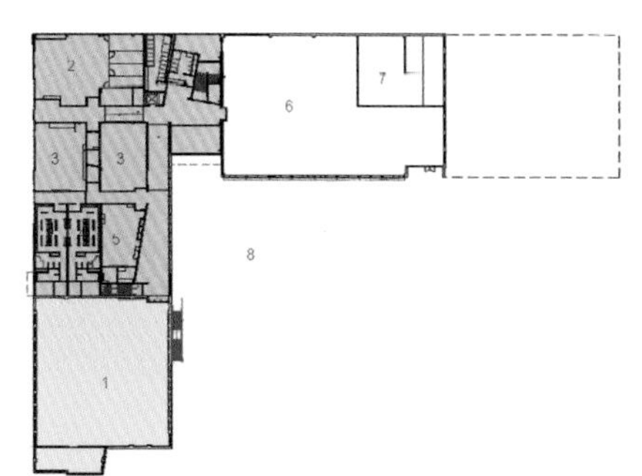

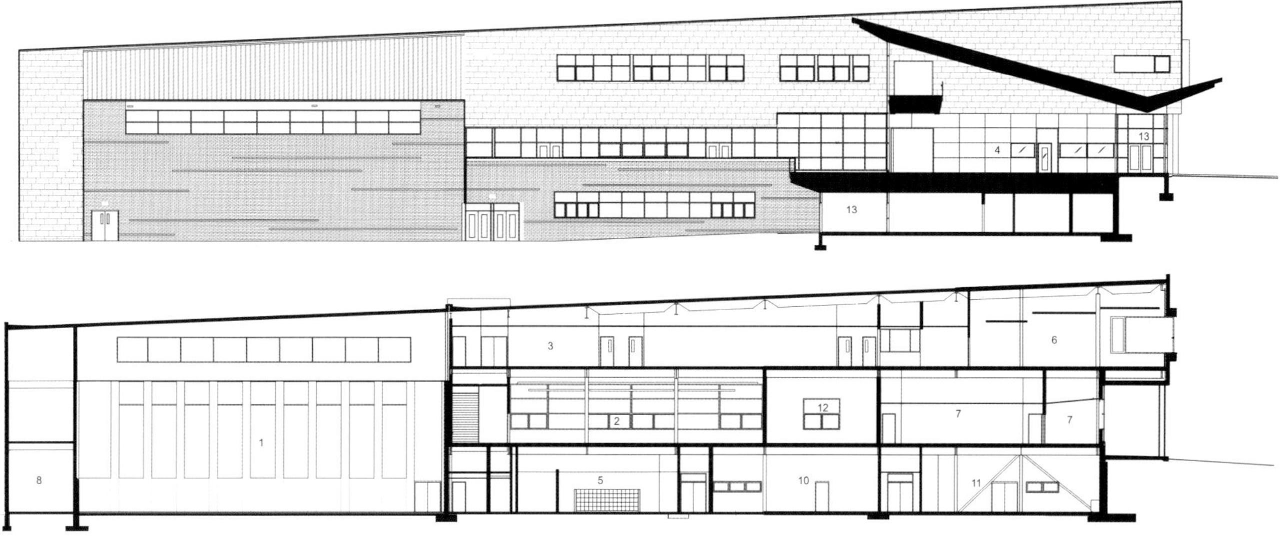

1_Gym
2_Cafeteria
3_Passage
4_Lobby
5_Locker Room
6_Library
7_Office
8_Storage
9_Conference
10_Industrial Arts
11_Outdoor Education
12_Food Prep
13_Vestibule

Carroll A. Campbell Jr. Graduate Engineering Center

卡罗尔 A. 坎贝尔 研究生工程研究中心

Location / 地点:
Greenville, USA
Date of Completion / 竣工时间:
2006
Area / 占地面积:
7896.75 m²
Architecture / 建筑设计:
Mack Scogin Merrill Elam Architects with The Facility Design Group
Interior Design / 室内设计:
Mack Scogin Merrill Elam Architects
Landscape / 景观设计:
Michael Van Valkenburgh Associates
Photography / 摄影:
Timothy Hursley
Client / 客户:
Clemson University

The Carroll A. Campbell Jr. Graduate Engineering Center is the first academic building of a new automotive engineering and research campus for Clemson University. Guidelines that served as the foundation for the development of the Graduate Engineering Center included: Satisfy the functional requirements of the program of research and the program of teaching. Empower the individual student. Sponsor both specialized and collective research. Satisfy the aspirations of the partnership between industry, academia, and the public. Encourage collaboration. Be emblematic and incorporate a unique integration of the automobile. Be environmentally responsible and sustainable. Have a defined plan for growth and expansion.

Research conducted at the Graduate Engineering Center focuses on systems integration with concentrations in Lightweight Design, Manufacturing, and Electronics with a chaired professorship for each. The specific program elements associated with each element concentration are clustered around the chaired professor's suite of offices and research labs. Because a multitude of individuals will work on projects related to or within the research coordinated by these chairs, there are a variety of flexible spaces. While each individual pursues his or her own work, there is collaboration, room for informal discussion, and a sense of community. A component of the research sector is the introduction of industry, both in the presence of individuals from the various fields as well as support of certain research. The results of the research being performed at the center benefit both the University and industry. The teaching component of the Graduate Engineering Center curriculum centers on mechanical engineering. The school is comprised of Masters Degree, Post-Doctoral, and Doctoral students as well as faculty members, visiting faculty, partners, and assistant faculty. The Center also houses complementary administrative functions. The school anticipates an initial annual enrollment of forty graduate students with the expectation of growth to over one hundred students per year. The third component of the building is the public function which includes classrooms, auditorium, café, library, and lobby/display spaces. While several portions of the building may not be physically accessible, many are visually accessible encouraging a broader exchange with the general public.

卡罗尔 A. 坎贝尔研究生工程研究中心是克莱姆森大学新成立的汽车工程研究学院的第一栋教学大楼。该中心发展的基本准则包括以下几个方面：满足研究项目和教学计划的功能需求；赋予学生个体以能力；支持专项研究和综合研究；加强产业界、学术界和公众之间的伙伴关系；鼓励协作研究；具有象征性，并集成独特的与汽车相关的元素；承担环境责任并注重可持续性；对发展和扩张有明确的规划。

该教学楼的设计和教学系统相结合，包括轻量化设计、加工制造和电子设备研究，其中每个分支各由一个主持教授负责。这里通常由多人进行需要相应教授指导的穿插实验，因此设计师在此设立了多种灵活的空间。每个人在进行自己工作的同时可以相互合作、随意交流。这个研究部门既有来自各领域人物的参与，又有对某些研究机构的支持。该中心的研究成果将使校方、工业方同时受益。工程研究中心的教学课程以机械工程为主。这所学院由硕士、博士后、博士生、教职员工、访问学者、合作伙伴和助理教师组成。该中心还设有与之功能相对应的行政单位。学院初步预计每年录取研究生40名，之后会增长到每年100名以上。大楼的第三个构成要素是公共设施，包括教室、礼堂、咖啡厅、图书馆以及大厅兼展示空间。除某些部分无法直达，大楼的多数空间都没有视觉障碍，使得相关人员之间进行更广泛的交流。

Floors: Concrete, Engineered Wood, Broadloom and Modular Carpet;
Interior Walls: Concrete, Colored Concrete, Steel, Painted Gypsum Board, Acoustic Wrapped Panels, Glazing;
Ceilings: Painted Concrete, Painted Gypsum Board, Acoustic Ceiling Panel, Custom Acoustic Ceiling Panel;
Roof: Thermoplastic Polyolefin Single-Ply Roofing Membrane;
Exterior Walls: Zinc, Brick, Colored Concrete, Glazing
Seating: Fixed Auditorium Seating;
Structural System: Concrete;
Equipment: ETS Lindgren Electromagnetic Compatibility Chamber, FEV 500 HP Engine Dyno Test Cell, Zeiss Pro T Compact Dual Column Full Vehicle CMM, MTS 320 Tire Coupled Road Simulator and Weiss Climate Test Chamber, Renk Labeco 4-Wheel 500 HP Chassis Dyno and Faist Semi-Anechoic Chamber.

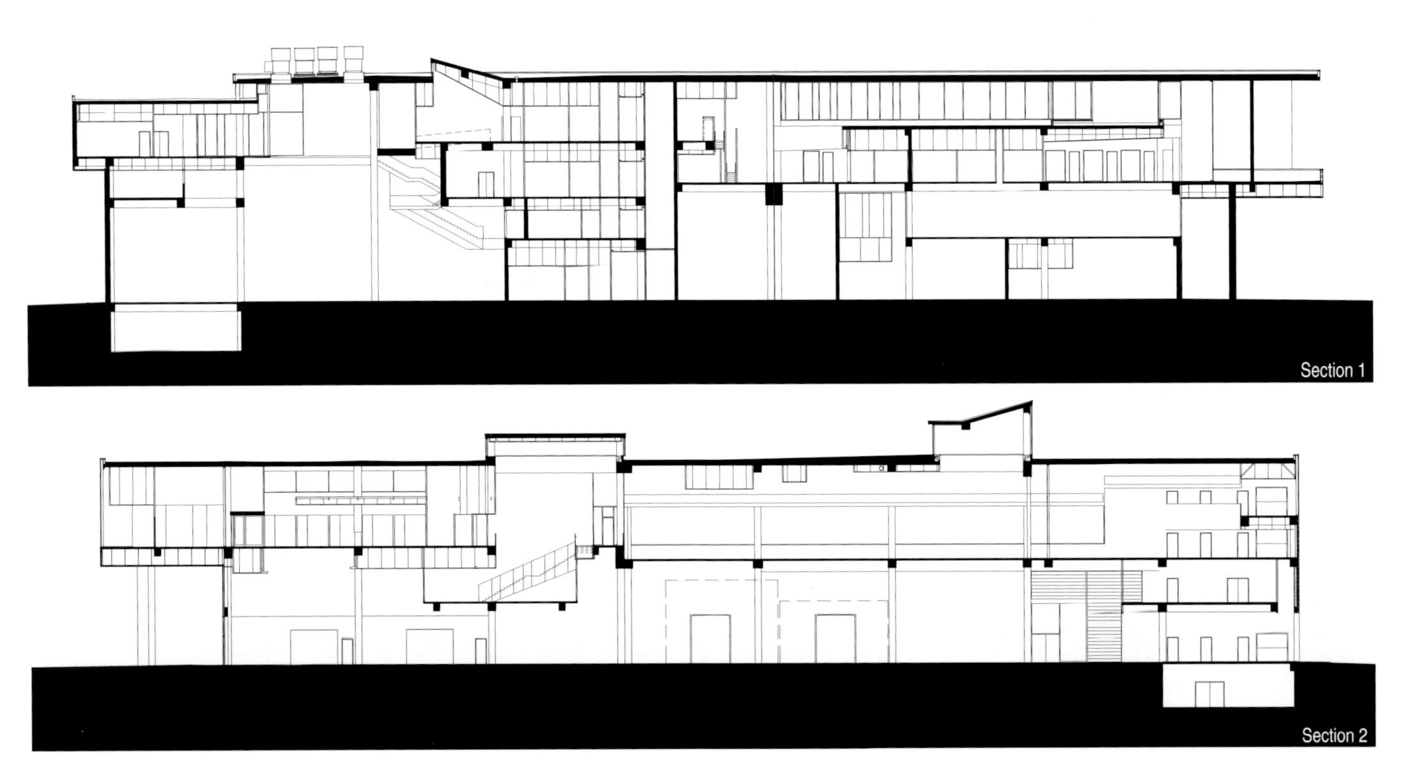
Section 1
Section 2

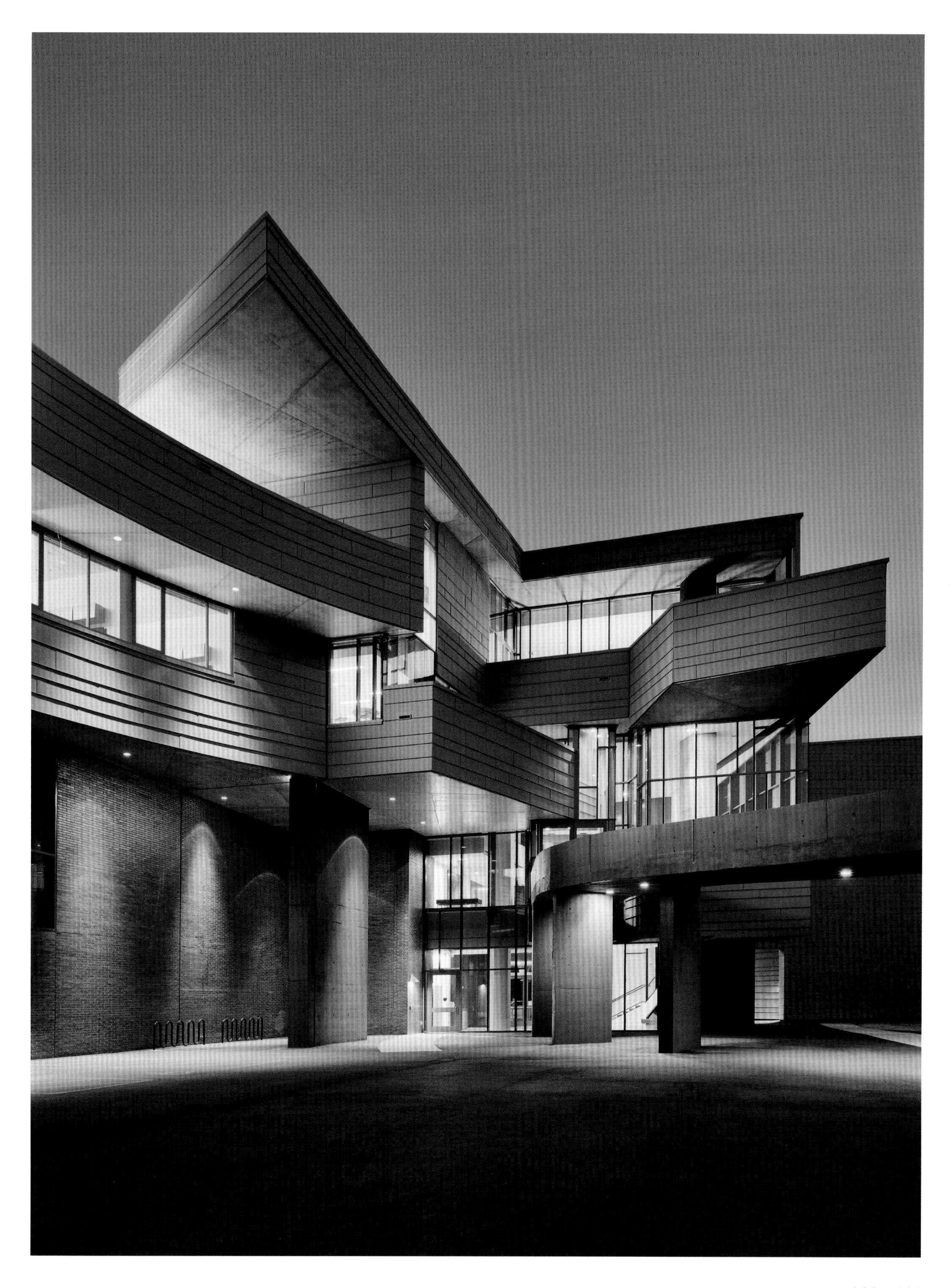

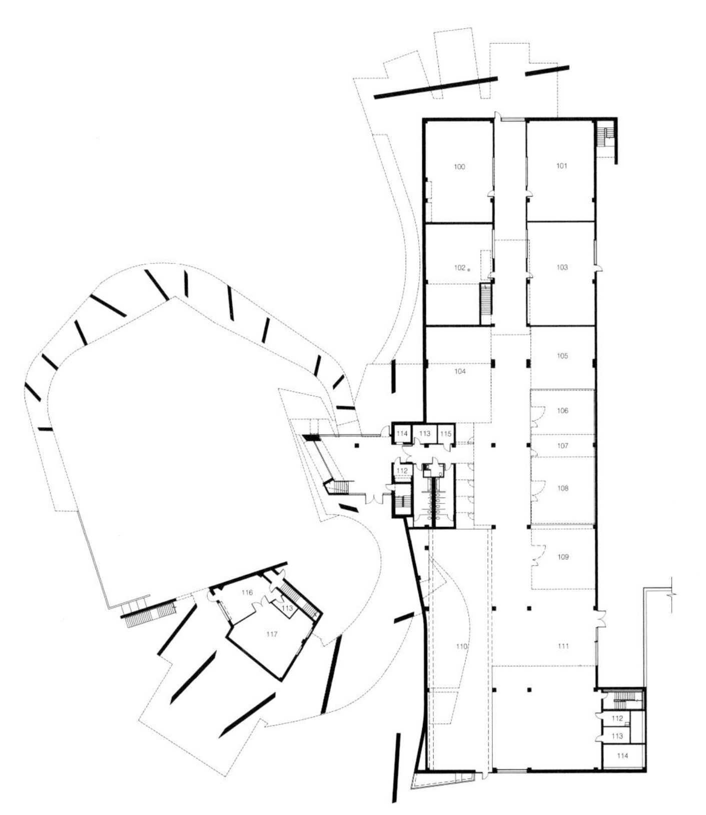

LEVEL 1 FLOOR PLAN

100 Systems Integration Lab 101 Lightweight Design Lab 102 Electronics Lab 103 Manufacturing Lab 104 Tool Shop 105 General Lab 106 Anechoic/Dynomometer Lab 107 Control Room 108 4-Post Shaker Lab 109 Full Vehicle Measurement Lab 110 General Labs West 111 General Labs East 112 Electrical Room 113 CIT Room 114 Elevator 115 Janitorial Closet 116 General Storage 117 Mechanical Room

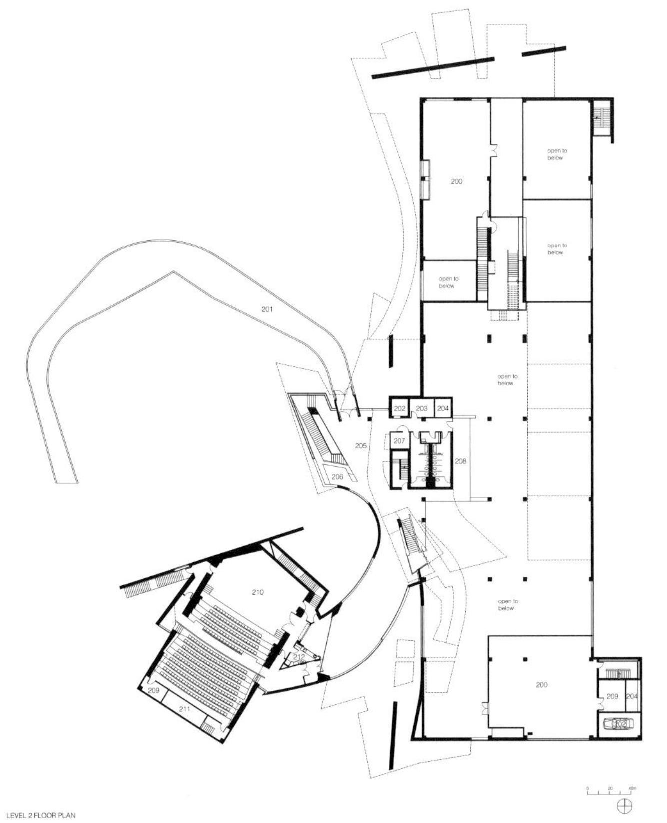

LEVEL 2 FLOOR PLAN

200 Mechanical Mezzanine 201 Pedestrian Bridge 202 Elevator 203 CIT Room 204 Chase 205 Lobby 206 Reception 207 Electrical Room 208 Viewing Platform 209 Storage Room 210 Auditorium 211 Projector Room 212 Refreshments Room

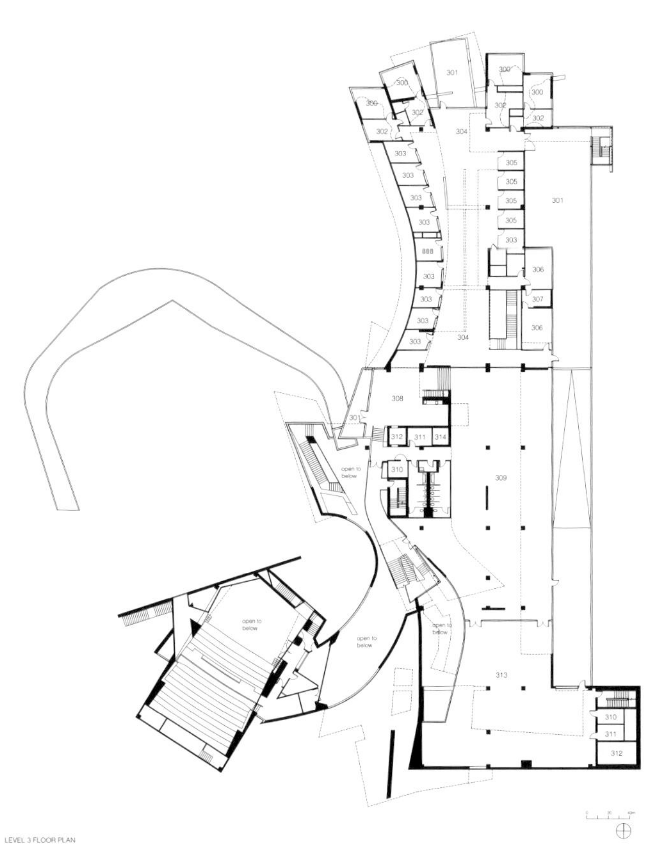

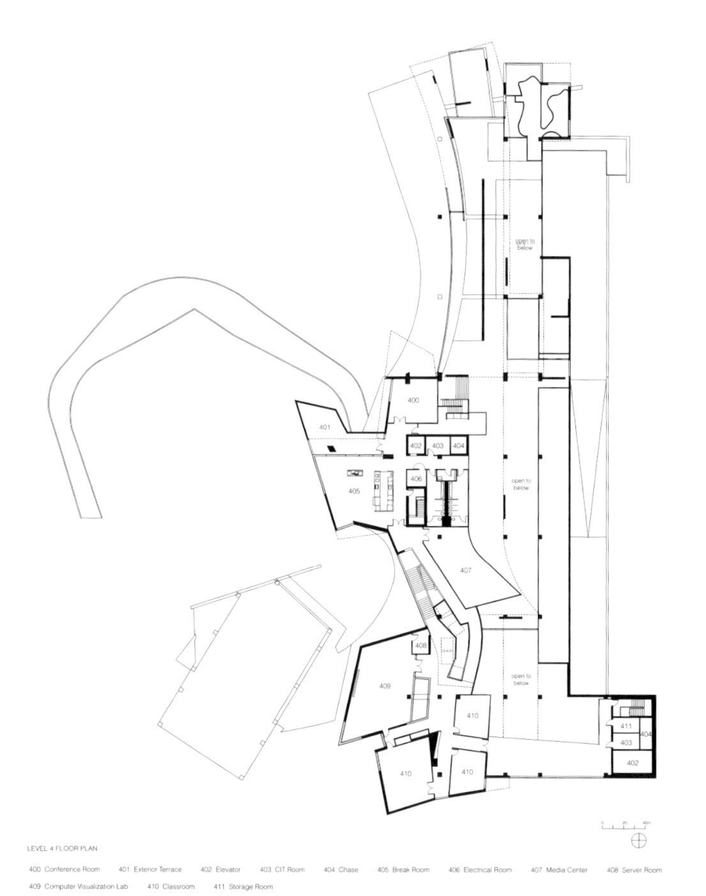

LEVEL 3 FLOOR PLAN

300 Chair Office 301 Exterior Terrace 302 Staff Office 303 Faculty Office 304 Common Area 305 Industry Office 306 Conference Room 307 Storage Room 308 Break Room 309 Student Area 310 Electrical Room 311 CIT Room 312 Elevator 313 Teaching Lab 314 Chase

LEVEL 4 FLOOR PLAN

400 Conference Room 401 Exterior Terrace 402 Elevator 403 CIT Room 404 Chase 405 Break Room 406 Electrical Room 407 Media Center 408 Server Room 409 Computer Visualization Lab 410 Classroom 411 Storage Room

La Bulle Enchantee - Nursery in Sarreguemines

魔法气泡——萨尔格米讷幼儿园

Location / 地点:
Sarreguemines, France
Date of Completion / 竣工时间:
2011
Area / 占地面积:
1,350 m²
Architecture / 建筑设计:
Paul Le Quernec & Michel Grasso
Interior Design / 室内设计:
Paul Le Quernec and Michel Grasso
Landscape / 景观设计:
Paul Le Quernec and Michel Grasso
Photography / 摄影:
Paul Le Quernec and Michel Grasso , Guillaume Duret
Client / 客户:
Communauté d'Agglomération Sarreguemines Confluences

The project is designed as a body cell, placing the nursery at the center of the layout as the nucleus, surrounding gardens as the cytoplasm, and a circumscribing enclosing wall as the membrane. A large outdoor playground is generated through a continuous curvilinear wall that shapes out the boundaries of the facility. The "vaginal" entrance is characterized by a concrete vault which is a continuation of the peripheral wall. Floating within its membrane, the nursery welcomes visitors into a round space at the center of the building, enlightened by a covered patio. All the children units radiate outwards from this central room. This concept of transitional space and centripetal walkway avoids all effects of corridor and all effects of start or end of movement. Thus, the playrooms spread out like a hand-held fan ideally oriented to the south where each one leads to its own playground. The curved lines of walls and ceilings underline the organic and uterin concept which provides safety and comfort within the building for children as for parents.

Double scaling: The ceiling height in the surrounding playrooms is 2.10 m to create a comfortable environment for the babies and children. Thus, to give a stronger spotlight on this idea, they combined it with a variation of ceilings heights up to 4.20 m, creating a large range of spaces and volumes. Thanks to this contrast effect, the units with low ceilings are perceived as protective recesses. The circulation space and the children units are lighted up by zenithal openings, creating the contrast between spaces with low ceiling that are moderately lightened and spaces that generously receive the day light. The transition between different ceiling heights is smoothly done by the curves, generating a various reflection of light on the ceilings.

Buildings fragmentation: Considering the children's perception of the building, they decided to turn it into a raw of variously dimensioned boxes coming out a bush of bamboos and capturing the day light. This idea does not only consist in an aesthetical function: thanks to its shade, the bamboo coat protects the frontages of the building from overheating while the hat shaped boxes regulate the energy intake depending on seasons.

该项目的外观被设计成一个人体细胞，位于中心的幼儿园的教室象征着细胞核，周围的花园象征着细胞质，封闭式的围墙则象征着细胞膜。连续的曲线墙体围合成一个大型室外运动场，同时勾勒出幼儿园场地的轮廓。入口呈“叶鞘”状，是一个混凝土拱顶结构。幼儿园教室漂浮在“细胞膜”内部，穿过“细胞膜”就会到达这里。这里是一处圆形的空间，宛如一个有顶的天井。所有的儿童活动单元采用从中心向外辐射的布局。这种过渡空间和向心步道的设计避免了走廊结构带来的弊端，行动起来不会有起点和终点的限制。同时，游戏室呈风扇状扩展开来，并朝向南方。墙壁和顶棚的弧形线条保证了孩子和父母在建筑物内部的安全性和舒适性。

双尺度：周围游戏室的顶棚高2.1m，为孩子们营造出一个舒适的环境。同时，为了强化这一设计手法，设计者将其与不同的顶棚高度相结合，最高至4.2m，创造出一个大尺度的空间和体量。顶棚较低的空间则呈现出防护式的凹形结构。流通空间和儿童单元通过顶部的天窗获得光照。曲线形构造使得不同高度的顶棚有个流畅的过渡，还使顶棚上产生各种不同的光反射效果。

建筑细部：考虑到孩子们对建筑的理解，设计者建立了不同大小的盒状结构，这些盒状结构如同一个个伸出竹林呼吸阳光的气孔。这种设计不仅实现了建筑的美学功能，而且具有新型现代建筑的节能环保功能。

Exterior finish: Roughcast;
Entrance vault: Concrete;
Floor: Red Linoleum;
Ceilings: Acoustic Plaster, Curved Plaster, MDF Blades;
Furniture: Painted MDF;
Illuminations: Fluorescent Lights, Floor Lights;
Plants: Bamboos, Cherry Blossom;
Other: Red Rubber, Black Asphalt.

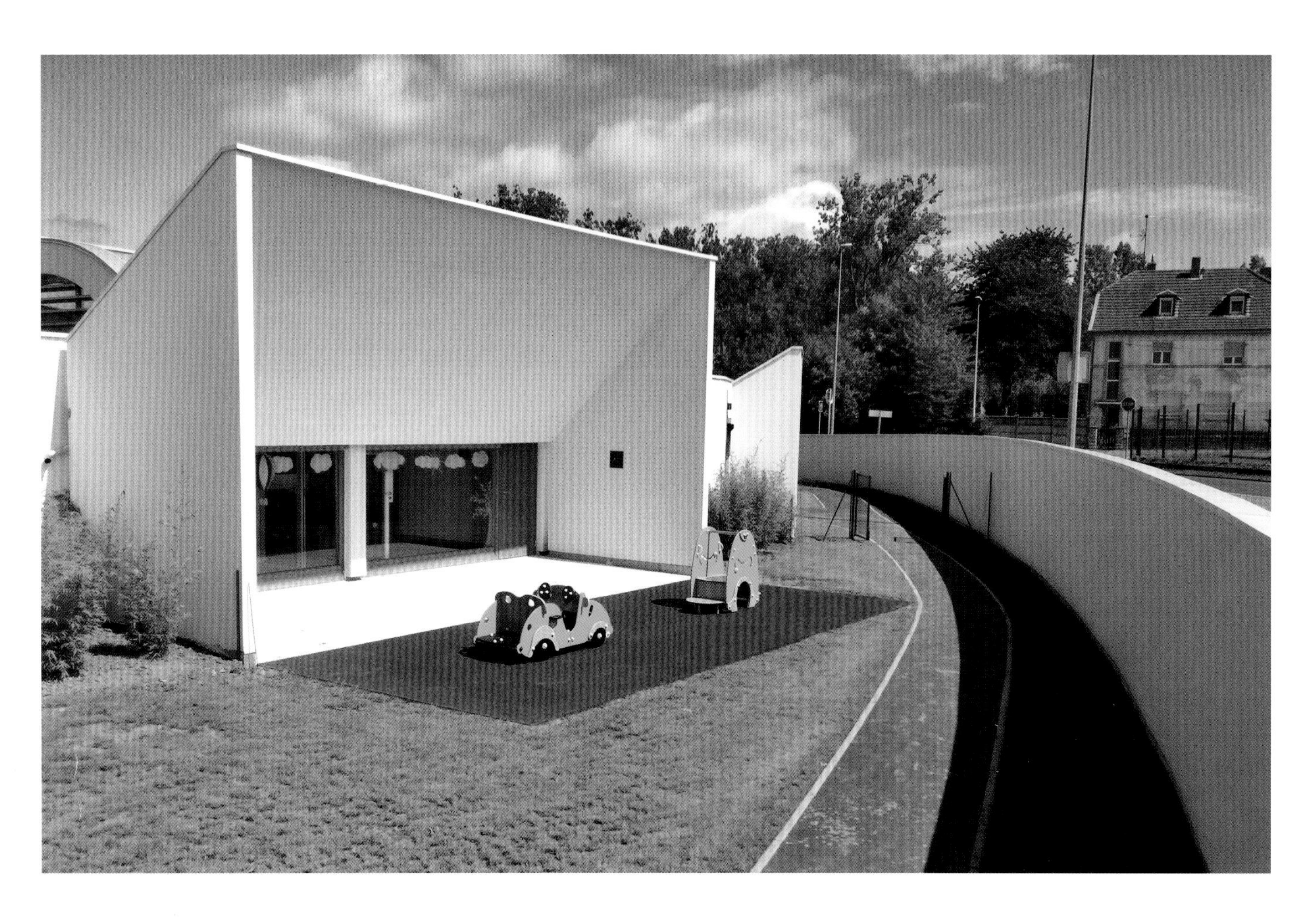

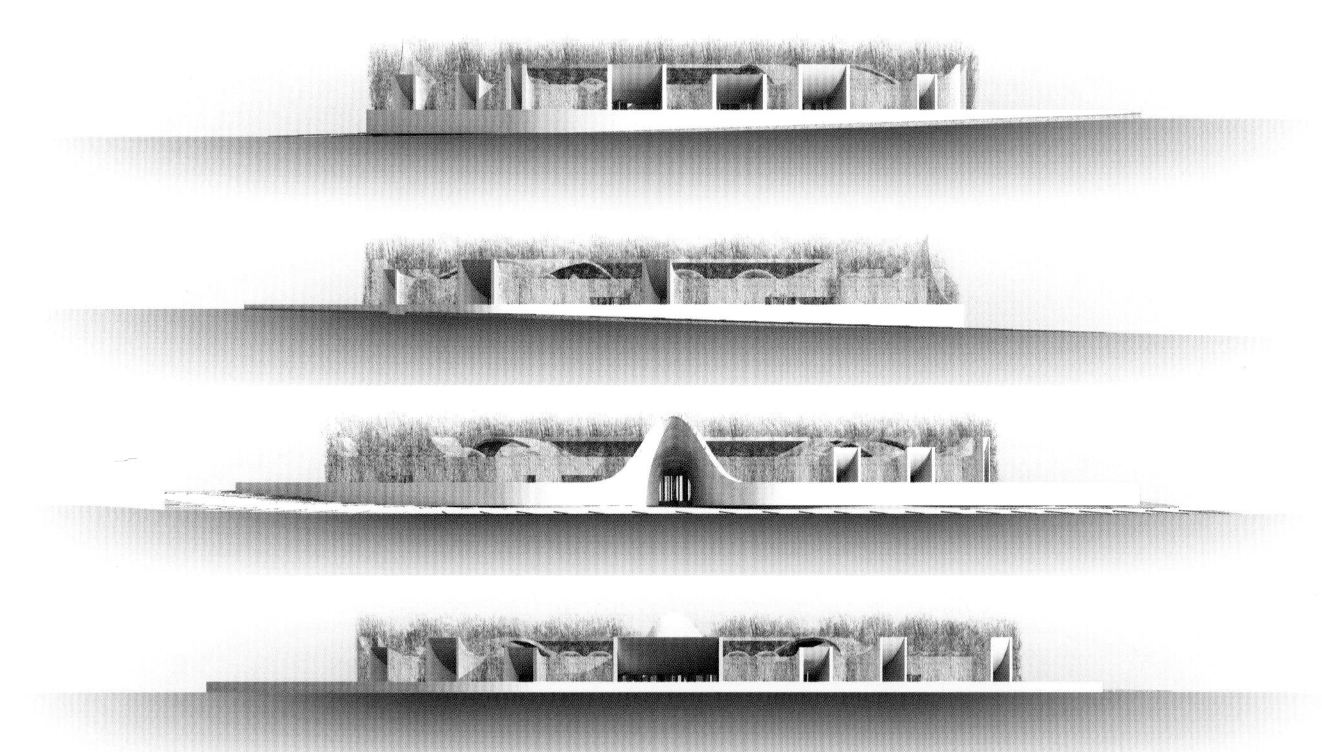

Facades

Master Plan

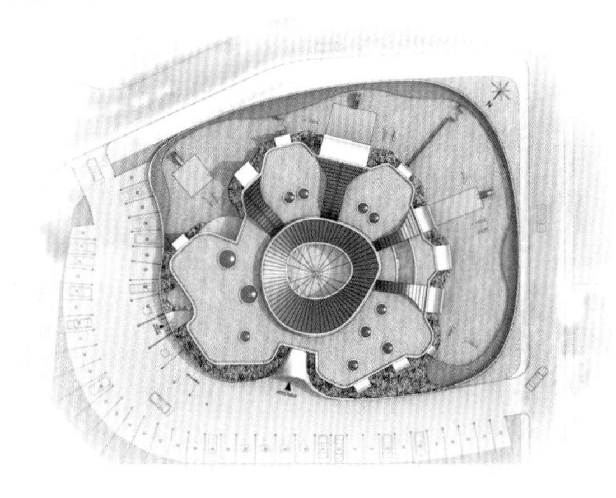

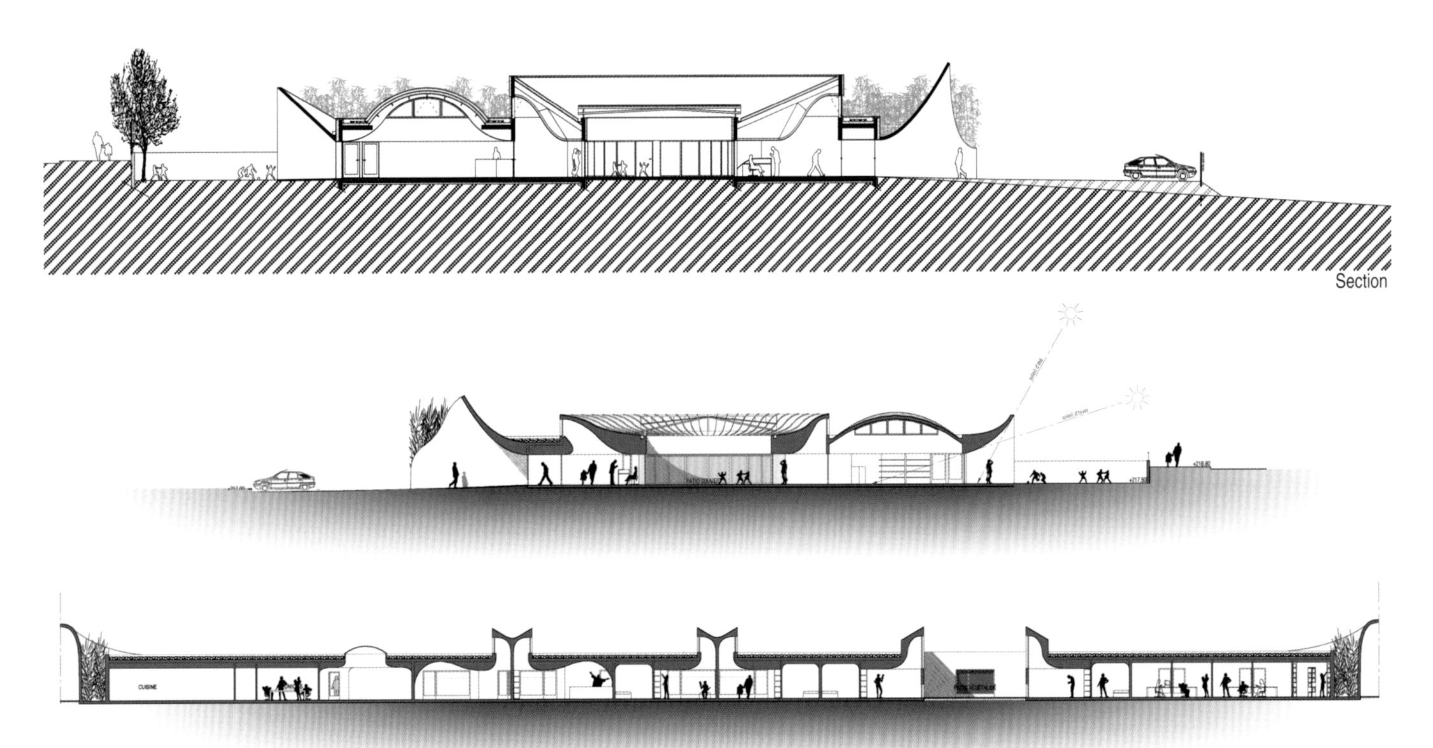

Section

Facades

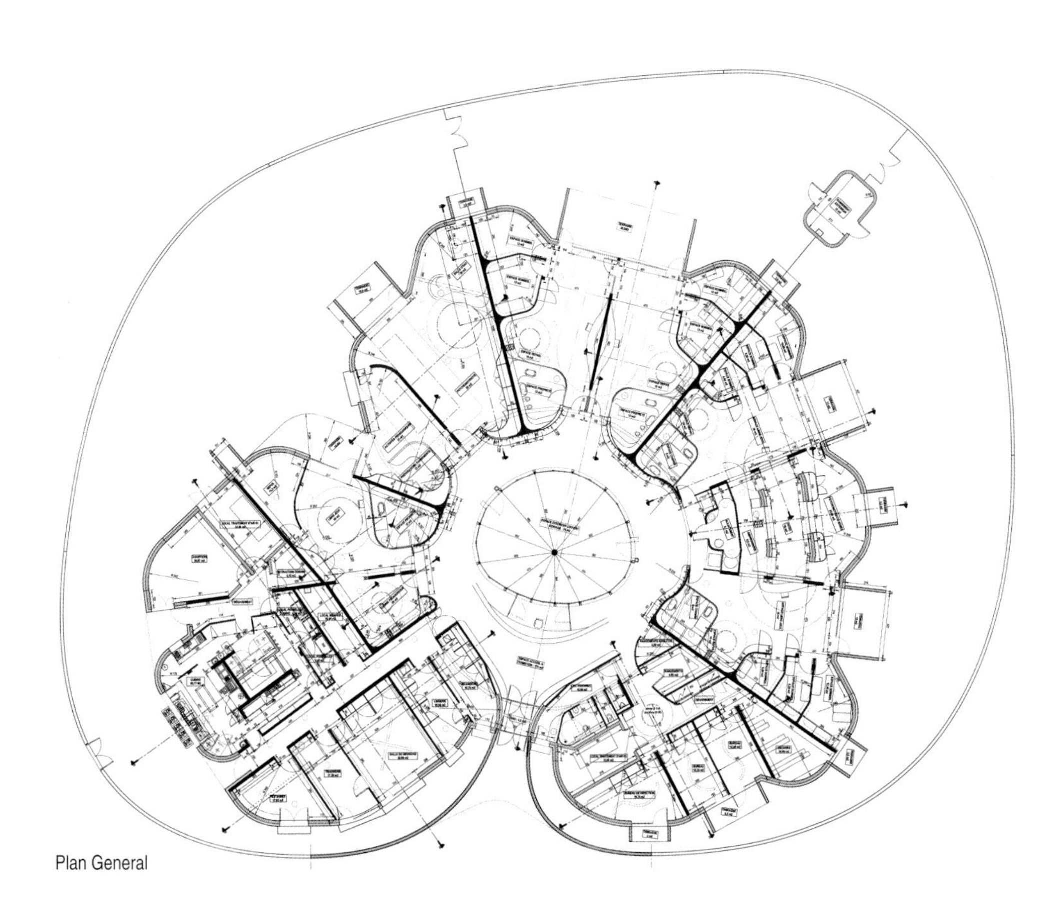

Plan General

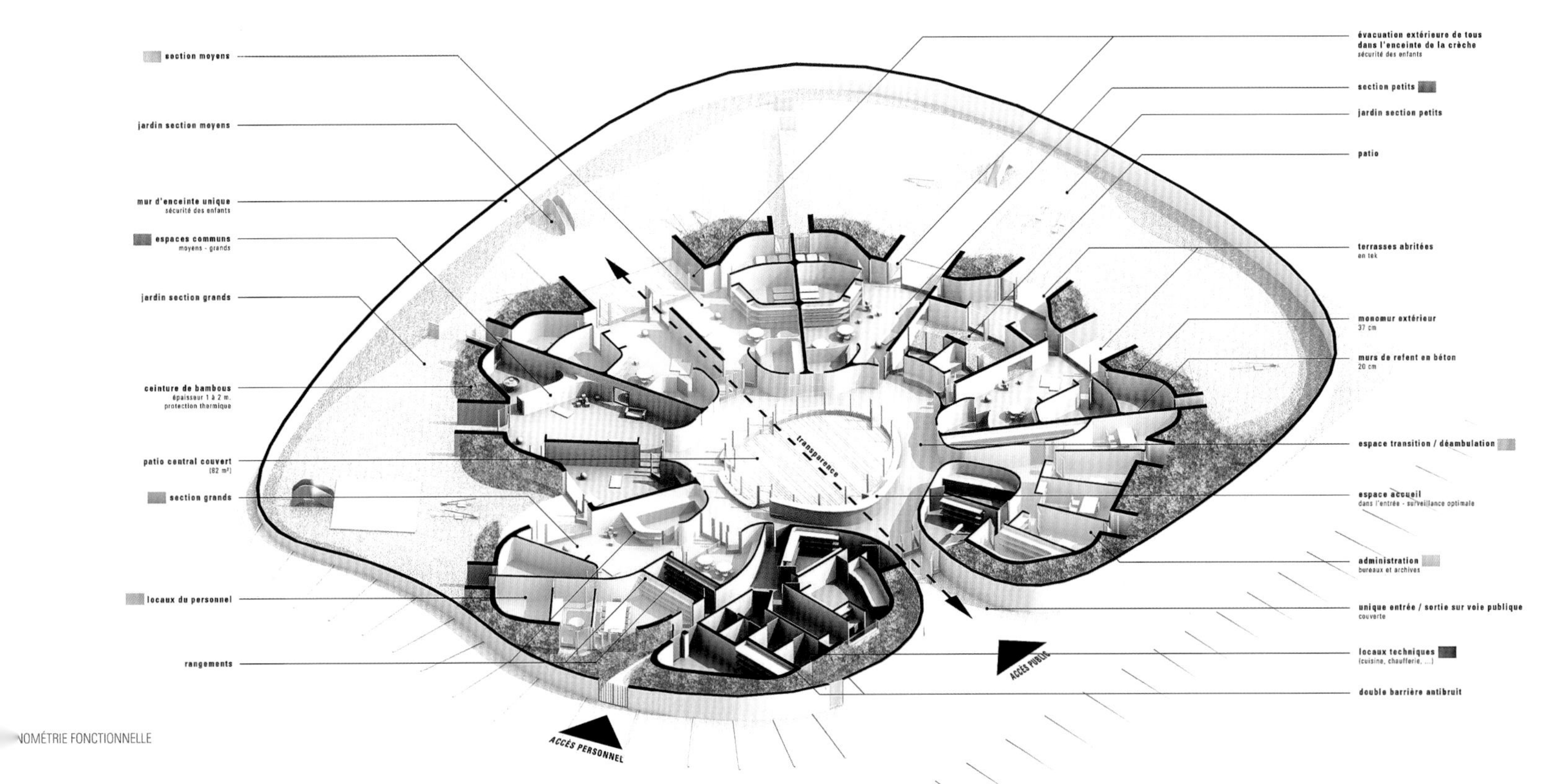

NOMÉTRIE FONCTIONNELLE

University of Wales Newport City Campus

威尔士大学纽波特分校

The University of Wales, Newport commissioned a masterplan to create a new campus for the schools of Art, Media, Design, Business and Technology on the waterfront in the heart of Newport. The University's vision was to create a new learning quarter integrated with the city centre. A former wharf, the site is at the heart of the regeneration vision for the city and its region; to revitalise the waterfront and to foster physical links between the water and the city centre. The new buildings create a new visual identity, and an inspirational place for learning. The new campus has achieved a 14.5% improvement on Part L and BREEAM Excellent without heavy investment in low or zero carbon technologies.

The main entrance is from the new piazza beside the dramatic masts of the pedestrian bridge. This will form the new civic heart of the city, re-engaging the existing city centre with the river. A generous cascading stair leads up to the 'learning resource' terraces which form a stepped plateau above the suite of lecture theatres and technical teaching space at ground level. The more intensively used spaces with potential public access such as the lecture theatres and performing arts space have been located at ground level, together with the more technical teaching facilities at the southern end of the building. Access to this suite of lecture rooms is via a glazed exhibition 'passeggiata' running along the west side of the building parallel to and looking over the new 'tree-lined' Usk way. This 'passeggiata' creates an exhibition space suitable for a variety of displays to showcase both the work of the University and to host external exhibitions. A small café and seating are located within this space. An important aim of the design is to allow passers-by to see into the building and to experience the creative buzz of University life.

The main performing arts studio also opens out onto a new 'public' space which can be used for external theatre performances on the waterfront. The 'passeggiata' also leads out onto this external riverside space forming both a second entrance to the building as well as a link to the 'technical' areas within the podium. By setting the library or 'learning resource' zone, café/restaurant and some seminar spaces on top of the podium, users will enjoy panoramic views along the river to the new waterfront square and to the city. The 'terraced' plateau creates a range of environments from quiet, study group spaces to convivial social spaces.

威尔士大学纽波特分校为其艺术、传媒、设计、商业和科学学院规划了一个新校区，其位于纽波特中心的河畔上。校方的目标是建立一个类似城市中心的新学园。这块场地曾是一个码头，如今作为城市区域振兴计划的核心项目，设计师力求使这个河滨区域重现生机，并强化水域和市中心的物质联系。新的建筑呈现出新的外观特征，建立了一处读书学习的佳所。新校区没有大量使用低碳或零碳技术就取得了高于“英国建筑节能设计标准”（Part L）14.5％和“建筑研究院环境评估法”（BREEAM）优异级认证的好成绩。

行人天桥是一个桅杆式建筑，位于新建的广场上，可以通向建筑的正门。这个方案将重新塑造这块城市心脏地带，重新构建当前的市中心与河流的联系。一个设计大方的串联式楼梯连接着台地式的自习室。这些自习室位于首层的演讲厅和技术培训室上方。演讲厅和艺术表演空间使用频繁，它与位于大楼南侧末端的技术培训室一起被安置在首层。演讲厅的入口处位于大楼西侧的玻璃装潢的长廊内，与“绿树成荫”的乌斯克街道平行。这个“长廊”构成了一个展示空间，既可陈列校园作品，也可举办外界展览。这里同时内设了一个小型咖啡厅和座位区。这项设计的一个重要目的是让路人可以一睹大楼内部，体验大学的创意生活。

主要的艺术表演室向外衍生出一个新的公共空间，可以用于举行室外水边戏剧表演。长廊设计扩展到外部河滨区域，构成大楼的二级入口，同时连接着楼内的功能空间。建筑的顶层设置了图书馆兼自习室、咖啡厅兼餐厅和研讨区，人们可以在这里欣赏滨水广场和城市的全景。台地结构营造出了安静的学习环境和愉悦的社交氛围。

Location / 地点:
Newport, UK
Date of Completion / 竣工时间:
2010
Area / 占地面积:
12,000 m^2
Architecture / 建筑设计:
BDP
Interior Design / 室内设计:
BDP
Landscape / 景观设计:
BDP
Photography / 摄影:
Sanna Fisher-Payne
Client / 客户:
University of Wales Newport

Walking Paths: Concrete, Feature Paths, Stenciled;
Turf: Drought Tolerant Natural Turf, Synthetic Turf in High Wear Areas;
Walls: Sandstone, Concrete Block;
Plants: Grevillea, Callistemon, "Bush Tucker" Gardens;
Drainage Gravel: Recycled Concrete.

City
Campus
Campws
Canol y Ddinas

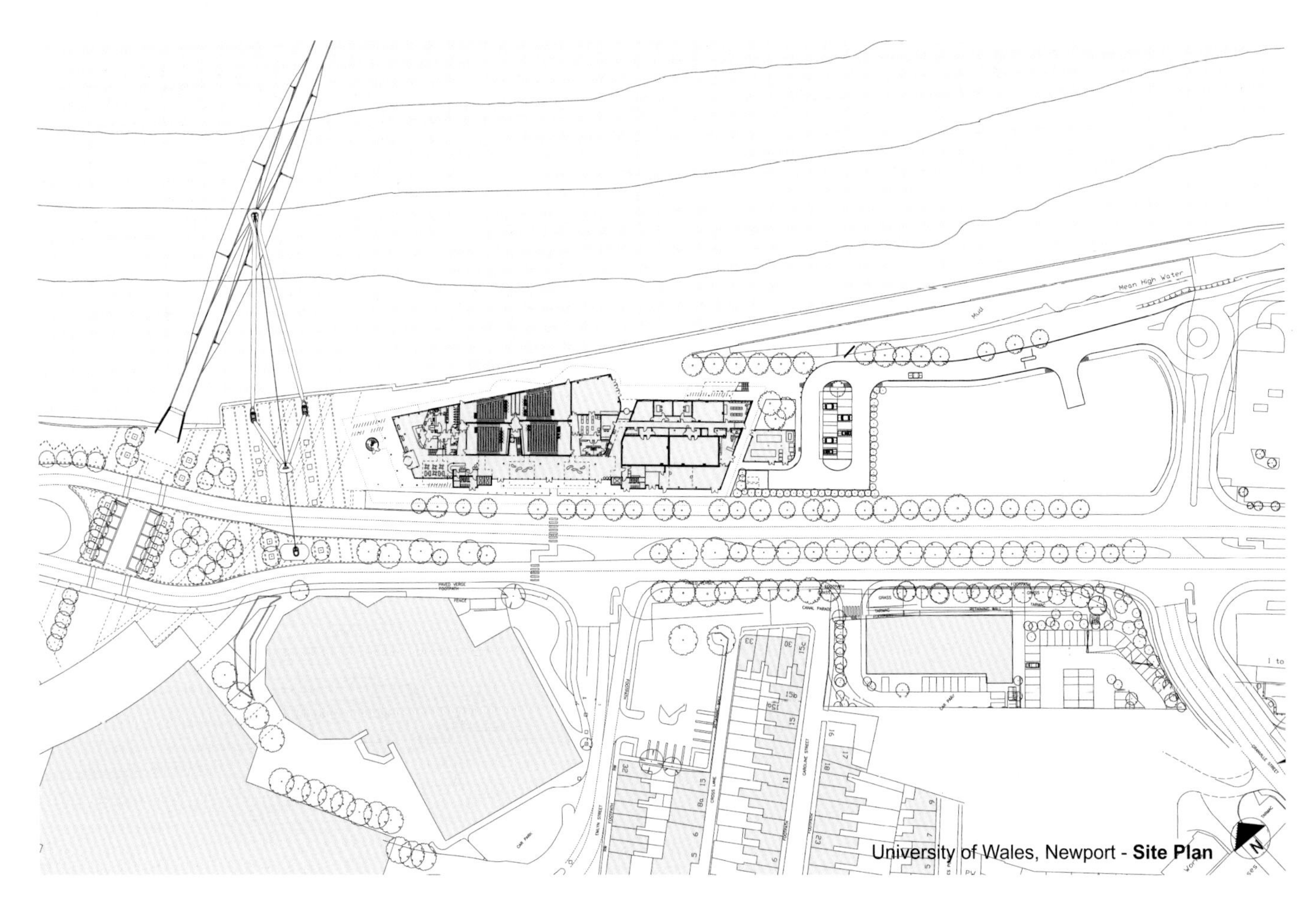

University of Wales, Newport - **Site Plan**

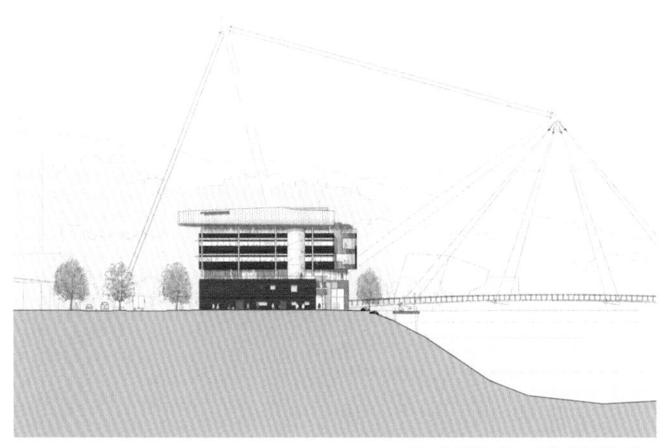

University of Wales, Newport - **South East Elevation**

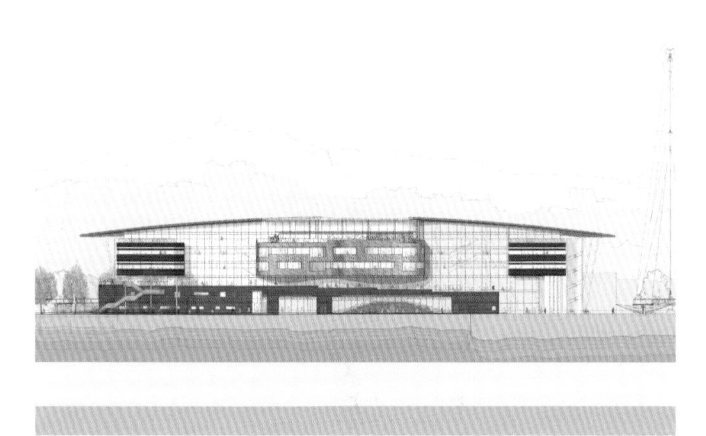

University of Wales, Newport - **North East Elevation**

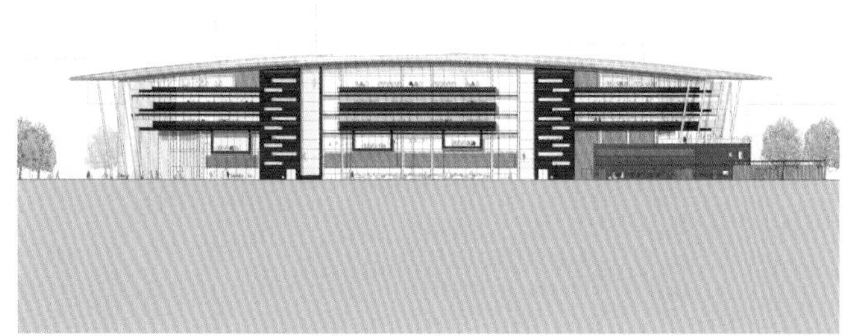

University of Wales, Newport - **South West Elevation**

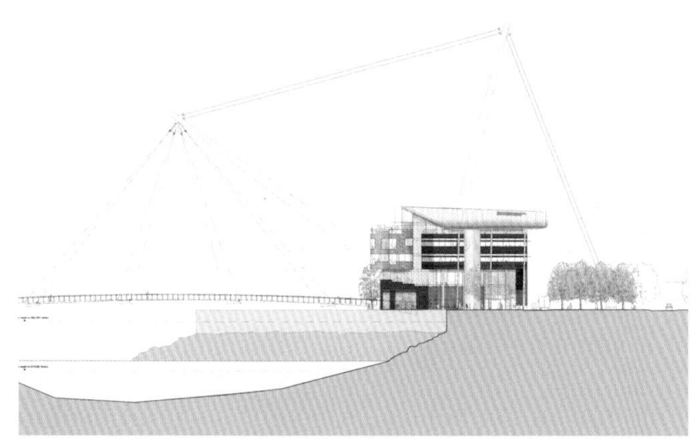

University of Wales, Newport - **North West Elevation**

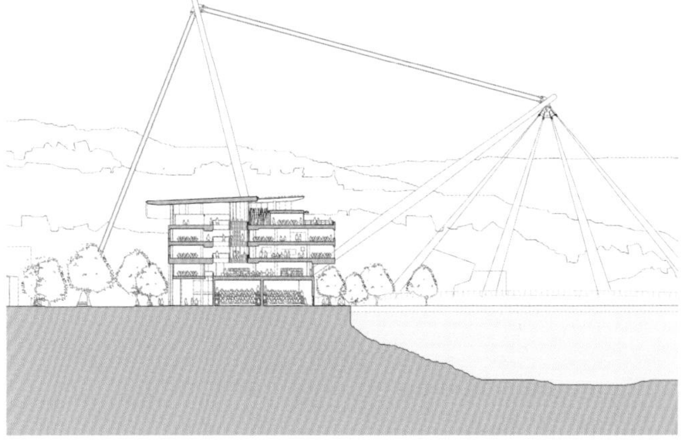

University of Wales, Newport - **Cross Section**

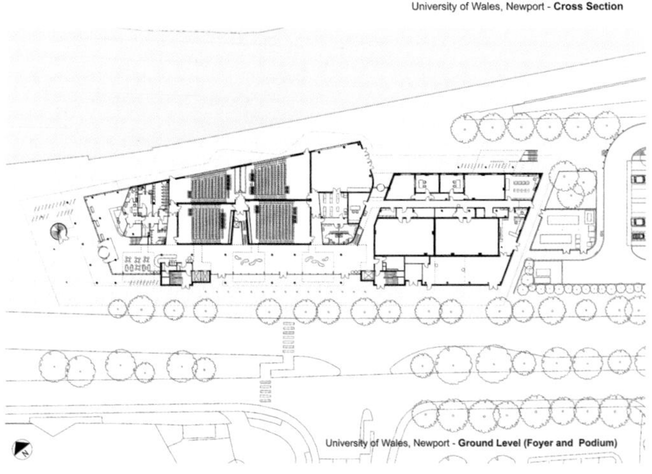

University of Wales, Newport - **Ground Level (Foyer and Podium)**

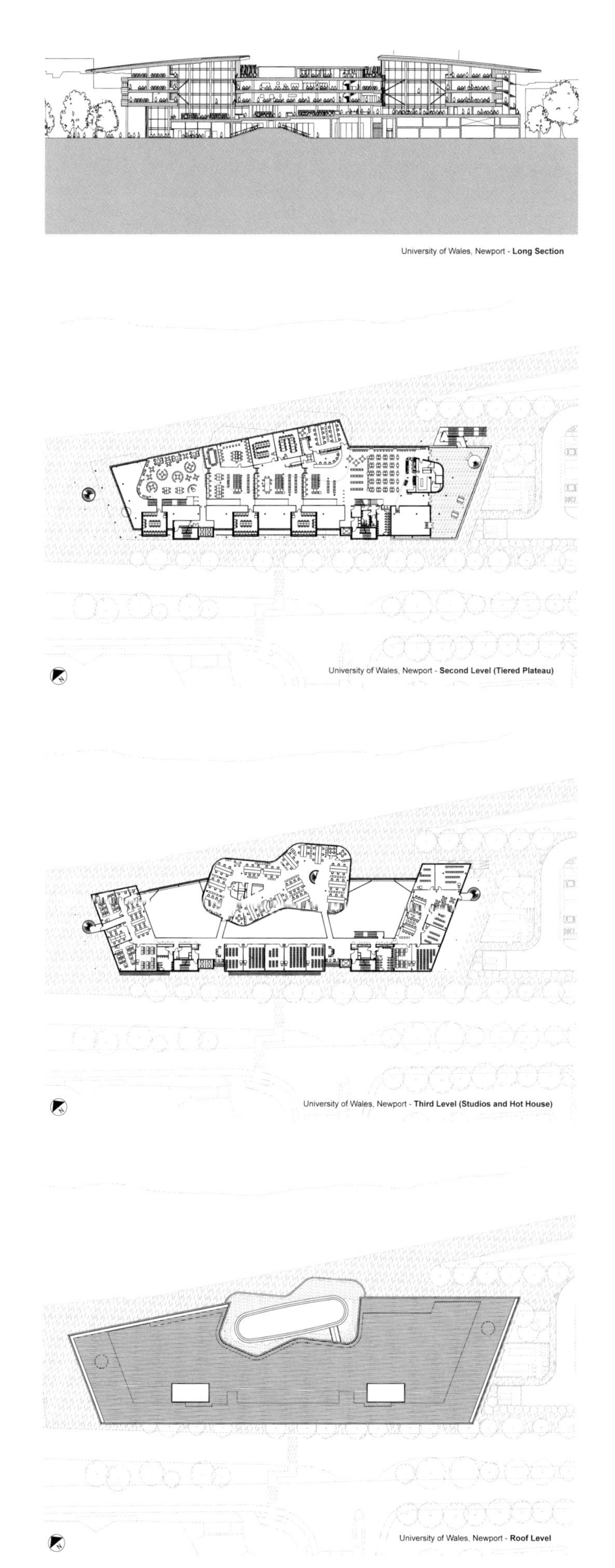

University of Wales, Newport - **Long Section**

University of Wales, Newport - **Second Level (Tiered Plateau)**

University of Wales, Newport - **Third Level (Studios and Hot House)**

University of Wales, Newport - **Roof Level**

University of Minnesota Duluth Civil Engineering Building

明尼苏达州大学德卢斯分校土木工程学院大楼

The University of Minnesota Duluth has instituted a new Bachelor of Science degree in Civil Engineering. The new Swenson Civil Engineering Building provides a home for this new program, containing approximately 3,280 m^2 gross to house classrooms, instructional and research laboratories and office space. The new facility incorporates the existing circulation patterns that are part of the UMD campus.

In designing the Swenson Civil Engineering Building, the project team was charged with the task of incorporating the numerous programmatic and equipment requirements while enhancing the educational function of the building. The program called for large pieces of equipment, including a strong wall and floor system, two 15 ton gantry cranes, and a hydraulic flume. In addition to these items, the design developed to incorporate three large 36' x 24' operable doors to facilitate the movement of the cranes through the building. The layout and openness of the main building spaces enhances the educational experience by providing visual connections to activities within and fostering interactions between students and faculty. The centrally located hydraulics laboratory serves as a main node of activity to which other spaces relate visually and functionally. The east wall of the second floor transportation laboratory incorporates a picture window that provides views into the hydraulics laboratory below. Other laboratory spaces wrap this central volume. Designed to display the building systems as a pedagogical tool, the building showcases the structural, and mechanical systems as well as stormwater management techniques. The building acts as a working classroom for the students using the space. Structurally, the building utilizes precast concrete walls, precast hollowcore floor slabs, and steel.

Having achieved a LEED Gold certification, the new building creates a healthy environment for the occupants through the use of integrated sustainable strategies. The building materials were selected to showcase the beauty of locally available raw, natural, unaltered materials that not only provide the basis for a sustainable building product, but also serve as a teaching tool for the students within the Civil Engineering Department. Through highlighting the properties of the materials in their natural state, very few 'finish' materials are needed or used on the project. The use of raw and locally available products resulted in over 20% of the total building materials being regionally harvested and manufactured, and over 30% of the materials being recycled.

明尼苏达州大学德卢斯分校新设了一所土木工程本科教学机构。该机构位于新落成的斯文森土木工程学院大楼，这里共有约3280m^{2}的空间，含有教室、研究实验室以及办公室。新设施改善了现有的校园流通模式。

在斯文森土木工程学院大楼的设计过程中，项目团队需要考虑总体规划和设备对空间大小的要求，同时要强化大楼的教学功能。该大楼内需设立大型的装备，包括坚固的墙壁、地板系统、两台重15吨的门式起重机和一个液压水槽。二楼交通实验室的东墙包含一个观景窗，透过窗口可以看到下方的液压实验室。中央结构的周围分布着其他实验室。大楼通过其结构体系、机械系统以及给排水系统体现了它服务于教学的宗旨。对于学生来说，这里是一处工作教室。建筑采用预制混凝土墙、预制中空型楼板和钢材。

新建筑已获得“能源与环境设计先锋奖”（LEED）金级认证，通过整合可持续性战略措施创造了一个健康的环境。建筑材料是本地原生、自然、未加工的材料，这不但实现了建设环保可持续的大楼的设计理念，还为土木工程系学生提供了课堂学习的案例。设计方案着重强调材料在自然状态下的性能，因此，项目中极少使用或需要加工后的材料。建筑建设中，超过20％的建筑材料都是从当地获取的，30％以上的材料被回收利用。

Location / 地点:
Duluth, USA
Date of Completion / 竣工时间:
2010
Area / 占地面积:
90,285 m^2
Architecture / 建筑设计:
Ross Barney Architects
Interior Design / 室内设计:
Ross Barney Architects
Landscape / 景观设计:
Oslund and Associates
Photography / 摄影:
Ross Barney Architects
Client / 客户:
University of Minnesota – Duluth

Walking Paths: Concrete;
Accent Path: Steel Plate;
Exterior Accent Surfaces: Taconite Pellets, Composite Pavers Composed of Recycled Materials;
Drainage Gravel: Rip Rap, Mine Waste Rock;
Plants: Native Forbs & Grasses;
Walls: Aluminum Curtainwalls, Exposed Precast Concrete, Weathered Steel Rainscreen Panels Attached to Concrete Masonry Unit Walls.

JAMES I. SWENSON
COLLEGE OF SCIENCE AND ENGINEERING
CIVIL ENGINEERING BUILDING

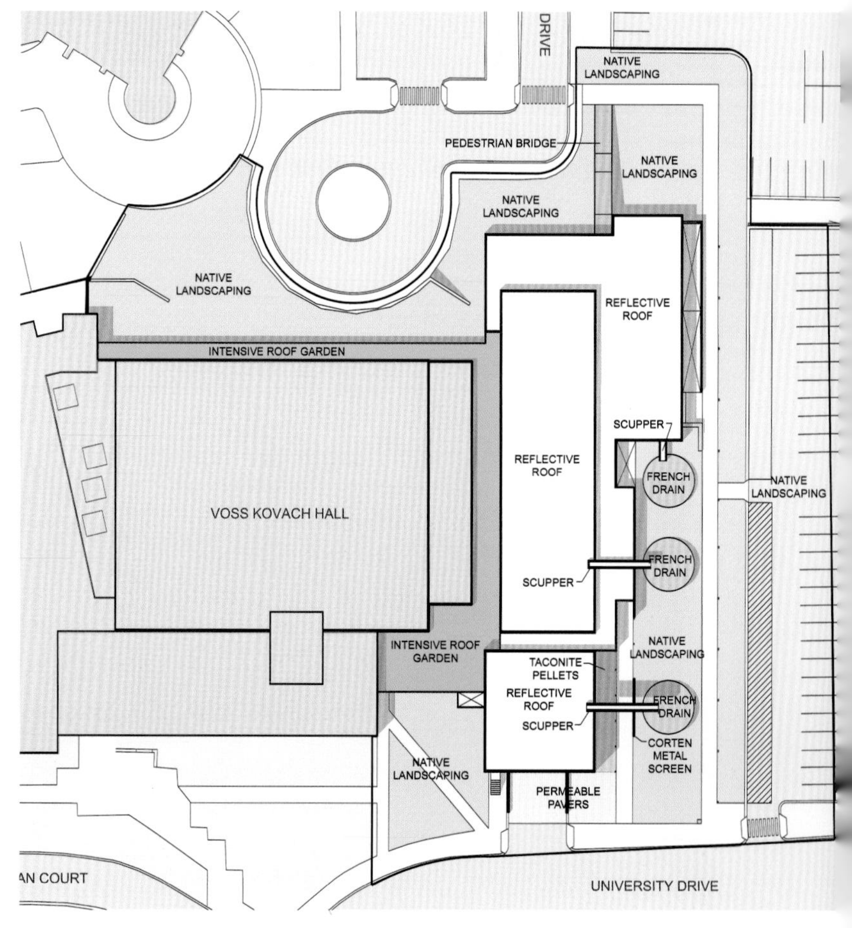
DRIVE
NATIVE LANDSCAPING
PEDESTRIAN BRIDGE
NATIVE LANDSCAPING
NATIVE LANDSCAPING
NATIVE LANDSCAPING
REFLECTIVE ROOF
INTENSIVE ROOF GARDEN
SCUPPER
REFLECTIVE ROOF
FRENCH DRAIN
NATIVE LANDSCAPING
VOSS KOVACH HALL
FRENCH DRAIN
SCUPPER
NATIVE LANDSCAPING
INTENSIVE ROOF GARDEN
TACONITE PELLETS
REFLECTIVE ROOF
SCUPPER
FRENCH DRAIN
CORTEN METAL SCREEN
NATIVE LANDSCAPING
PERMEABLE PAVERS
AN COURT
UNIVERSITY DRIVE

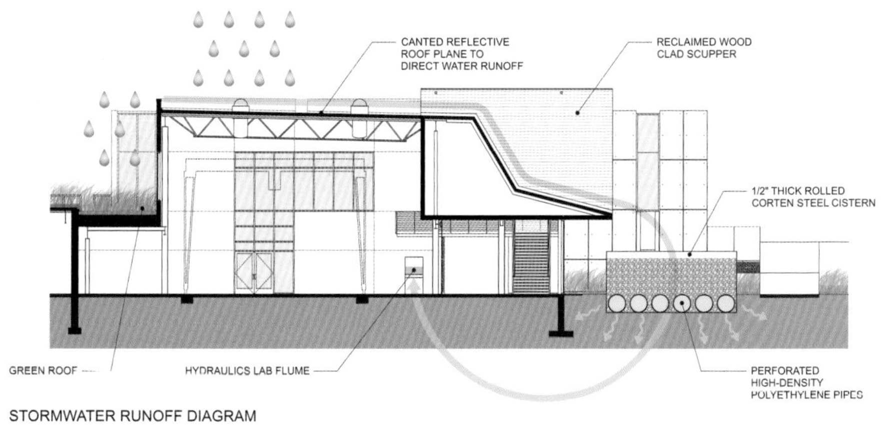

STORMWATER RUNOFF DIAGRAM

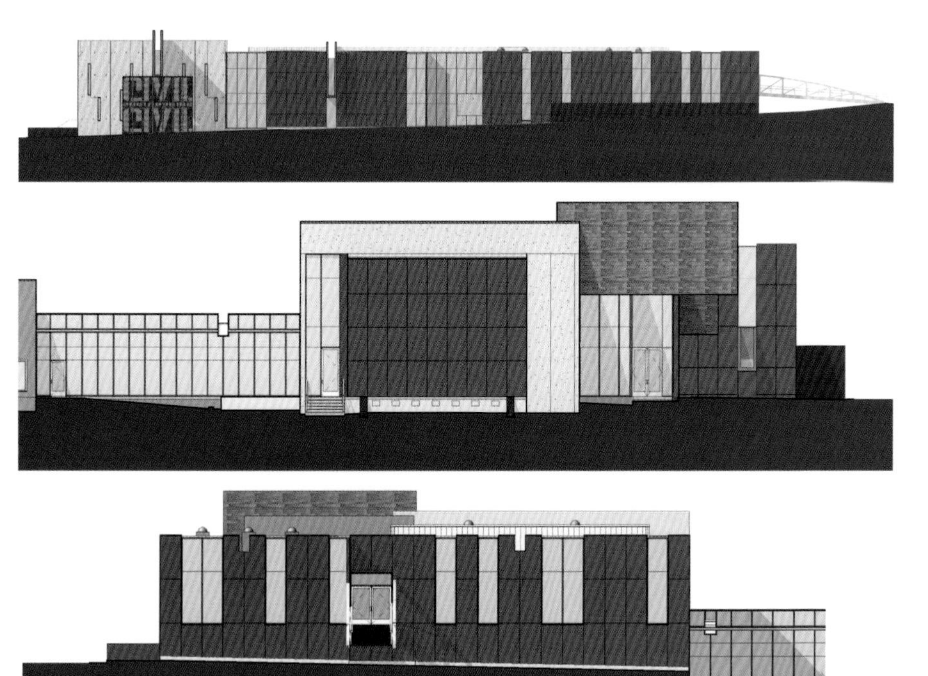

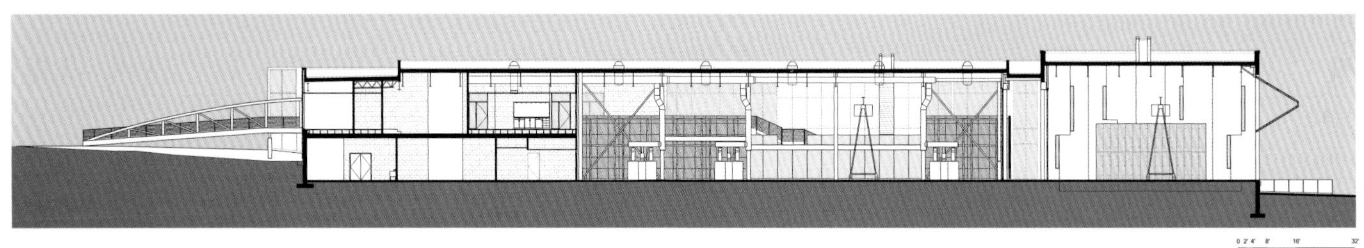

BUILDING SECTION

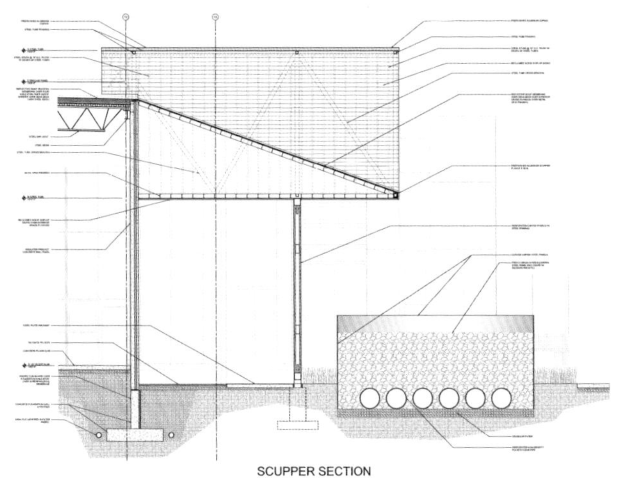

SCUPPER SECTION

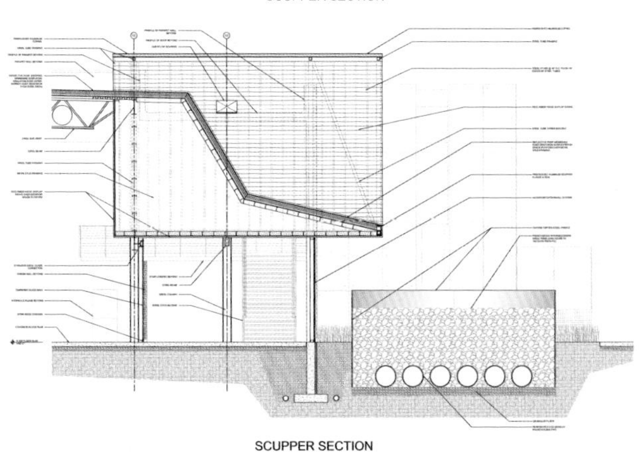

SCUPPER SECTION

15 TON

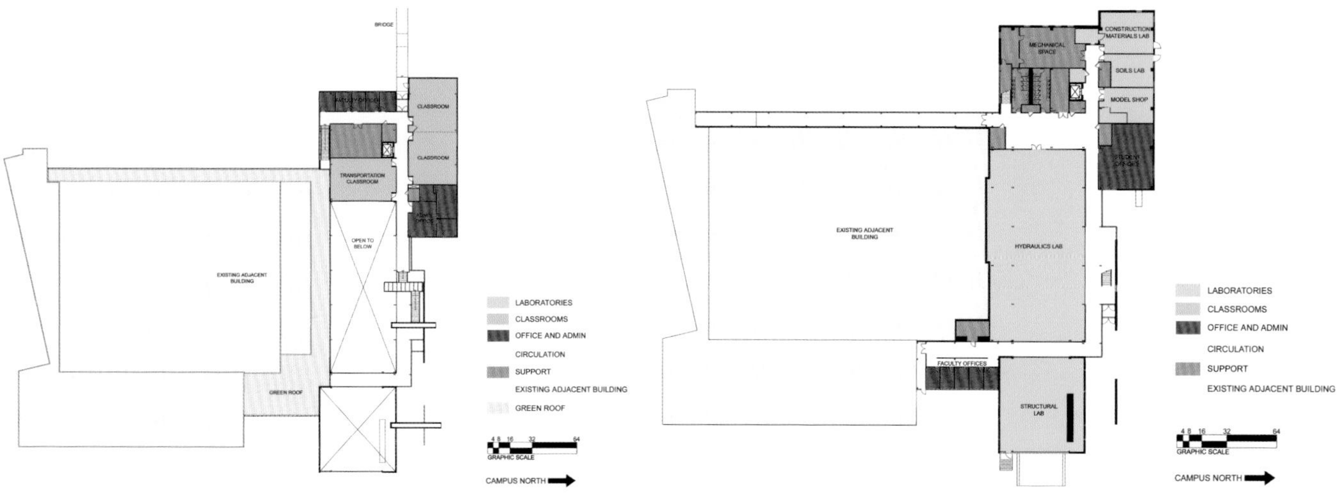
BRIDGE
CLASSROOM
CLASSROOM
TRANSPORTATION CLASSROOM
OPEN TO BELOW
EXISTING ADJACENT BUILDING
GREEN ROOF
LABORATORIES
CLASSROOMS
OFFICE AND ADMIN
CIRCULATION
SUPPORT
EXISTING ADJACENT BUILDING
GREEN ROOF
4 8 16 32 64
GRAPHIC SCALE
CAMPUS NORTH
MECHANICAL SPACE
CONSTRUCTION MATERIALS LAB
SOILS LAB
MODEL SHOP
EXISTING ADJACENT BUILDING
HYDRAULICS LAB
FACULTY OFFICES
STRUCTURAL LAB
LABORATORIES
CLASSROOMS
OFFICE AND ADMIN
CIRCULATION
SUPPORT
EXISTING ADJACENT BUILDING
4 8 16 32 64
GRAPHIC SCALE
CAMPUS NORTH

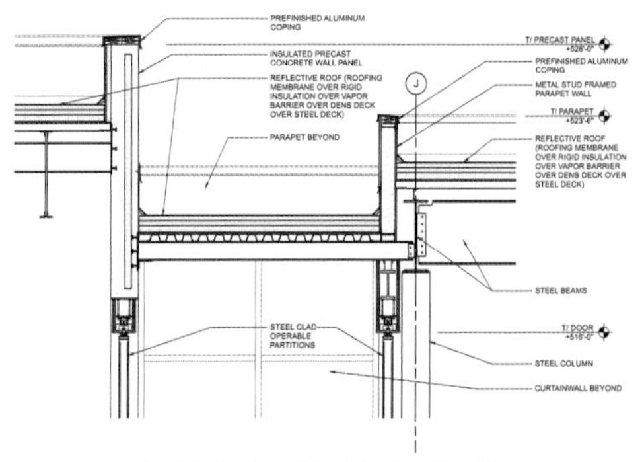

SECTION THRU INTERIOR OPERABLE PARTITIONS

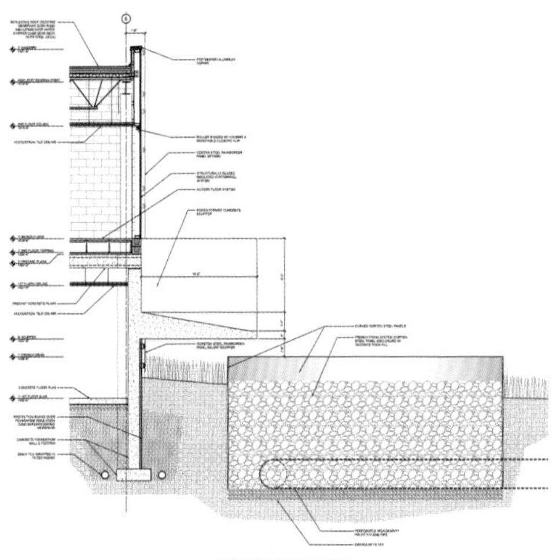

SCUPPER SECTION

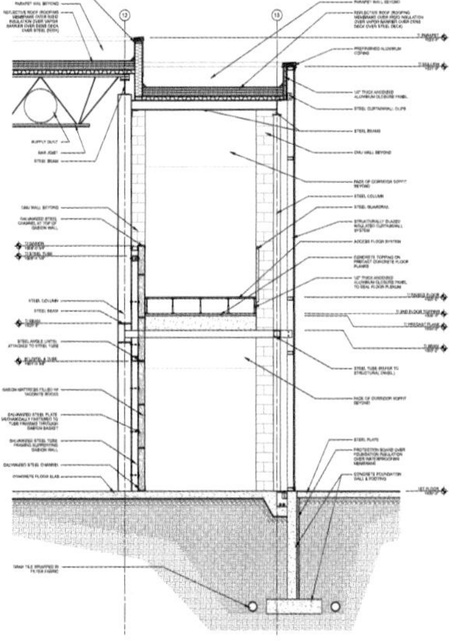

SECTION THRU GABION WALL

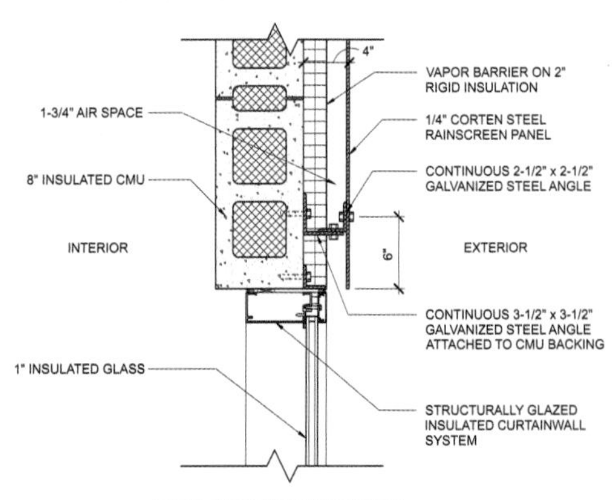

DETAIL AT CORTEN RAINSCREEN

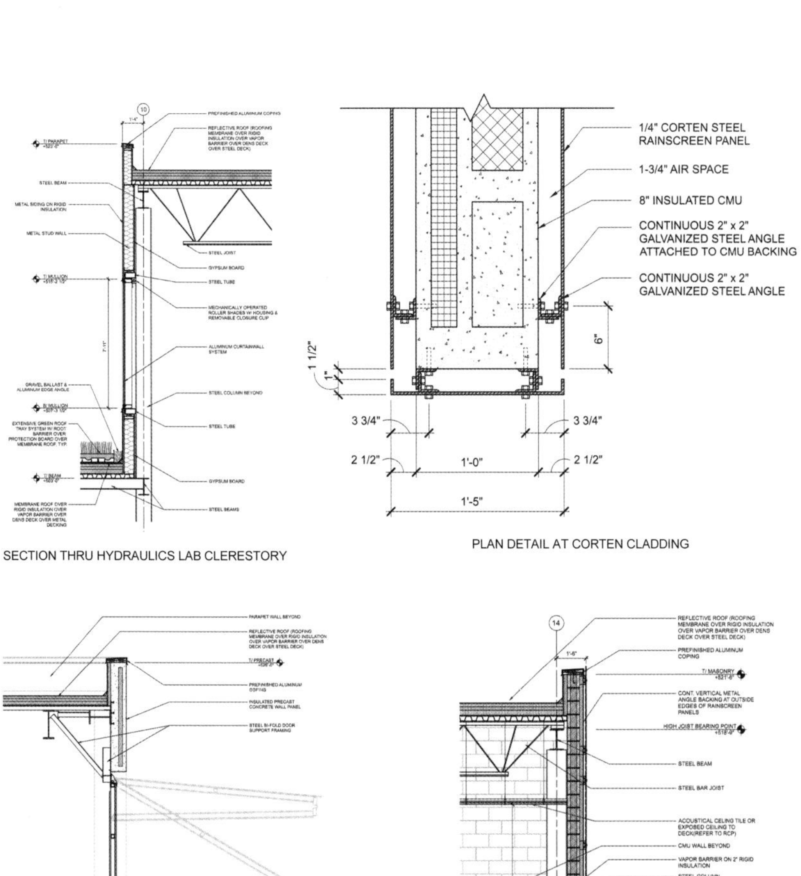

SECTION THRU HYDRAULICS LAB CLERESTORY

PLAN DETAIL AT CORTEN CLADDING

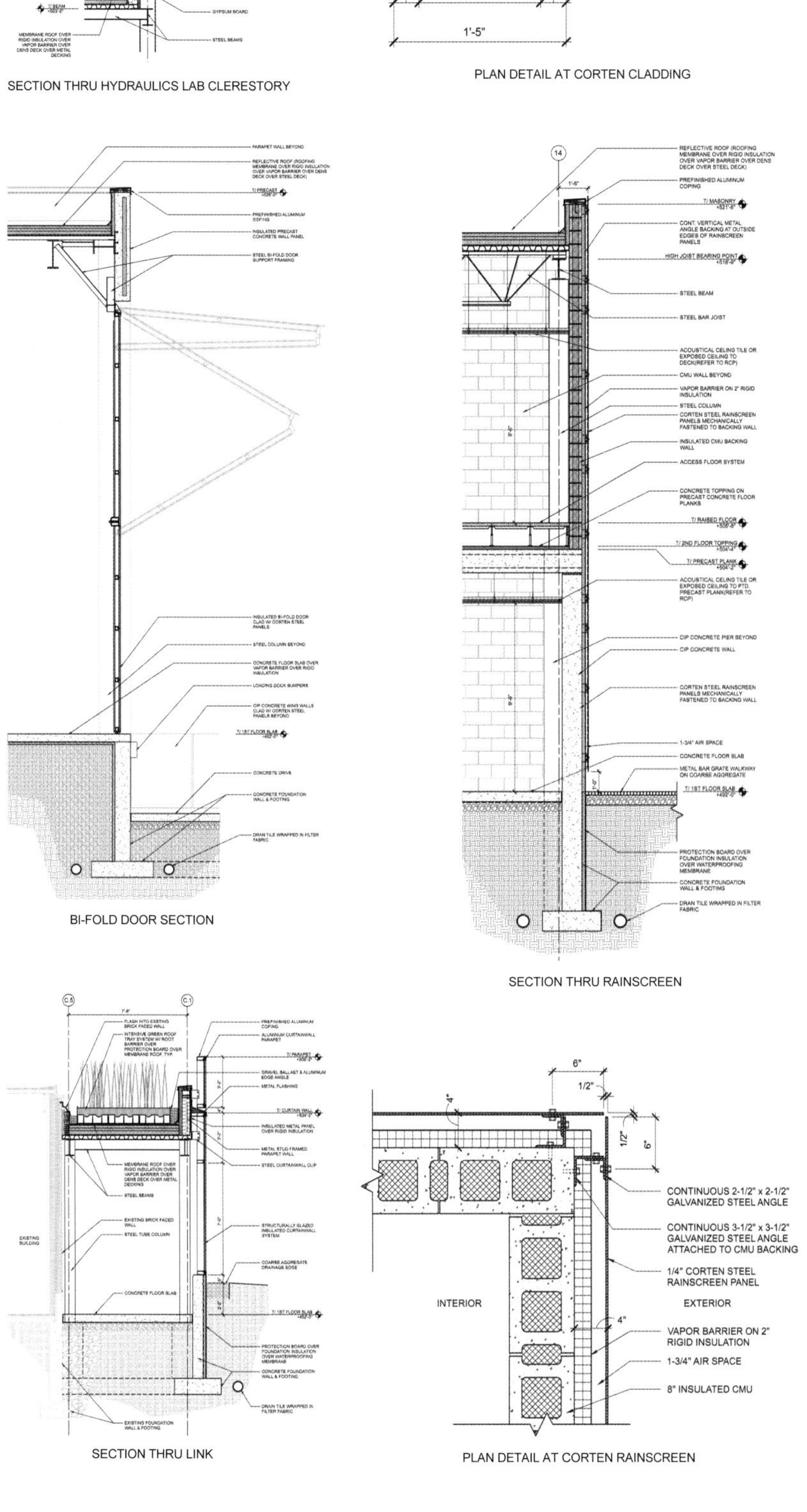

BI-FOLD DOOR SECTION

SECTION THRU RAINSCREEN

SECTION THRU LINK

PLAN DETAIL AT CORTEN RAINSCREEN

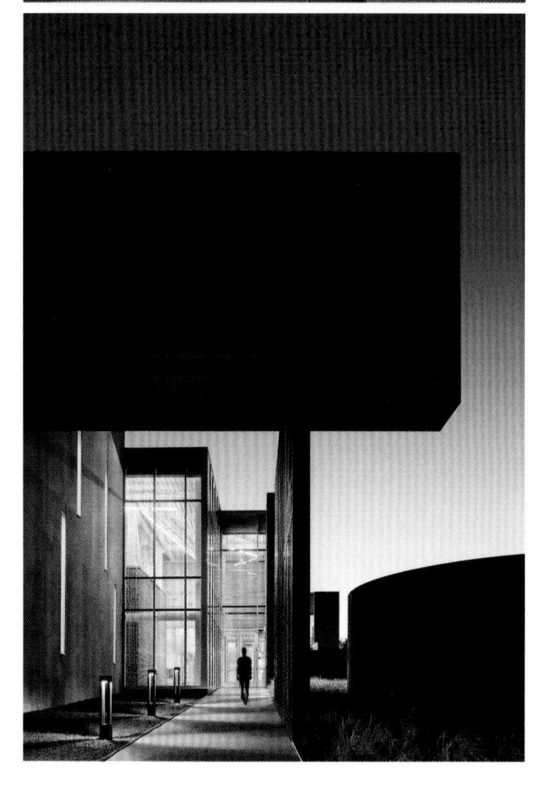

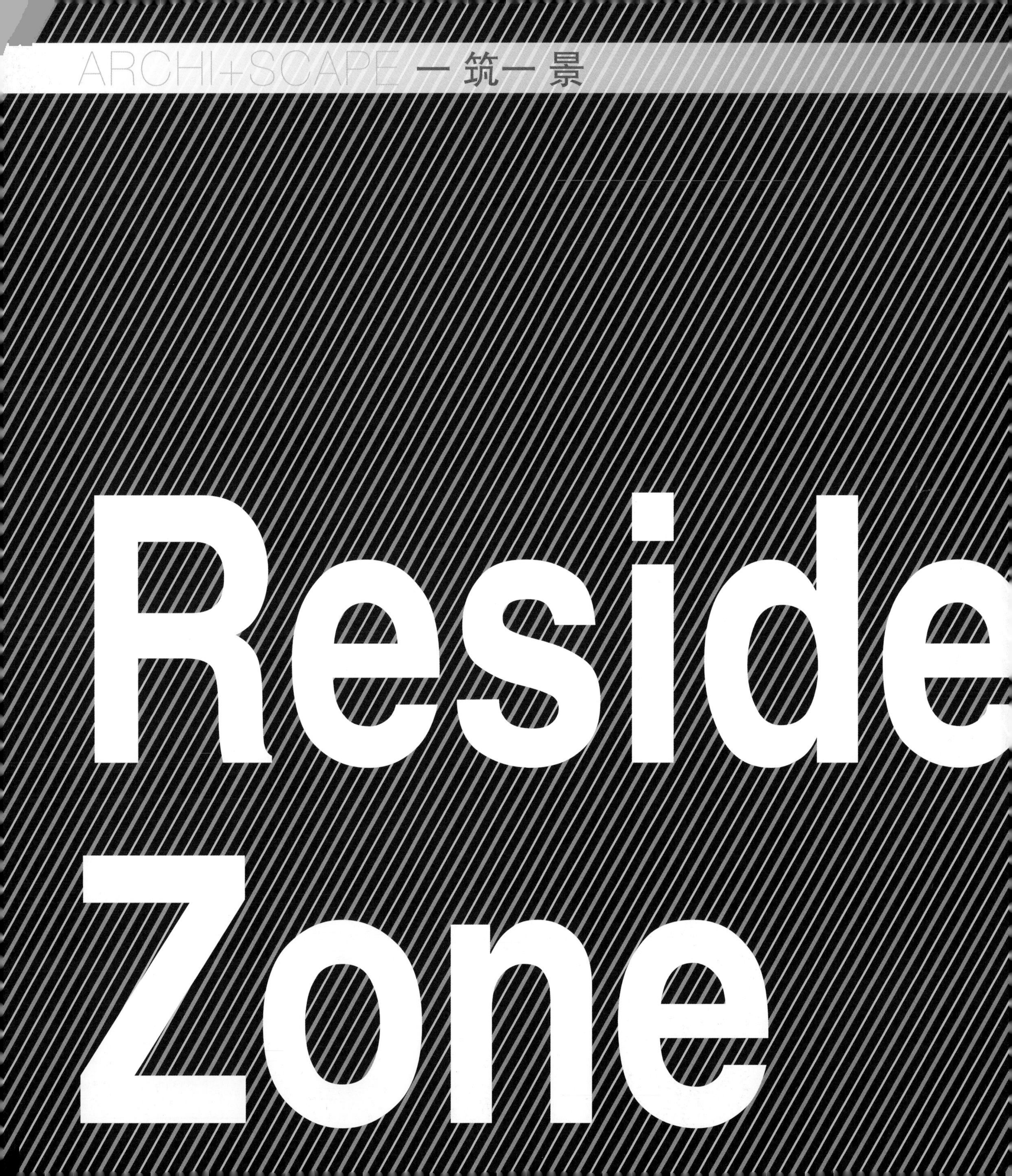
Reside
Zone

ntial

住宅区

NO.5 204-311

Dakar Sow in Senegal

塞内加尔 Dakar Sow住宅

Situated on a cliffside overlooking the Atlantic Ocean, Villa Sow in Dakar, Senegal was designed for a Senegalese businessman and his family. Built on the site of an old World War Two bunker and on the edge of a cliff, Villa Sow maximises its commanding position to create a house that is not only dramatic but with the incorporation of historical elements quite magical and mysterious.

Part of the old bunker has been retained and a portion of it now houses an underground cinema that opens up into a water courtyard or moat that runs along the boundary creating a water feature at the gateway to the property. It is connected back to the house via a timber panelled walkway leading to a spiral staircase that runs from the lower ground through to the first floor and second floor levels of the villa. The ground floor of the house, designed to facilitate seamless indoor and outdoor living and entertainment, is arranged in an L-shape around the pool, the pool terrace and the garden. The formal living and dining spaces cantilever over the cliff and hang over the Atlantic Ocean enjoying panoramic sea views as well as views back to the house. The Kitchen is made up of an open kitchen, a separate traditional kitchen and the garage and staff facilities run along the east west axis and along the northern side of the boundary. From the entrance one moves past the sculptural circular stair to the entertainment room and the double volume family lounge which connects up with a floating stair to the upper level pyjama lounge. The main and the two children's bedrooms are placed on this upper level.

One of the features of the house is the spiral staircase, clad in stainless steel, while the treads are clad in white granite. To add to the sense of continuity between the levels the 20 mm in diameter stainless steel rods run from the first floor handrail to the lower ground floor, thus making the stairwell look like a sculptural steel cylinder. A skylight above the stairwell as well as floor to ceiling glazing in the lounges adds to the sense of transparency. The main bedroom suite opens up onto a large terrace which is the roof of the more formal living wing of the house and the element which projects over to the ocean. The Main Bathroom opens into the private garden and outdoor shower situated over the garages. The study or office sits in a separate block and is joined to the main house by a hallway running along the spine of the building. Under the study/office is a separate fully contained guest room, alongside which is a private gym and a reflecting pond.

Location / 地点:
Dakar, Senegal
Date of Completion / 竣工时间:
2011
Area / 占地面积:
2,643 m²
Architecture / 建筑设计:
SAOTA - Stefan Antoni Olmesdahl Truen Architects
Interior Design / 室内设计:
ANTONI ASSOCIATES
Landscape / 景观设计:
N/A
Photography / 摄影:
SAOTA

该项目是为一个塞内加尔商人及其家人建造的住宅，位于悬崖边缘地带，此地可俯瞰大西洋。建筑所处的场地是二战时期一个古老地堡的旧址，紧邻悬崖边。设计方案充分利用了场地居高临下的地势，不仅塑造出令人叹为观止的景观，还融入了极富传奇和神秘色彩的历史元素。

设计师保留了旧地堡的某些局部结构，并用其中一部分开辟了一个地下家庭影院。影院面向一个水景庭院或护城河，它沿着边界处展开，在入口处形成具有特色的水景。水景庭院通过一个木栈道与住宅相连。栈道通向一个旋梯，从底部一直延伸至住宅的一楼和二楼。底楼的布局呈L形，围绕着泳池和花园而建，为室内外生活和娱乐提供全面的设施。客厅和用餐区依崖傍海，使业主和来客既可享受大西洋无与伦比的全景，又可回望豪华的居所。厨房区沿着东西轴向和北部边界设置，由一个开放式厨房、一个独立的传统厨房以及车库和设备存储间构成。从入口处进入，穿过具有雕塑风格的旋梯，便可到达娱乐室和家庭休息室，休息室的一个悬梯连接着上层的卧室区。上层设置了主卧室和两间儿童房。

旋梯是该住宅的一项特色设计，它的外层镀有不锈钢，而台阶则以白色花岗岩饰面。为了增加各层之间的连续感，设计师还在一楼到底层的扶手处设计了直径20mm的不锈钢体，这使楼梯井的雕塑感油然而生。楼梯井上方的天窗以及休息室的落地玻璃使得房屋的通透感倍增。主卧室套房敞向一个大型露台，这里是主要的居住空间，也是探向大海的结构部分。主浴室面向一个私人花园和车库上方的室外淋浴装置。书房或办公室位于一个单独的区域，通过一个沿建筑脊柱设立的走廊与住宅相连。书房兼办公室的下方是一个独立的客房，客房旁边设有一个私人健身房和一处倒影池。

Materials: Marble, Granite, Stainless Steel, Timber.

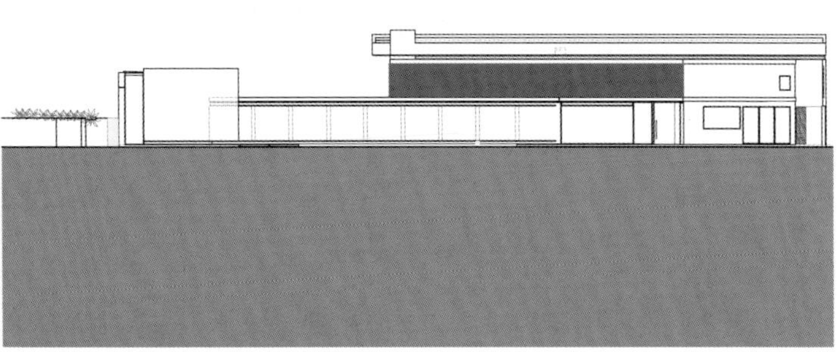

NORTH EAST ELEVATION

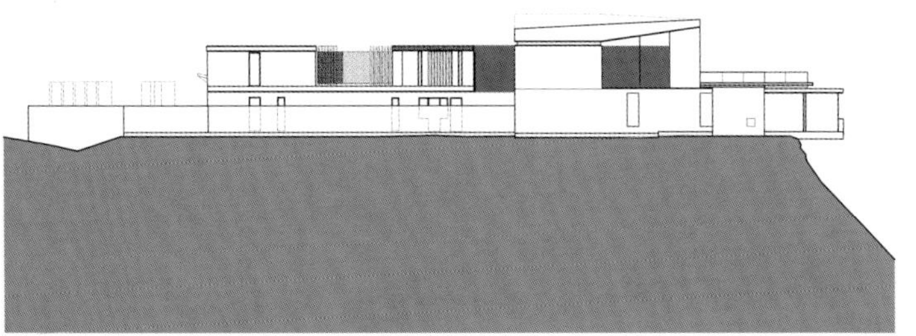

NORTH WEST ELEVATION

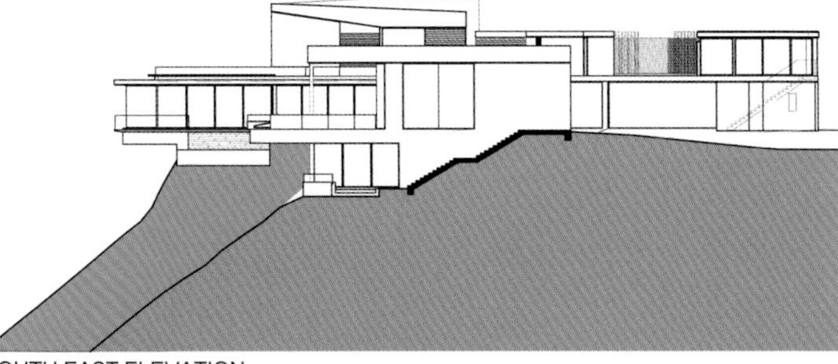

SOUTH EAST ELEVATION

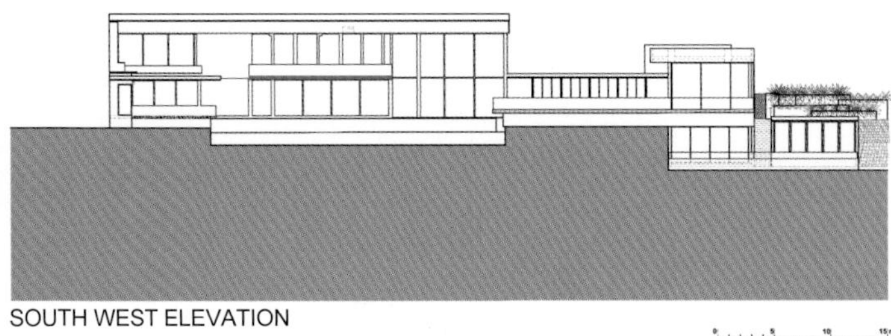

SOUTH WEST ELEVATION

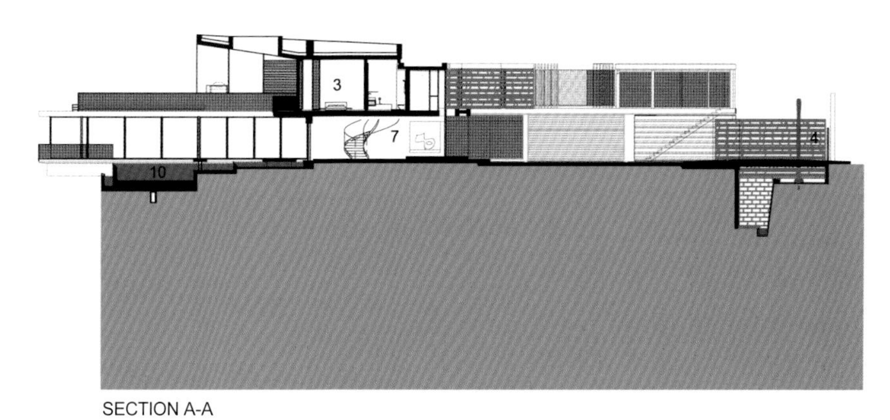

SECTION A-A

SECTION B-B

0 5 10 15m

LEGEND

1. CINEMA
2. SERVICE AREA
3. BEDROOMS
4. POND
5. GYM
6. KITCHEN
7. ENTRANCE
8. LOUNGE
9. STUDY
10. POOL
11. TERRACE
12. GARAGE
13. ENTRANCE GATE
14. GATE HOUSE

LEGEND

1. CINEMA
2. SERVICE AREA
3. BEDROOMS
4. POND
5. GYM
6. KITCHEN
7. ENTRANCE
8. LOUNGE
9. STUDY
10. POOL
11. TERRACE
12. GARAGE
13. ENTRANCE GATE
14. GATE HOUSE

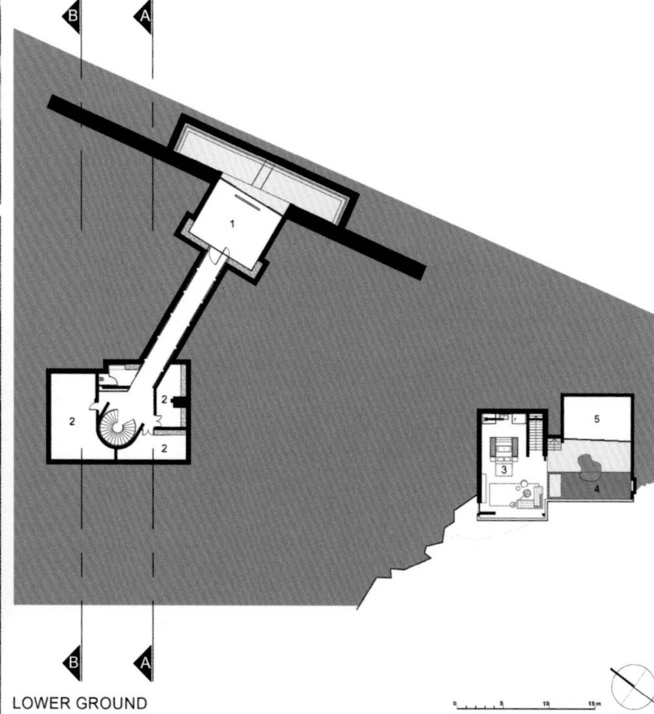

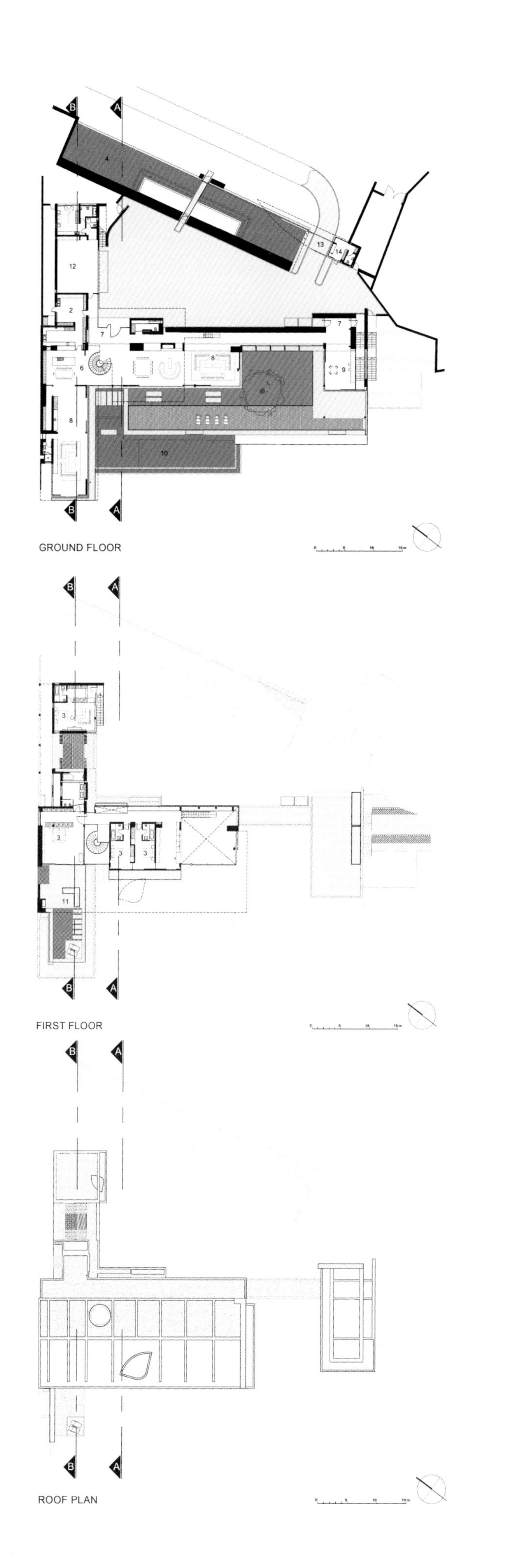

LEGEND

1. CINEMA
2. SERVICE AREA
3. BEDROOMS
4. POND
5. GYM
6. KITCHEN
7. ENTRANCE
8. LOUNGE
9. STUDY
10. POOL
11. TERRACE
12. GARAGE
13. ENTRANCE GATE
14. GATE HOUSE

LEGEND

1. CINEMA
2. SERVICE AREA
3. BEDROOMS
4. POND
5. GYM
6. KITCHEN
7. ENTRANCE
8. LOUNGE
9. STUDY
10. POOL
11. TERRACE
12. GARAGE
13. ENTRANCE GATE
14. GATE HOUSE

LEGEND

1. CINEMA
2. SERVICE AREA
3. BEDROOMS
4. POND
5. GYM
6. KITCHEN
7. ENTRANCE
8. LOUNGE
9. STUDY
10. POOL
11. TERRACE
12. GARAGE
13. ENTRANCE GATE
14. GATE HOUSE

FM-House in Nova Lima

新利马FM-House

The FM-House is the result of the solution of three main contradictions.

The request of a social plan of use in a single level, as largest as possible, on an uphill topography with a slope larger than 20 m, searching to diminish the visual impact of this operation at the landscape. To achieve this, the topography was redesigned, creating a gardened plateau at the bottom, in the cutting area, with slopes of small height that integrate the landscaping; the house is stepped in two levels. The main access is defined by a gentle ramp that starts at the highest street level and achieves an arrival covered square. A sequence of varied spatial routes and vertical articulations – ramp, square, curved staircase, the upper balcony – minimizes the perception of the vertical distance.

The possibility of openning the house to enjoy a variety of views – mountain and city – while ensuring the privacy of interior spaces. The 'showcase effect' is avoided through a clear differentiation between social and intimate areas (with venetian blinds); the elevation of the main volume, that avoids frontal views to the glass pavillion that houses the social areas; and the placement of leisure areas on the back, so they cannot be seen from the street. The social spaces were turned to the day view of the mountains; the master bedroom, with more privacy, was oriented to the night view of the city lights. The service spaces, concentrated on a lateral core, comprise an opaque barrier to the future neighbors.

The request of amplifying the integration between interior and exterior while ensuring an enviromental control, protecting internal spaces from strong winds, sun, cold and the mountain fog. To minimize strong wind impact, the social and leisure areas are organized around a central courtyard, that organizes also the entrance transition. An elevated glazing works as windbreaker along one side of the pool. The generous cantilevered slabs help to keep the rain away and conform more efficient sun protection, allowing large openings that integrate interior and exterior spaces to the landscape. The roof slabs undulation echoes the geometry of the plot, oblique, and the direction of the surrounding landscape views.

FM-House的设计方案解决了三大矛盾。

首先，该项目需要由单一楼层来提供社交用途，体量要尽可能大，而且考虑到场地位于20多米长的斜坡上，设计方案须努力减少建筑对自然景观的视觉破坏。为达到这一目标，设计师重新塑造了地形，在底部和基垫部分创建了一块带有小型斜坡的园林化高地，使建筑场地与周围景观相融合。同时，房屋采用两层结构。主入口由一个起始于最高街道水平的缓坡形成，并设置影壁对道路进行遮挡。坡道、广场、弧形楼梯和上方的阳台构成有序的布局，以多变的空间流线和垂直联系将垂直距离感降至最低。

其次，设计方案须使房屋拥有山景、城市景观等多种视野，同时又要保证室内空间的私密性。避免“展示效应”的措施包括以下几点：社交和私人空间通过软百叶帘被清晰地间隔开来；主体立面的设计使得玻璃馆结构（其内为社交空间）避开正面视野；休闲区被设置在后方，从而无法从街道上看到。社交空间面向日光下的山景而设，而主人的私密空间则面向城市灯光下的夜景，更具私密性。起居室被集中于一个位于旁侧的核心结构，同将来可能存在的邻居之间形成一道不透明的屏障。

最后，内部和外部环境之间的整合需要被强化，同时，又要对环境因素进行有效控制，使内部空间免于强风、阳光、寒冷和山雾的侵扰。为尽量减少强风的影响，社交和休闲区围绕一个中心庭院而建，这种布局也同入口之间形成一种过渡。沿池塘一侧的高架玻璃可以起到挡风的作用。一个大方的悬挑板有助于使住宅免受雨淋，提供高效的防晒功能，同时构成更大的开口结构，将内部和外部空间同自然景观融为一体。房屋的顶板呈起伏状，同场地的斜面几何形态和周围景观的导向相呼应。

Location / 地点:
Nova Lima, Brazil
Date of Completion / 竣工时间:
2008
Area / 占地面积:
710 m^2
Architecture / 建筑设计:
Carlos Alberto Maciel
Interior Design / 室内设计:
Carlos Alberto Maciel
Landscape / 景观设计:
Carlos Alberto Maciel
Photography / 摄影:
Eduardo Eckenfels

Structure: Pos-Tensioned Concrete Structure Associated with Thin Steel Tube Columns;
Handrails: Glass, Steel;
Walls: Stucco Painted White, Ceramic Tiles;
Floors: Wood, Stone.

PHM

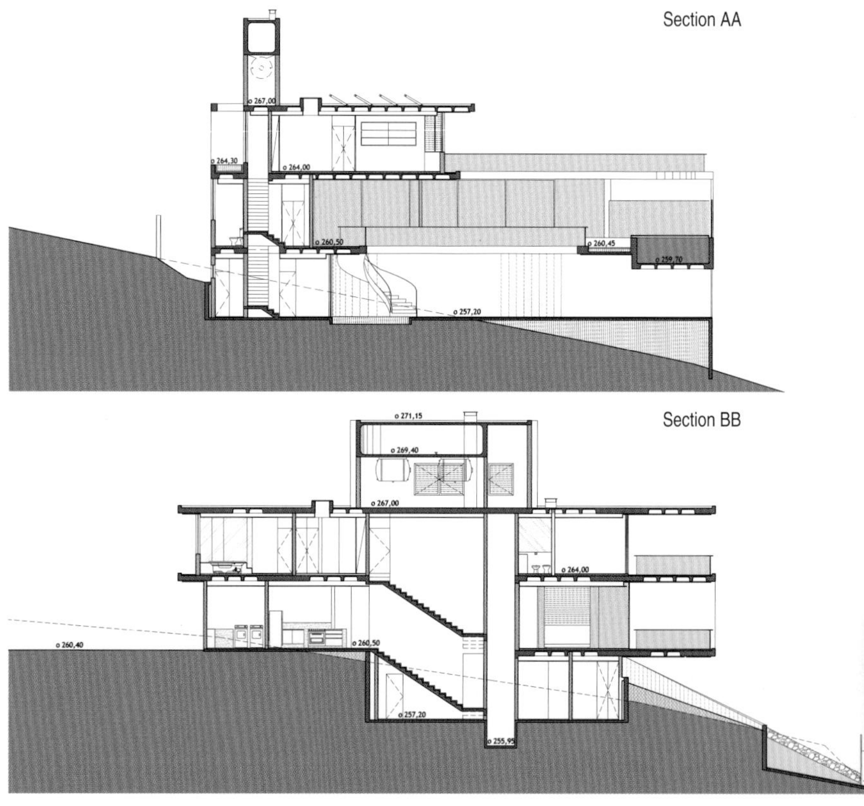
Section AA
Section BB

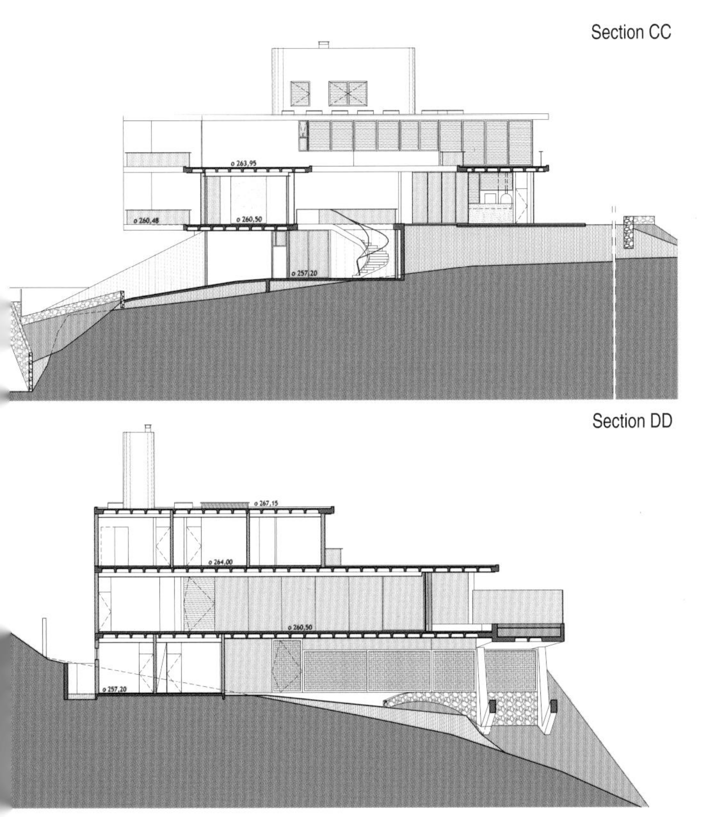

Section CC

Section DD

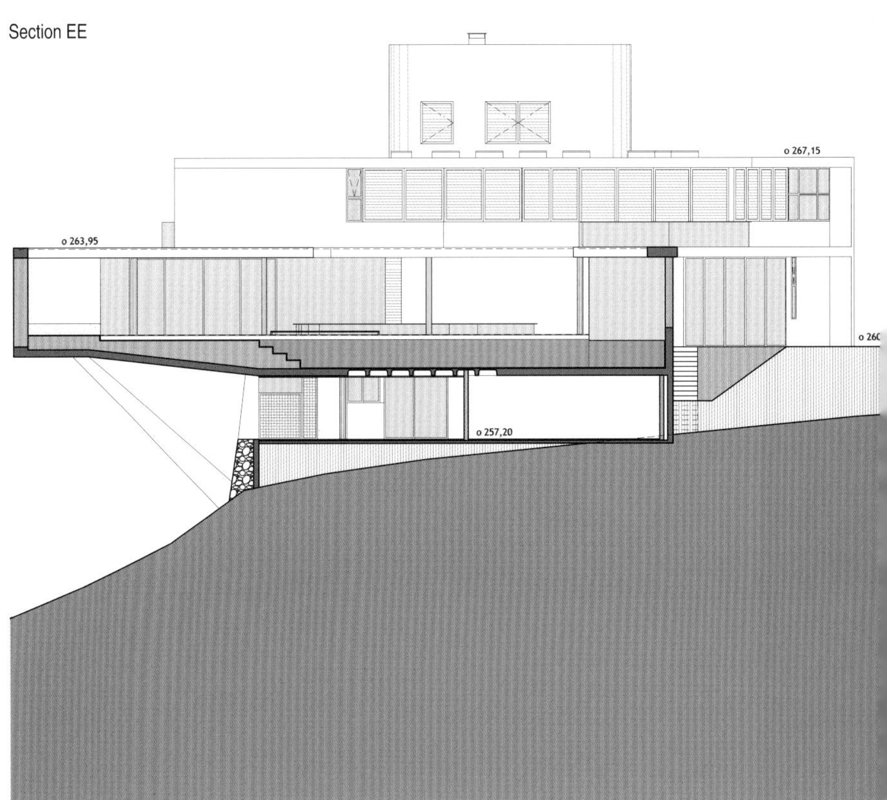

Section EE

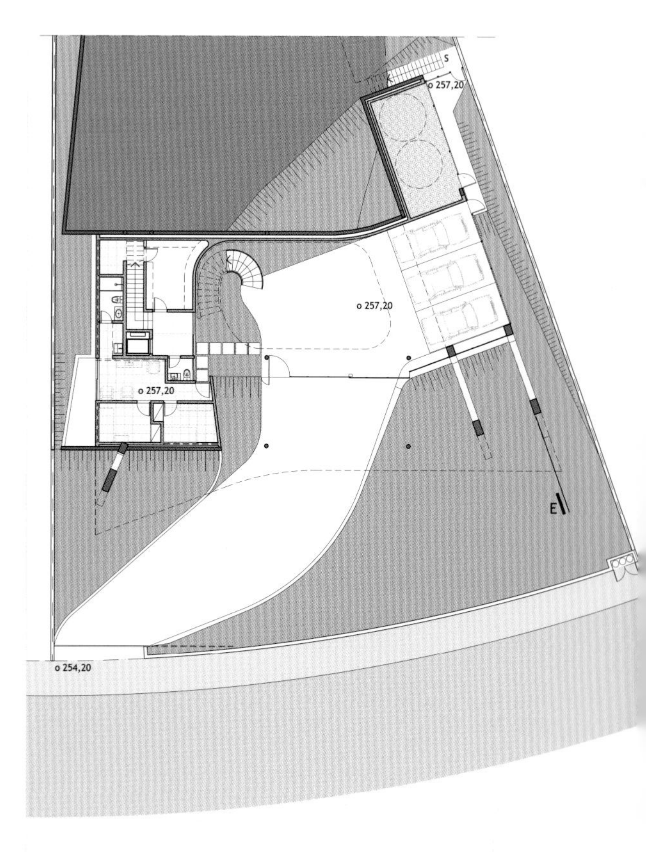

Plan Level 1

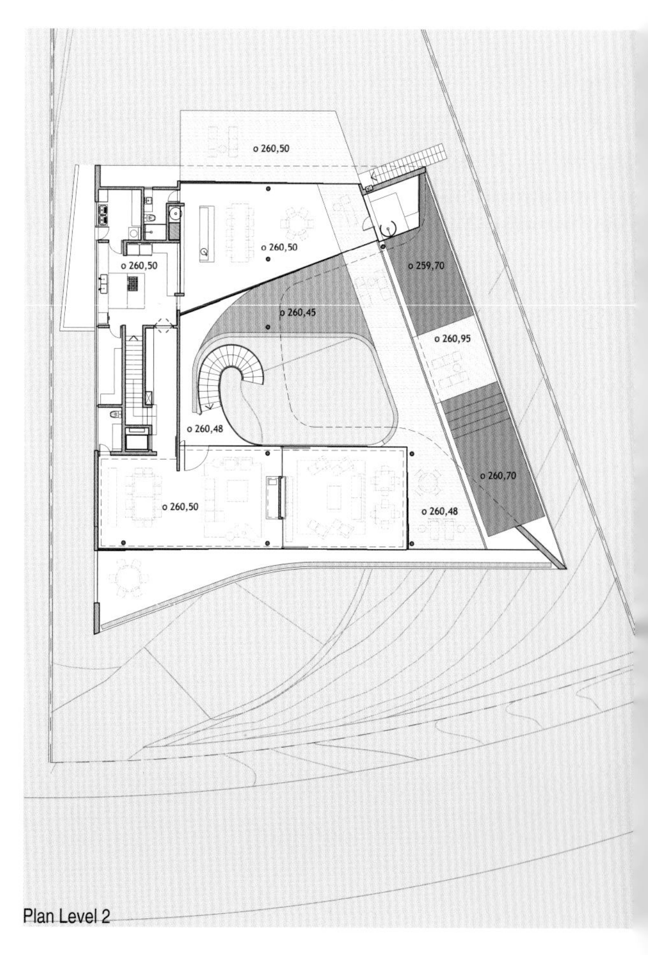

Plan Level 2

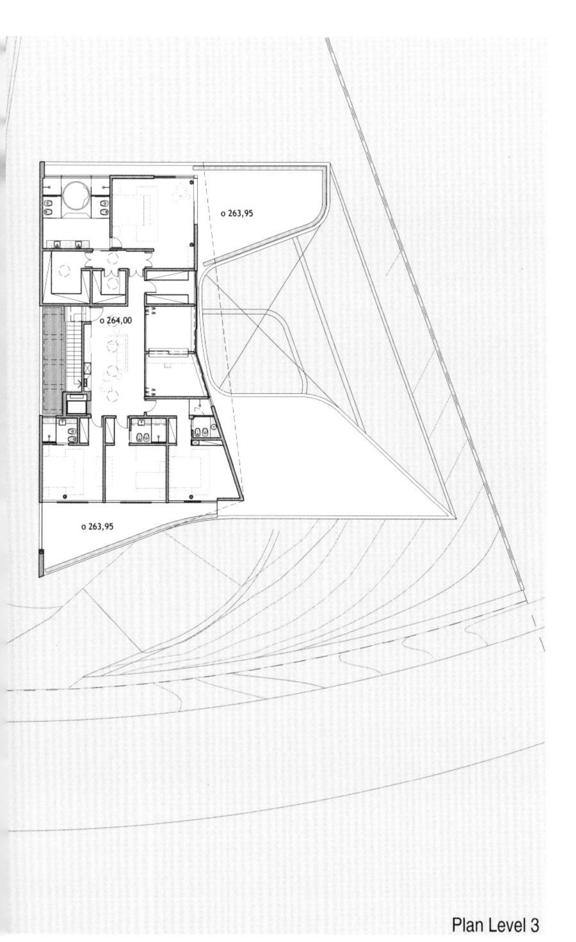

Plan Level 3

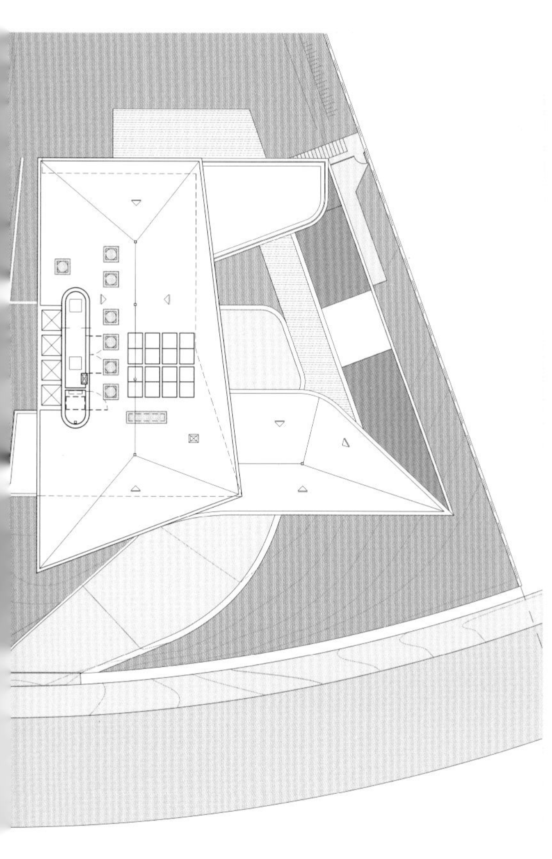

Roof Plan

Grid House in Brazil

巴西网格楼

In an area of 532,400 m^2, only 65,000 m^2 are not covered by the lush native intact forest that is permanently protected. In this area of accidented topography, where large stones are surrounded by Araucaria trees, a small valley was chosen, protected from the winds and close to the forest. This is where the natural walking paths cross: the site where people who arrive at the plot of land go to, access to the paths leading to the heart of the forest and to the top of the hill where one sees an impressive view.

Three main issues have guided the project conception: the demand for a single-story house, the wish to establish a direct relationship with the land and nature and also the need to provide privacy to the members of the family, with the main area located in one single building. Another important factor that architects took into consideration was the region's high humidity levels, which led them to suggest a house above the ground. A structural grid in wood, with 5.5x5.5x3 m modules, suspended above this nucleus for accesses, connects the existing paths and creates new ones. The program in the grid is made up of a nucleus, with washing area, social area, guest room and the owner's apartment, and three isolate modules, with two bedrooms each, for the children.

Suspended above the valley and merging into the hills, the house becomes the land and the land becomes the house, creating a new landscape. The built-up empty spaces, simultaneously inside and outside, allow people to see, under the grid, stones and garden of the native forest, the surrounding trees and the supporting stones, where the house is immersed. Three levels of landscape intervention have been defined. The idea is to rebuild the margins of the forest and create a transition between the open field and the closed forest by using native species, compatible with the region. At the same time, in the remainder of the open area, the park is taken up by paths with resting areas in the areas where the best views can be observed. Finally, at the sites close to the buildings there is a garden prior to the architecture work. At the top floor, which is a continuation of the plot, there is a linear water mirror that avoids the use of body protector and relates to the large water mirror located in the lower garden, around which is the largest stone at the site.

Location / 地点:
St.Paul, Brazil
Date of Completion / 竣工时间:
2007
Area / 占地面积:
532,400 m^2
Architecture / 建筑设计:
FGMF Arquitetos
Landscape / 景观设计:
CAP, Fernando Chacel, Sidney Linhares
Photography / 摄影:
Ale Shneider

该项目占地面积达532400m^2，其中仅6.5hm^{2}的空间没有被永久保护的茂密植被所覆盖。这里的地形起伏不平，南洋杉树群分布在巨大的石头周围。建筑场地位于一个小山谷中，遮风良好，并被森林环绕。房前仅有一条天然小路，其成为人们到达该地区后的必经之地。它和通往森林和山顶的小路相连，拥有极好的景观视野。

设计方案主要围绕以下要求展开：住宅采用单层结构；在场地和自然景观之间建立直接的联系；将主要功能区域集于一个楼房的结构中，为家庭成员提供私密空间。另外一个影响设计的重要因素就是该地区的高湿度，为此，房屋采用高出地面的悬空式设计。5.5m×5.5m×3m的组件单元构成一个木质的网格结构，它悬于一个核心结构上方，下方成为通道系统，在连接现有路径的同时亦创造出新的路径。网格结构中的设施由一个核心结构和三个隔离的组件单元构成。其中，核心结构设有盥洗区、客房以及业主的房间；三个隔离的组件单元为孩子所用，内部各设有两间卧室。

房屋悬于山谷上，同山地融为一体，呈现出“房屋即山地，山地即房屋”的别样景观。建筑结构围建成的空间让人们无论从内部还是外部，都能在网格、石景和花园下看到与房屋浑然一体的原生森林和周围的树木。方案清晰地表现出三个层面的景观介入。这样做是为了重建森林的边缘，并通过使用本地区的树种在空地和森林间形成过渡区域。与此同时，在其余的空地中，公园随小路和周边区域上升至拥有最佳视野的位置。最后，在接近房屋的场地上有一个建筑开建前既有的花园。房屋的顶层构成场地的延续部分，这里设有一个线形镜面水系，作为屋顶安全防护装置的替代方案。与该水系相呼应的是下方花园中的大型水面，环绕水面四周的是场地中最大的石景。

Materials: Concrete, Steel, Metal, Wood, Stone, Araucaria Trees.

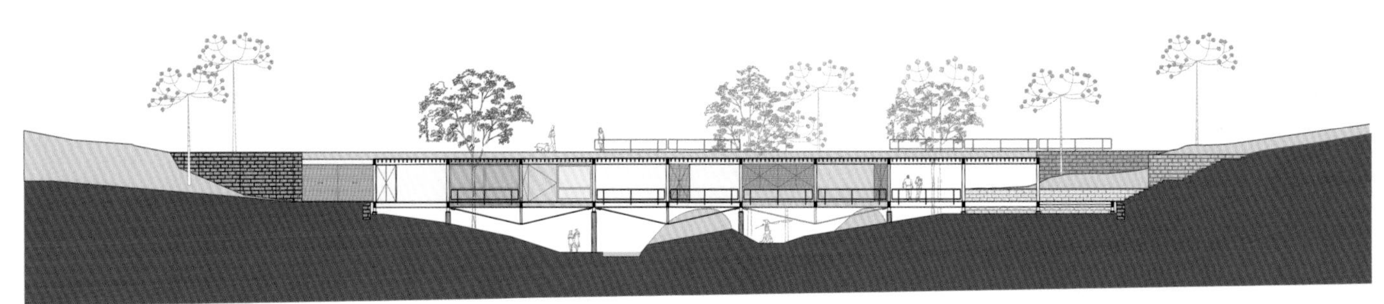

LONGITUDINAL SECTION

TRANSVERSE SECTION

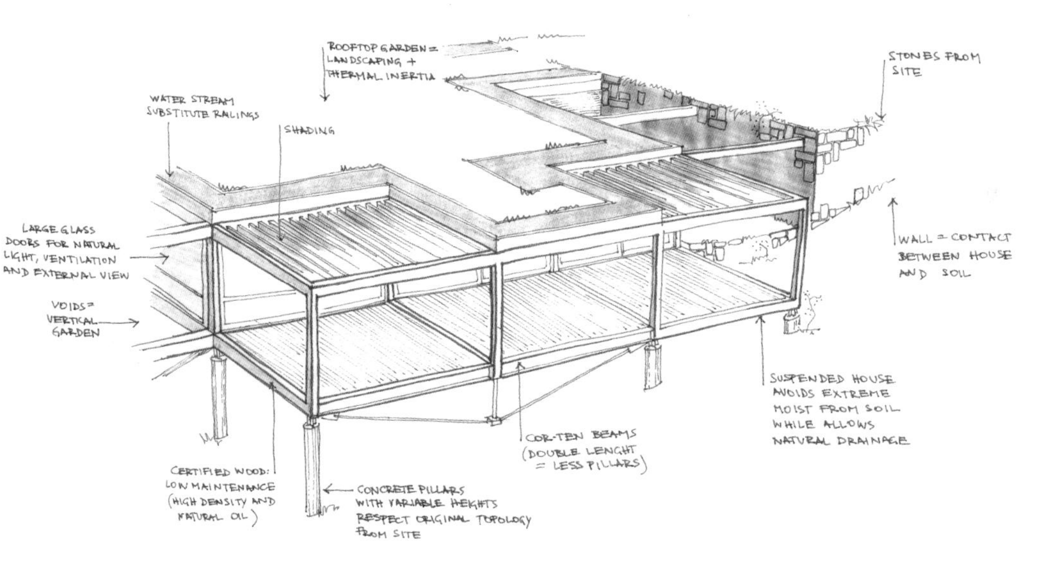
ROOFTOP GARDEN = LANDSCAPING + THERMAL INERTIA
STONES FROM SITE
WATER STREAM SUBSTITUTE RAILINGS
SHADING
LARGE GLASS DOORS FOR NATURAL LIGHT, VENTILATION AND EXTERNAL VIEW
VOIDS= VERTICAL GARDEN
WALL = CONTACT BETWEEN HOUSE AND SOIL
SUSPENDED HOUSE AVOIDS EXTREME MOIST FROM SOIL WHILE ALLOWS NATURAL DRAINAGE
COR-TEN BEAMS (DOUBLE LENGHT = LESS PILLARS)
CERTIFIED WOOD: LOW MAINTENANCE (HIGH DENSITY AND NATURAL OIL)
CONCRETE PILLARS WITH VARIABLE HEIGHTS RESPECT ORIGINAL TOPOLOGY FROM SITE

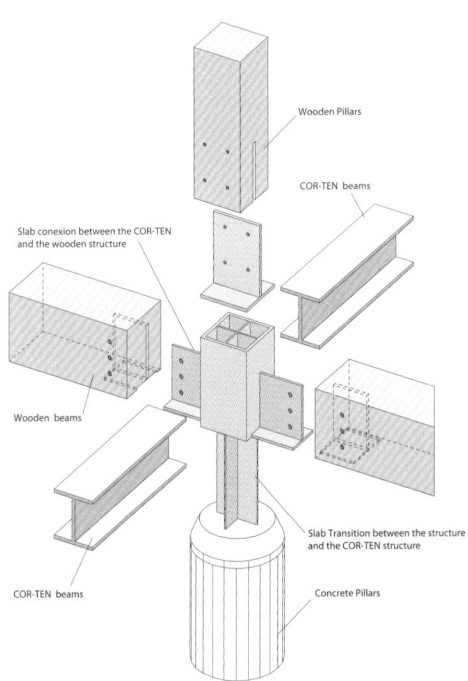
Wooden Pillars
COR-TEN beams
Slab conexion between the COR-TEN and the wooden structure
Wooden beams
Slab Transition between the structure and the COR-TEN structure
COR-TEN beams
Concrete Pillars

Site Plan

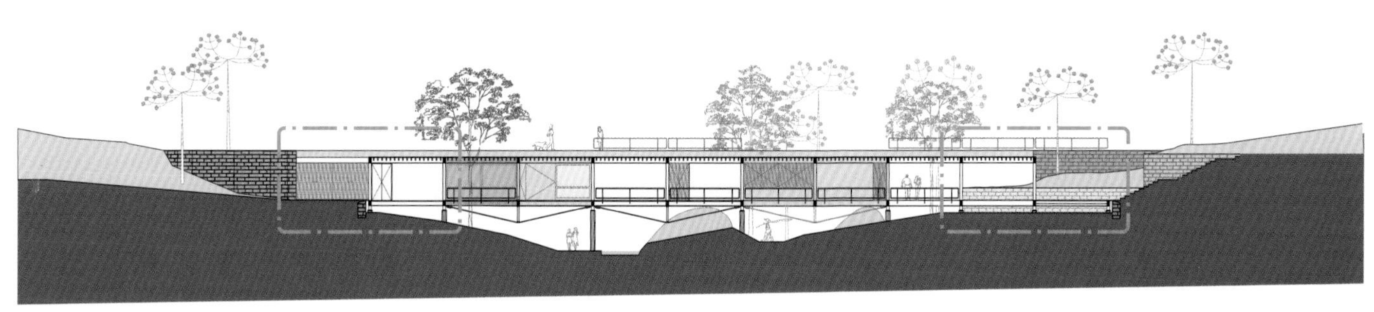

LONGITUDINAL SECTION

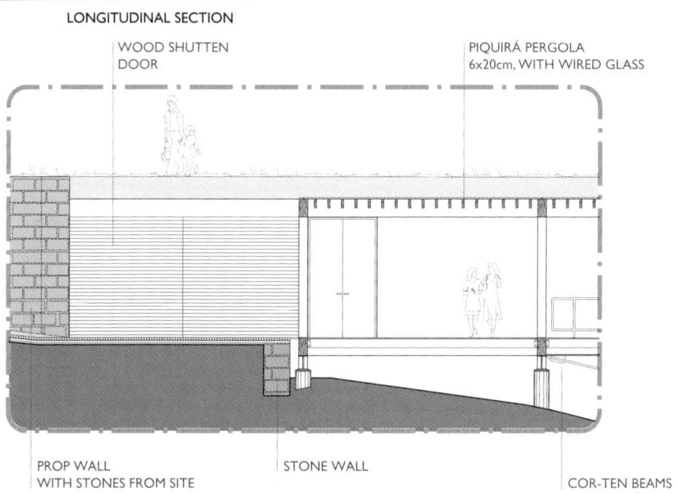

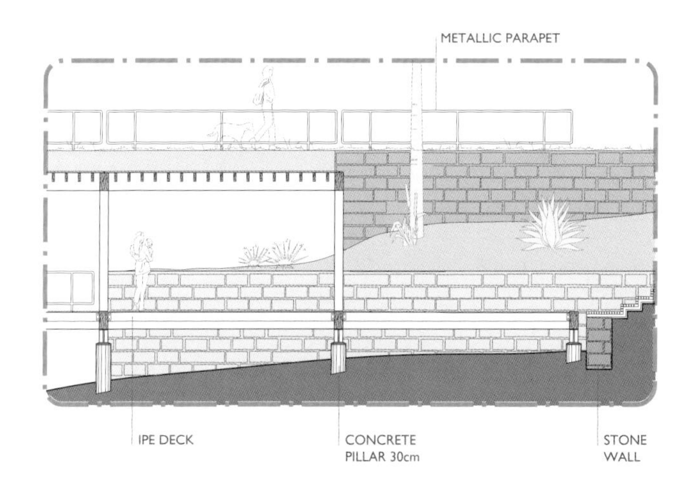

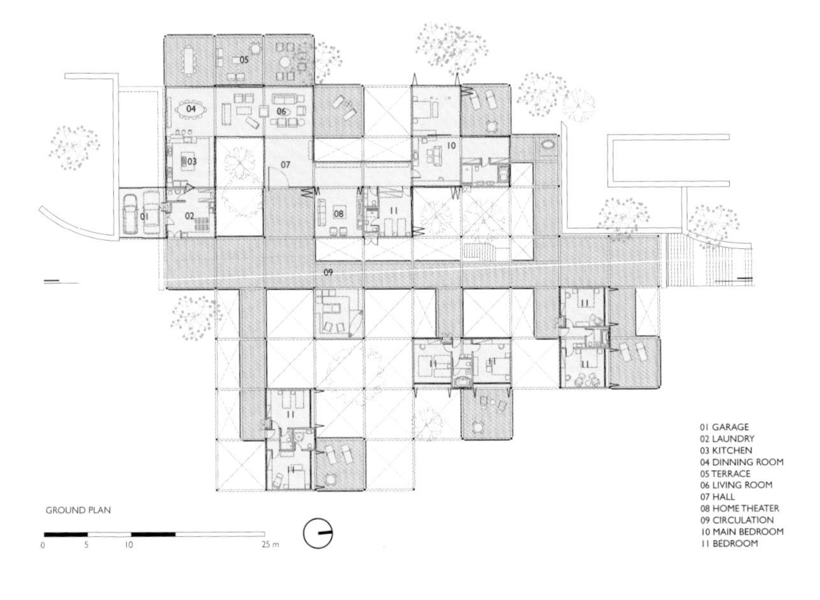
GROUND PLAN
0
5
10
25 m
01 GARAGE
02 LAUNDRY
03 KITCHEN
04 DINNING ROOM
05 TERRACE
06 LIVING ROOM
07 HALL
08 HOME THEATER
09 CIRCULATION
10 MAIN BEDROOM
11 BEDROOM

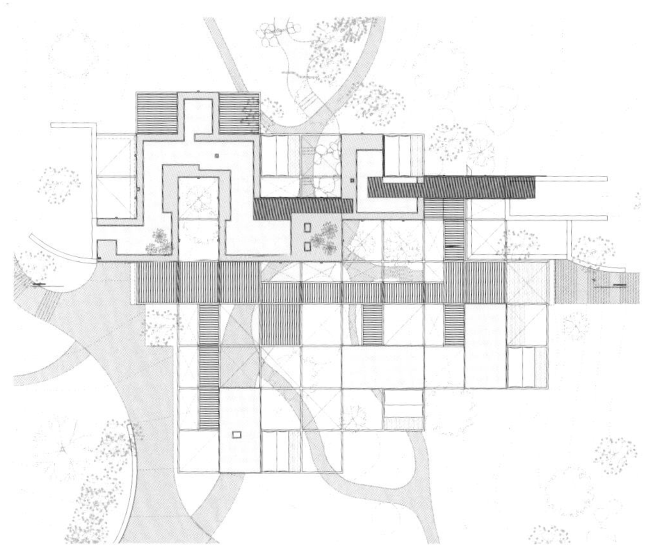

Housing in Toulouse

图卢兹房屋

The architects' aim here was not to work against a landscape, to impose upon it a specific residential model, but to suggest a model of a specific setting. The houses unfold in two sinuous curves that follow the natural lie of the land; it is the site itself that generates the design. This responsive manner of integration also means that each individual home is located in such a way as to enjoy wide-open views of the landscape beyond. In effect, the aim here was to slightly de-intensify the development of the site so as to create a better quality of life, with the lower level of the project being occupied by simplex accommodation and most of the upper level being given over to duplexes. The roof terrace gardens thus blend in with the landscape.

The work here was predicated upon a precise and concise architectural language, with unity of materials and finishing going together with a variety in types and sizes of housing. All the homes, even the two-room apartments extend through the depth of the building, thus guaranteeing good ventilation and above all a fine view of the valley. The relation between the accommodation and the landscape is articulated by the wide wooden terraces, which function as a sort of extension of the living-room areas; in effect, each is a veritable living-room in its own right, measuring from 9 to 20 m^2. The ground-level areas are preserved as a single piece of landscape, providing a garden that is open to all the residents. Arbours of chestnut trees mark out the boundaries.

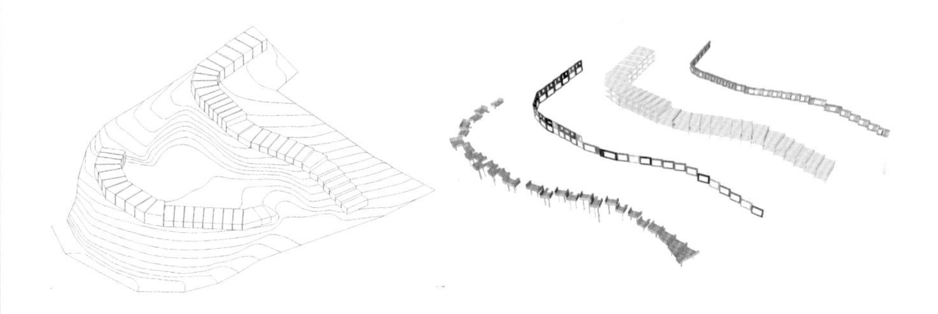

设计师的方案并非逆景观而行，将住宅生硬地置于场地中，而是根据这里的特殊环境量身定制一个建筑模式。这些住宅沿着地势，以两条蜿蜒的曲线展开，可以说是场地本身成就了这项设计。房屋通过这种方式结合在一起，使得在每个单独的住宅中都能够欣赏到远处开阔的景观。实际上，这项设计的目的是为建筑的结构发展达到一种缓冲式效果，建筑较低的部分为单一的住宅，上方大部分为复式住宅，从而营造出更好的生活环境。这也使屋顶露台花园与景观融为一体。

设计方案依据精确和简洁的建筑形式，将主体材料和装饰材料相互搭配，形成多种类型和大小不一的房屋。所有的住宅，甚至包括那些双房间的套房在内，都随着建筑的深度延伸开来，从而保证房屋拥有良好的通风，更重要的是，人们可以在房内或露台上一览山谷的优美景色。 住宅和景观之间的联系通过宽敞的木制露台展现出来。这些木质露台是客厅区的延伸，而实际上其本身都是一个客厅，其面积为9~20m^2。建筑周边的区域作为单独的自然景观被保留下来，供居民欣赏。其中，板栗树的树荫巧妙地勾勒出住宅区的界限。

Location / 地点:
Toulouse, France
Date of Completion / 竣工时间:
2011
Area / 占地面积:
13,500 m^2
Architecture / 建筑设计:
ECDM Architectes
Landscape / 景观设计:
ECDM Architectes
Photography / 摄影:
Vincent Piquet

Construction Materials: Wood, Concrete, Stone;
Plants: Chestnut Trees.

Elevation

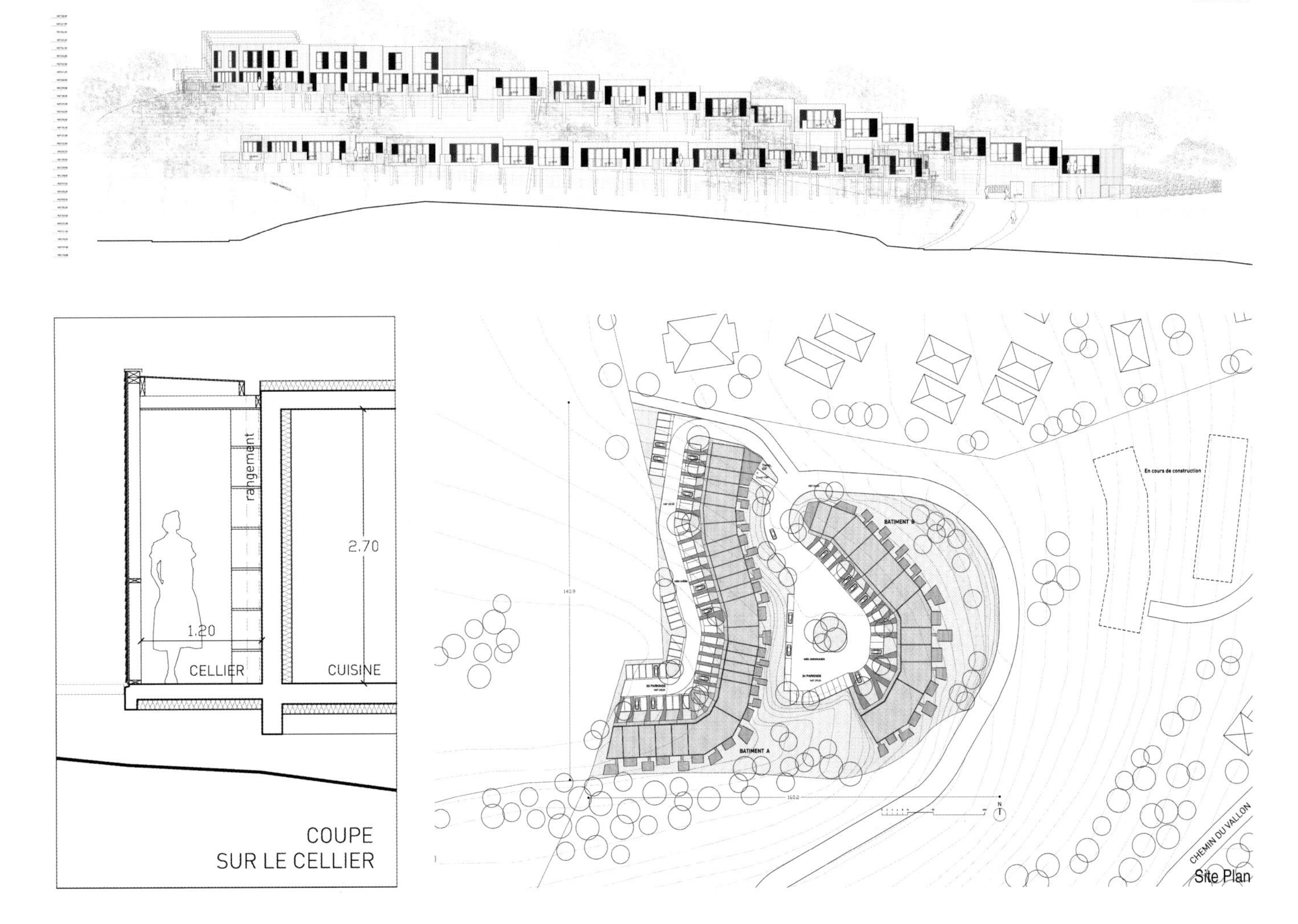

Site Plan

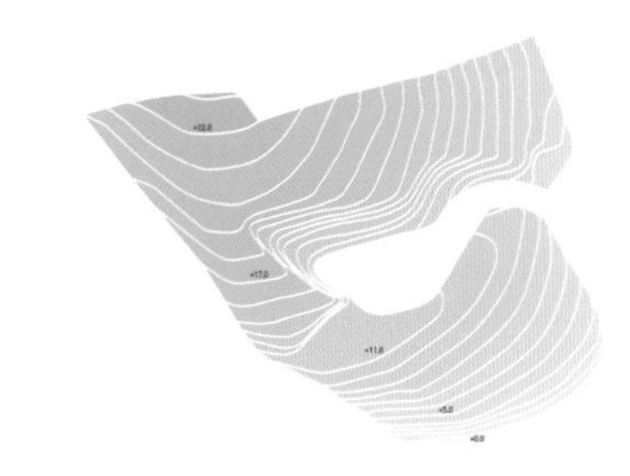

site area : 13 489 m²
height difference : 22 m

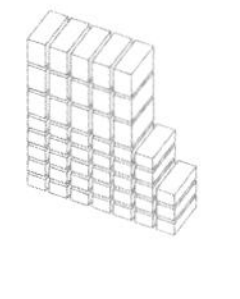

preservation of the big commun garden

program : 33 simplex and 15 duplex dwellings

pilotis houses following the topography

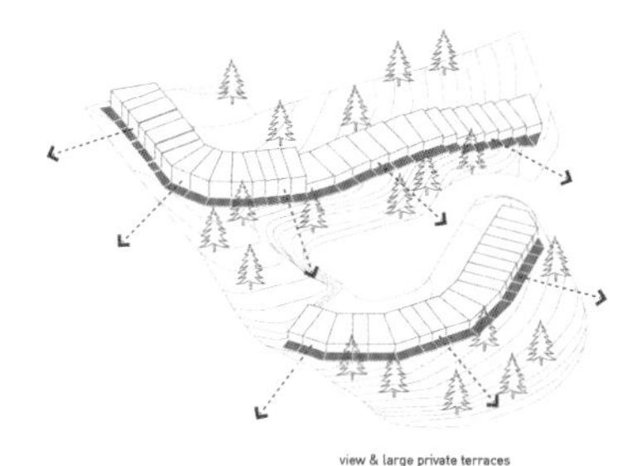

view & large private terraces

living landscape

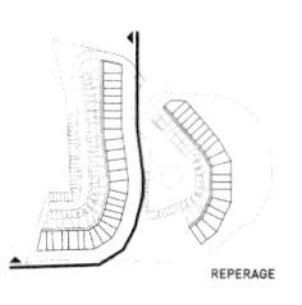

REPERAGE

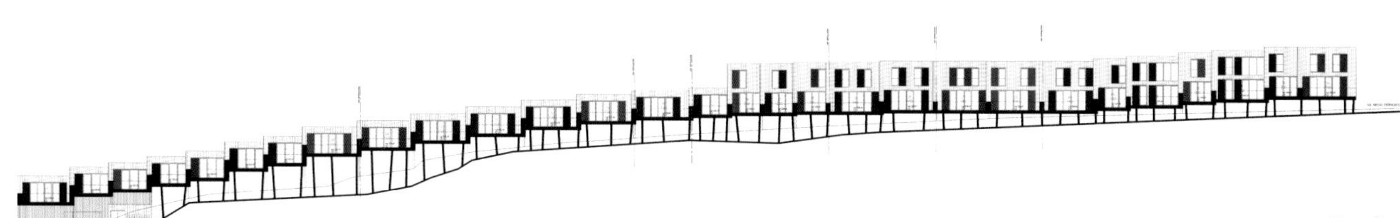

Elevation

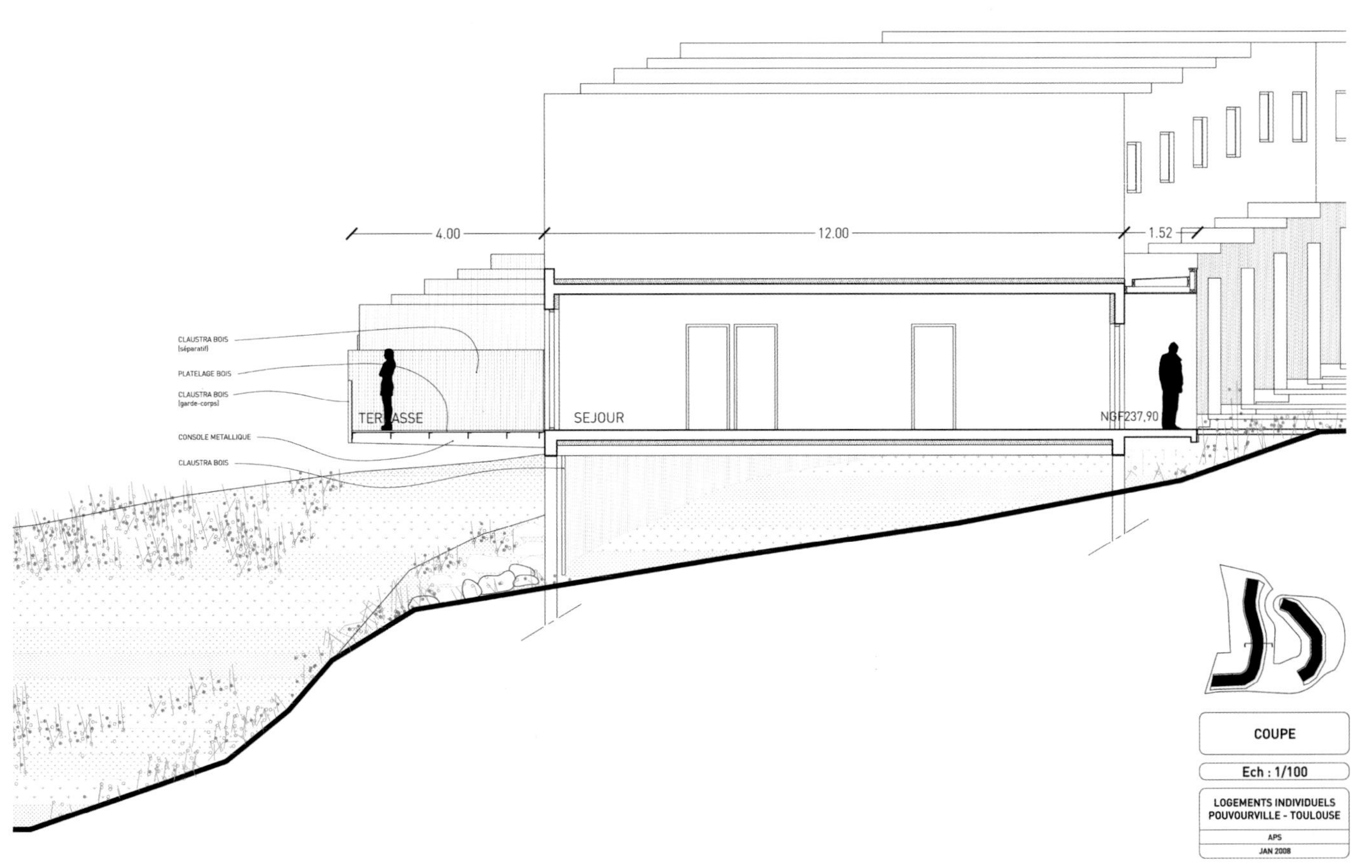
4.00
12.00
1.52
CLAUSTRA BOIS
(séparatif)
PLATELAGE BOIS
CLAUSTRA BOIS
(garde-corps)
CONSOLE METALLIQUE
CLAUSTRA BOIS
TER ASSE
SEJOUR
NGF237.90
COUPE
Ech : 1/100
LOGEMENTS INDIVIDUELS
POUVOURVILLE - TOULOUSE
APS
JAN 2008

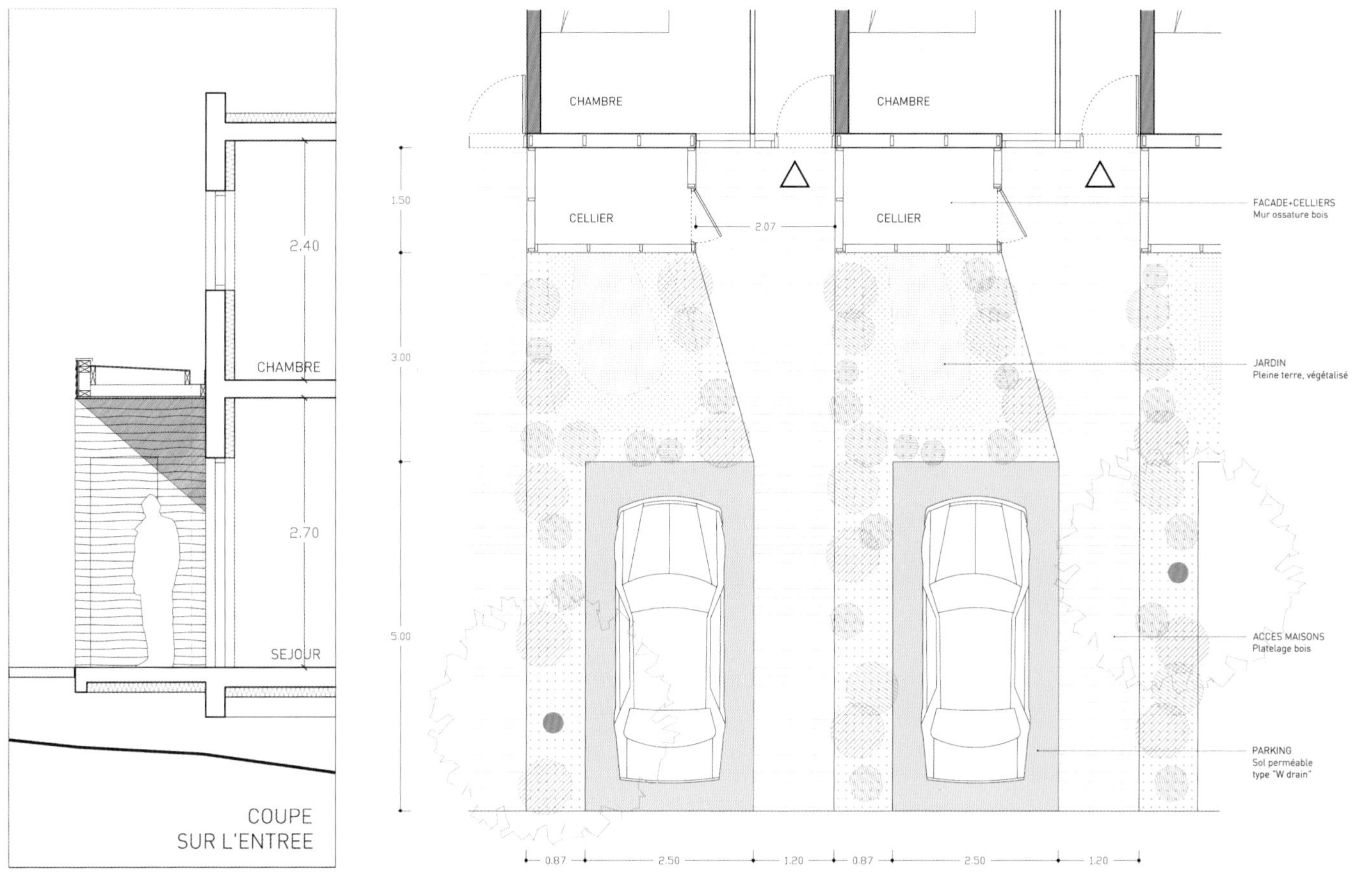
2.40
CHAMBRE
2.70
SEJOUR
COUPE
SUR L'ENTREE
CHAMBRE
CHAMBRE
CELLIER
CELLIER
1.50
2.07
3.00
5.00
FACADE+CELLIERS
Mur ossature bois
JARDIN
Pleine terre, végétalisé
ACCES MAISONS
Platelage bois
PARKING
Sol perméable
type "W drain"
0.87
2.50
1.20
0.87
2.50
1.20

Island House in Korea

韩国“小岛之家”

Location / 地点:
Gapyunggun, Korea
Date of Completion / 竣工时间:
2008
Area / 占地面积:
630 m^2
Architecture / 建筑设计:
IROJE KHM Architects, HyoMan Kim
Landscape / 景观设计:
IROJE KHM Architects, HyoMan Kim
Photography / 摄影:
KimJongOh

This site, where is floating on river and confronted the graceful landscape, was strongly recognized to people as a part of nature. "The architectural nature" is visualized as a place of recreation. While maximizing the efficiency of landuse, the leaner concrete mass, that cherishes the courtyard which is filled with the water and the greenery, was laid out on this site along the irregular formed site line. This courtyard is "the architectural nature" and a central recreation space as extended river that communicate the river and architecture.

Continuous circulation of stepped roof garden - creation of new green land in this site. The whole part of the step typed roofs, which is moving upward with various levels, are directly linked to the bedrooms in upper floor. Finally, these stepped roof gardens are linked to both sides of the inner court where swimming pool is. The inner court where is filled with water, flowers and fruits, and the whole of the roof gardens are circulated as the continuous landscape place and that is the place as "architectural nature" in concept.

Naturally, all of the rooms inside this site-shaped-mass are laid toward the picturesque landscape to enjoy the graceful scenery surrounding this site. The huge panoramic view framed with sloped ceiling line that is composed of the lines of stepped roof gardens and the bottom line of the inner court, is the major impression of inner space of this house.

Promenade inner space of skipfloor made the promenade roof garden-space of skipfloor. Skip floor plan of inside of this house produced various dramatic spaces. Floating boxes with bamboo garden - dynamic, unrealistic sequence of interior space. As "The architectural nature", floating white polyhedral masses that have the built-in bamboo gardens, produced the various stories of vertical space.

The shape of the mountain type composed of irregular polygonal shaped concrete mass and metal mesh was designed to harmonize with the context as "the architectural mountain". There was the intention to be a part of the surrounding context that is consisted of the river and the mountain. As a result, this house was to be "the island house" as an "architectural island".

该居所静静地盘踞在江边，与秀丽的景观对望，融入大自然的怀抱之中。这片“建筑化自然”被构想为一个休憩场所。该项目最大限度地提高土地利用效率，精简的混凝土体量沿着不规则的地形建立，守视着由水景和绿地修饰的庭院。庭院部分为“建筑化自然”的过渡区域，并作为中央休憩空间，以及沟通河流和建筑的纽带。

连续而循环的屋顶花园为场地创造出新的绿地。多级台阶盘桓而上，最终构成了整个建筑物的屋顶，与顶楼卧室直接相连，沿台阶而下直接通往泳池。设有水景、花卉、果树的内庭和整个屋顶花园构成一个循环结构的连续景观，共同形成了一片“建筑化自然”。

在顺应场地形态的结构内，所有房间都朝向如画的景观，人们可以享受景观无穷的魅力。屋顶花园和内庭组成一个框架，容纳了一幅巨大的全景视图，这构成了住宅内部空间的主要特色。

房屋内部的跃层结构形成了多种引人注目的空间。带有竹花园的悬厢设计让室内空间显得动力十足、亦真亦幻。这种悬厢体现了“建筑化自然”的理念，是一种悬空的白色多面体结构，内置竹花园，为垂直空间增添了一份情趣。

建筑犹如一座山峰，以不规则多边形的混凝土结构和金属丝网为主体轮廓，旨在通过“建筑式山峰”的设计与其山体背景取得协调。设计方案力求将建筑融入到周围的群山与河流中，因此，这个作为“建筑之岛”的“小岛之家”便应运而生。

Exterior Finishing : Exposed Concrete, Stainless Mesh;
Interior Finishing : Exposed Concrete, Lacquer.

Elevation 1

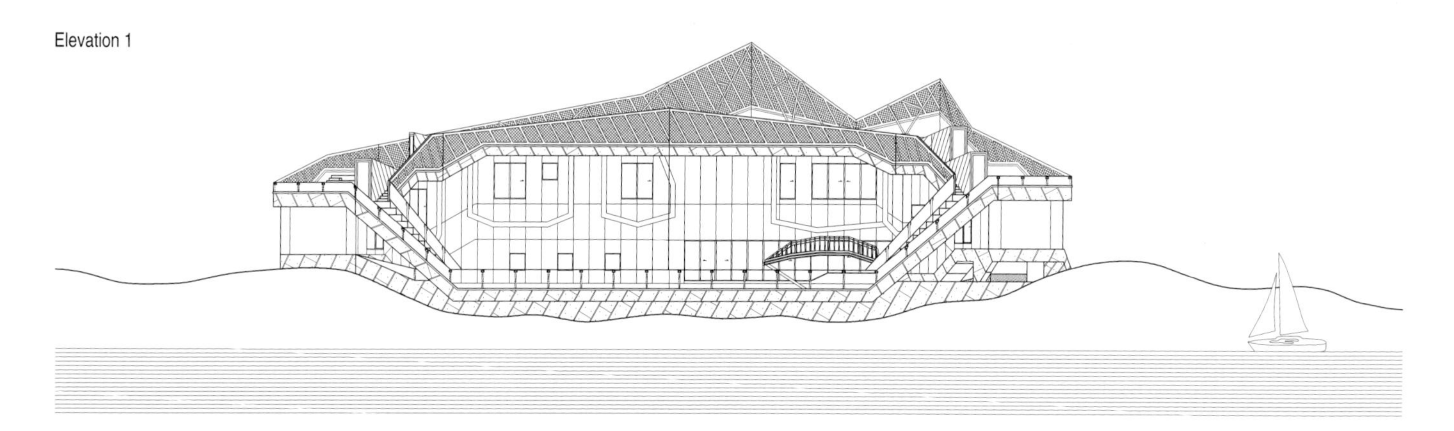

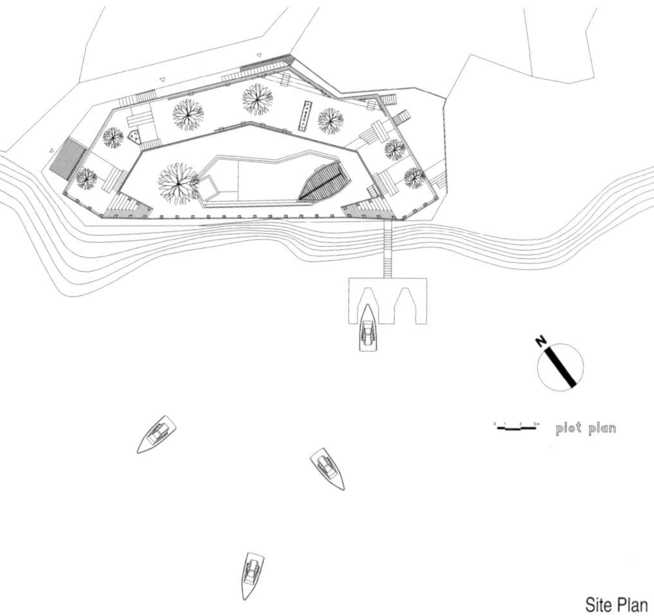

Site Plan

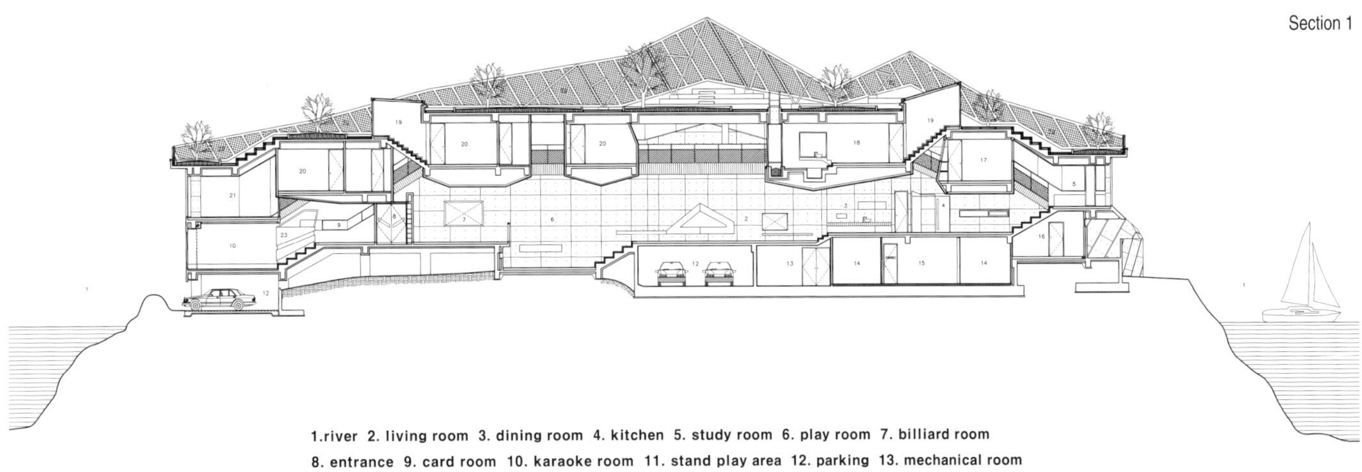

1.river 2. living room 3. dining room 4. kitchen 5. study room 6. play room 7. billiard room
8. entrance 9. card room 10. karaoke room 11. stand play area 12. parking 13. mechanical room
14. manager's bed room 15. manager's living dining 16. maid's room 17. child bed room 18. master bed room
19. stair case 20. guest bed room 21. fitness room 22. roof garden 23. stand karaoke area

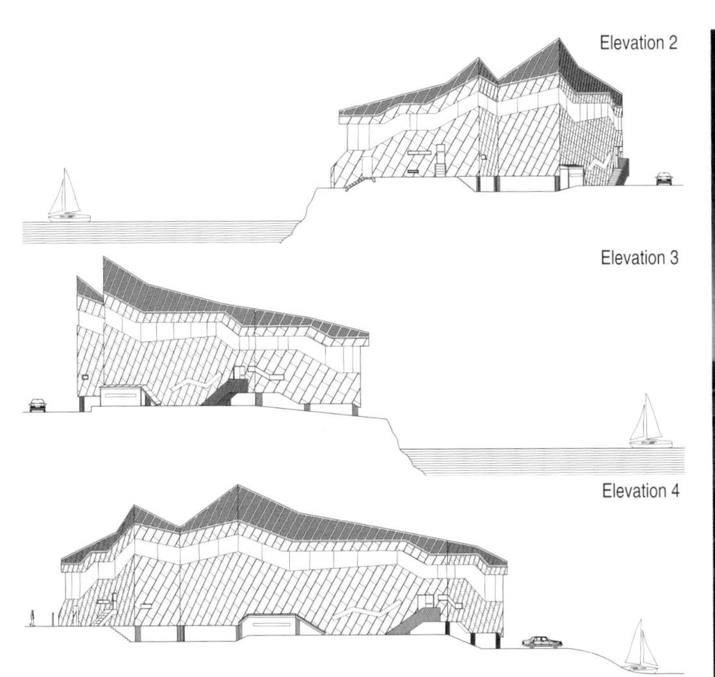

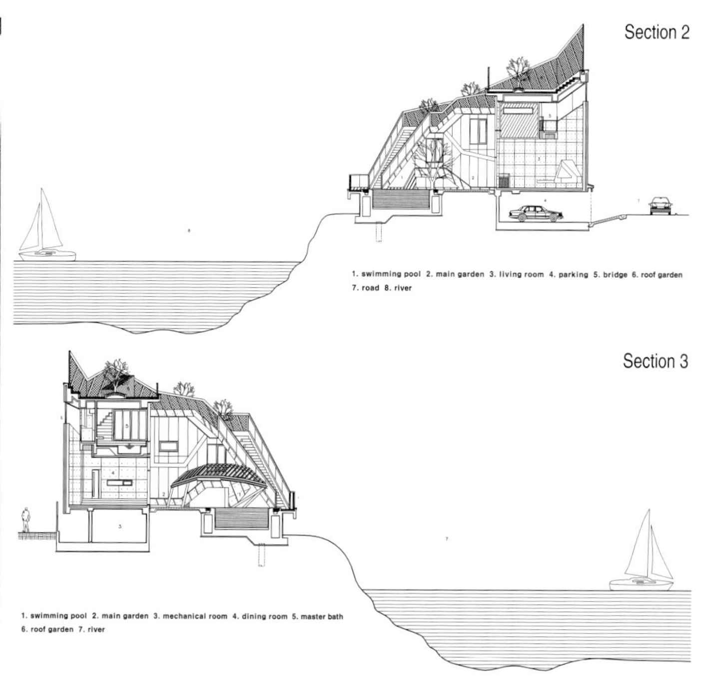
Section 2
1. swimming pool 2. main garden 3. living room 4. parking 5. bridge 6. roof garden
7. road 8. river
Section 3
1. swimming pool 2. main garden 3. mechanical room 4. dining room 5. master bath
6. roof garden 7. river

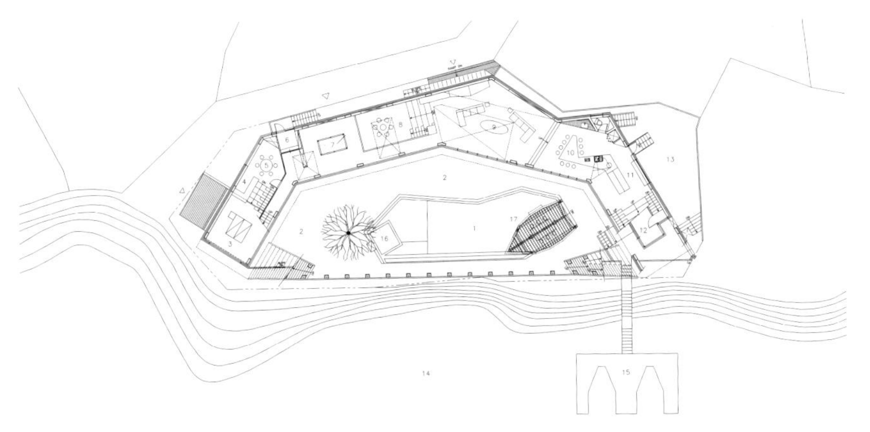

1. swimming pool 2. main garden 3. karaoke room 4. bar 5. card room
6. entrance 7. billiard room 8. play room 9. living room 10. dining room 11. kitchen 12. maid's room
13. service yard 14. river 15. boat deck 16. spa 17. pavilion

Plan Level 1

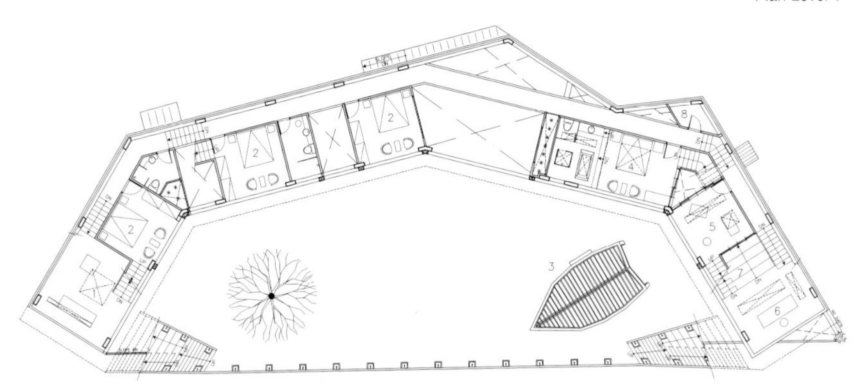

1. fitness room 2. guest bed room 3. roof of pavilion 4. master bed room 5. child bed room
6. study room 7. stand study area 8. boiler room

Plan Level 2

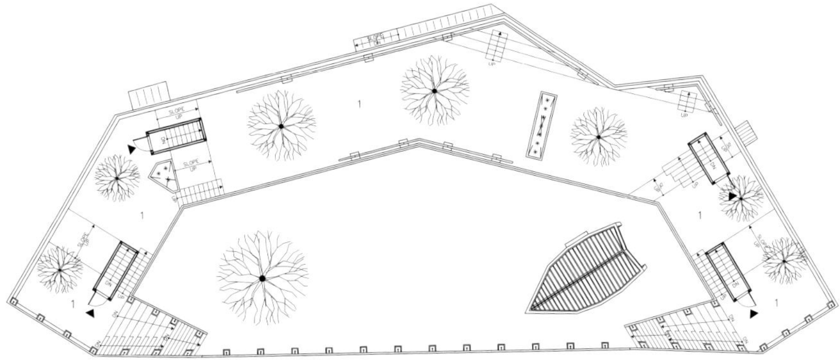

1. roof garden

Plan Level 3

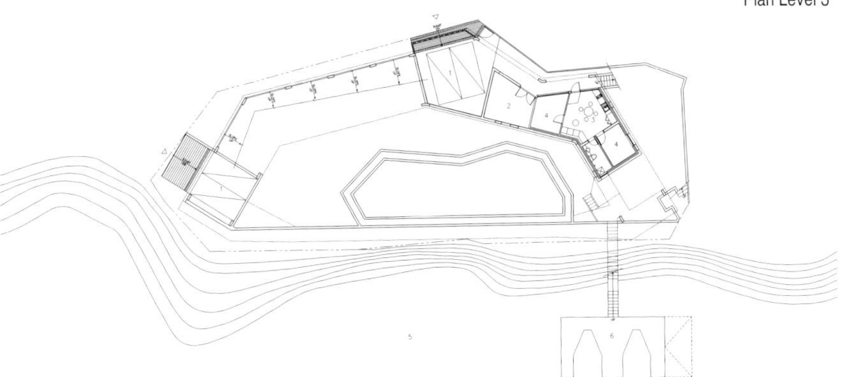

1. parking 2. mechanical room 3. manager's living dining 4. manager's bed 5. river
6. boat deck

Plan Level 4

Sao Francisco Xavier House

圣·弗朗西斯科沙维尔宫

The challenge of this Project is to invent a new and unique space for human beings, in an isolated location with impressive and exuberant nature and remarkable geography. The region is like a "sea of mountains" with deep valleys, flowing rivers and waterfalls. Also the sky has a strong presence in the landscape composing with the rest of the nature a classic view.

The terrain is a small hill, including the valley bottom and the hill top. It has a total area of 24,000 m and contains many types of vegetations; native vegetation nearby the river in the valley bottom, weald and high eucalyptus trees spread randomly.

The architects decided that the house should be placed on the edge of the hill with a main volume and 2 attachments. The main building is a long, narrow, rectilinear volume with a transparent facade facing the view and the valley. It contains 4 suites, a kitchen, a living room and a veranda. Almost all ambiences are facing the view, including bathrooms and kitchen.

The service areas (technical area, guest room, laundry and deposit) and the sauna are attached, cross-linked, to the main building, on the flat part of the terrain, in the hill top. The distribution of the service areas conform an open, but protected space, delimited by 3 sides, like a courtyard. This cozy space promotes the meeting of the people on the open air and also distributes the fluxes.

The house has two distinct spaces defined by the edge of the hill; a main volume with a long glassy facade facing the view, and a meeting space, protected and delimited, facing the interior of the terrain.

该项目所面临的挑战是在一个自然气息浓厚、地势特征显著的偏远地区建立一处清新、独特的空间。该地区宛如一片“山海”，并伴有幽谷、河流和瀑布。同时，天空以不凡的气势同其他自然元素构成一幅绚丽的风景画。

项目所在地是一个小山丘，包括谷底和山顶。其总面积达2.4hm^2，植被丰富。谷底河流附近的原生植被、旷野处高耸的桉树自由地蔓延开来。

建筑师决定将房屋建在山丘边缘，包括一个主体结构和两个附属结构。主体建筑为一个狭长的直线形结构，带有一个透明的朝向景观和山谷的立面。它包含四间套房、一个厨房，一个客厅和一个阳台。几乎所有的空间，包括卫生间和厨房在内，都面向自然景观。

服务区（包括技术区、客房、洗衣间和保管处）和桑拿浴室同位于山顶平坦处，与主体建筑相连。服务区的分布形成一个半开放的围合空间，建筑分布在三面，形成一个庭院。这个舒适的居所提升了人们在室外的聚会体验，并将流通空间散布开来。

本项目在山丘边缘勾勒出两个截然不同的空间：一个是房屋主体，拥有面向景观的玻璃立面；另一个是聚会空间，拥有防护结构和清晰的界限，并朝向地形内部。

Location / 地点:
St. Paul, Brazil
Date of Completion / 竣工时间:
2010
Area / 占地面积:
24,000 m^2
Architecture / 建筑设计:
Nitsche Arquitetos Associados
Landscape / 景观设计:
Nitsche Arquitetos Associados
Photography / 摄影:
Nelson Kon
Client / 客户:
Antonio Carlos Molina e Fulvia Maria Lucia Ribeiro Molina

Floor: Demolition Wood and Ceramic Floor;
Walking Paths: Portuguese Mosaic;
Illumination: Project Made by Reka Iluminação;
Plants: Existing Trees.

site plan
0
50m

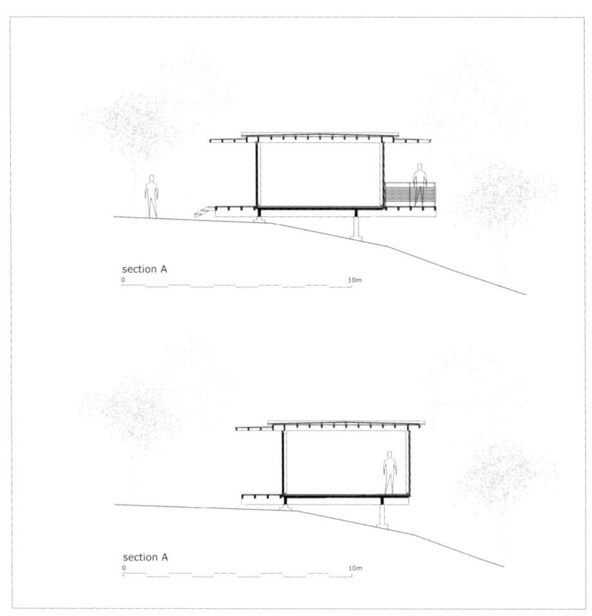
section A
0
10m
section A
0
10m

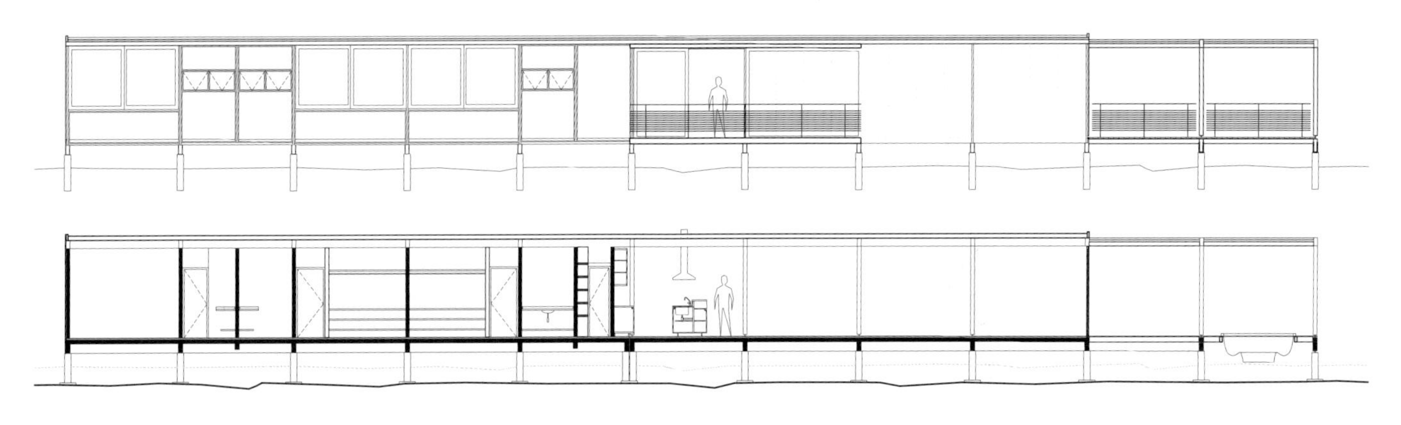

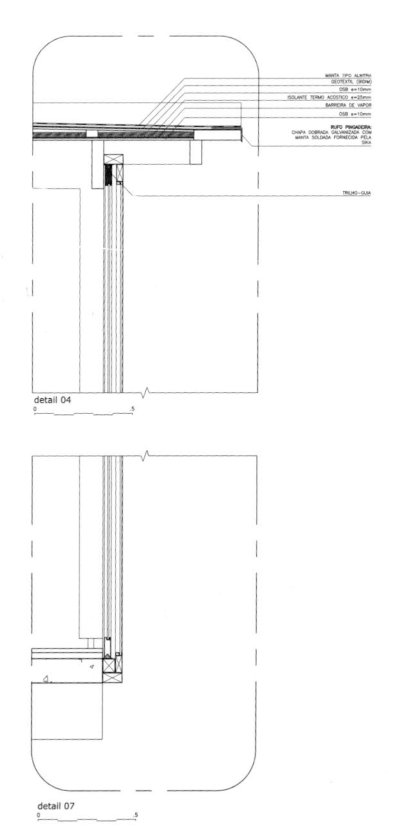
MANTA TIPO ALWITRA
GEOTEXTIL (BIDIM)
OSB e=10mm
ISOLANTE TERMO ACÚSTICO e=25mm
BARREIRA DE VAPOR
OSB e=10mm
RUFO PINGADEIRA:
CHAPA DOBRADA GALVANIZADA COM
MANTA SOLDADA FORNECIDA PELA
SIKA
TRILHO-GUIA
detail 04
detail 07

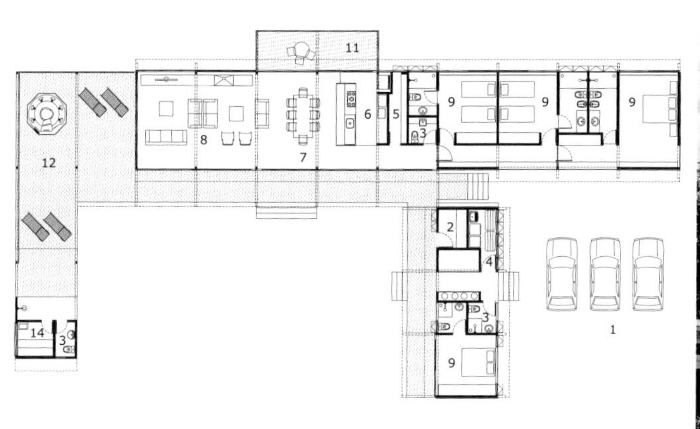

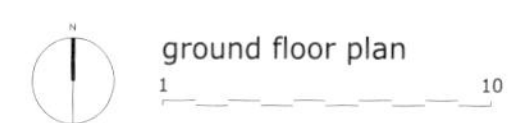

1 - Garage
2 - Deposit
3 - W.C.
4 - Utility room
5 - Spence
6 - Kitchen
7 - Dinning room
8 - Living room
9 - Suite
10 - Bedroom
11 - Veranda
12 - Deck
13 - Guest house
14 - Sauna

Casa Diaz in Mexico

墨西哥卡萨·迪亚兹

Location / 地点:
Valle de Bravo, Mexico
Date of Completion / 竣工时间:
2010
Area / 占地面积:
450 m^2
Architecture / 建筑设计:
PRODUCTORA
Interior Design / 室内设计:
PRODUCTORA
Landscape / 景观设计:
PRODUCTORA
Photography / 摄影:
Paul Czitrom, Rafael Gamo
Client / 客户:
Díaz Family

This property adjoins a large lake in a small town situated a few hours from Mexico City. To take full advantage of the relationship with the surroundings, a system of elongated rectangular volumes was used, with one side of each completely open toward the lake. The sloping plot and the amount of surface to be realized led to the creation of three volumes stacked in a zigzag pattern, generating spacious open terraces and irregular, sheltered patios between them.

From the street, the residence looks like a traditional construction; the use of roof tiles, wood, natural stone, and the plastered facade with small openings, grants it the regional character that is required by urban planning requirements. From the lake, the house is perceived as a composition of rectangular elements with large glass surfaces; a series of typical modernist volumes, stacked in a dynamic configuration.

这栋住宅位于距墨西哥城几个小时车程的一个小镇。为了充分利用建筑与环境的关系，将建筑设计成细长的长方形体块系统，每个体块都朝向湖景。倾斜的场地上可利用的区域共建立了3个体块，这些体块以“Z”字形分布，形成宽敞的开放式露台和不规则、有遮挡的避风庭院。

从街道角度看去，住宅呈现出传统建筑的外观，使用的屋顶瓦片、木材、天然石材以及带有开口结构的石膏外墙赋予房屋以区域特征，并且顺应了城市规划的要求。从湖边的角度观赏住宅，它呈现出的是带有大型玻璃立面的长方形体块组合体，这些体块具有独特的现代主义风格，布局方式颇具活力。

Floor: Quarried Stone (Volcanic);
Walking Paths: Flagstone;
Interior walls: Plaster with Smooth Finish Painted White, and Tzalan Wood;
Exterior walls: Flagstone;
Fixed Furniture: Brick masonry, Concrete and Tzalan Wood;
Illumination: Ceiling Lamps;
Roof: Ceramic Tiles;
Carpentry: Aluminum Frames.

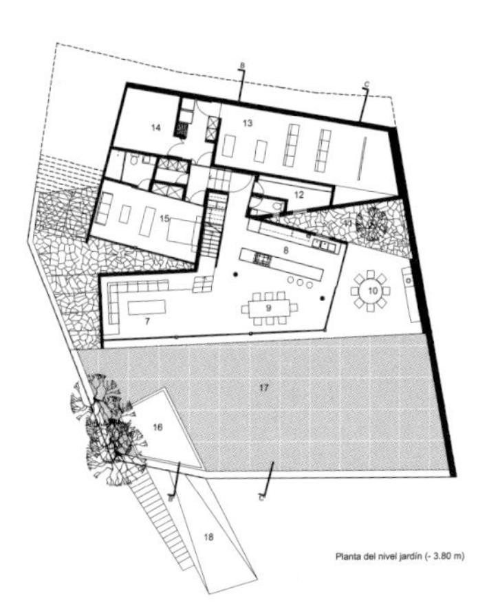

Planta del nivel jardín (- 3.80 m)

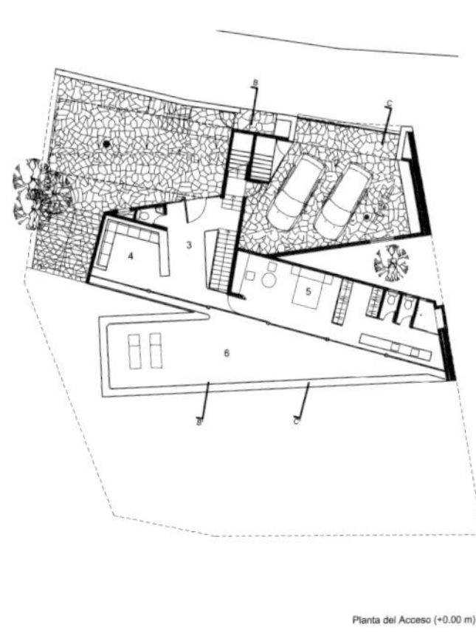

Planta del Acceso (+0.00 m)

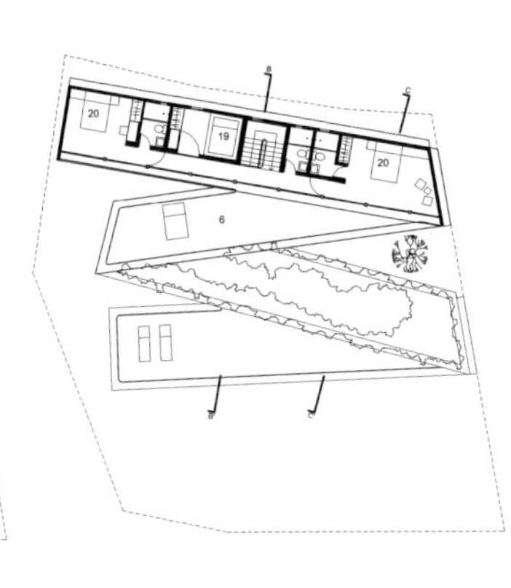

Planta superior (+3.00 m)

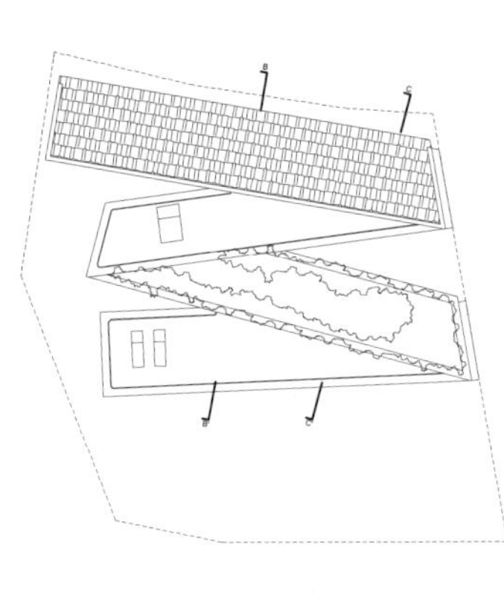

Planta de Techos

1. Acceso /
2. Cochera
3. Vestíbulo
4. Sala de TV
5. Dormitorio Principal
6. Terraza
7. Sala
8. Cocina
9. Comedor
10. Galería
11. Patio
12. Bodega
13. Sala de Proyecciones
14. Cuarto de Máquinas
15. Dormitorio de Servicio
16. Alberca
17. Jardín
18. Rampa de Acceso al Lc
19. Cuarto de Visitas
20. Dormitorio

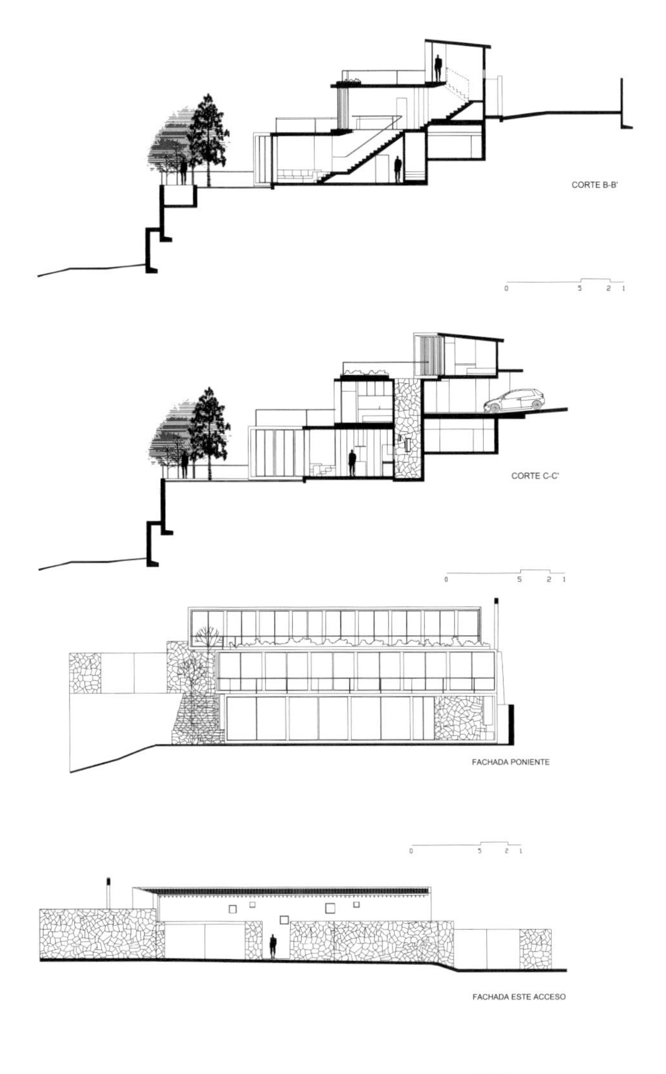
CORTE B-B'
0 5 2 1
CORTE C-C'
0 5 2 1
FACHADA PONIENTE
0 5 2 1

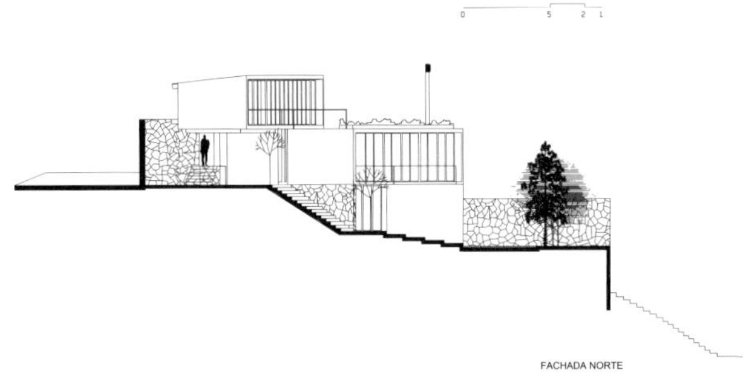
FACHADA ESTE ACCESO
0 5 2 1
FACHADA NORTE

Sow Geneva Switzerland

瑞士日内瓦居所

The owner commissioned SAOTA – Stefan Antoni Olmesdahl Truen Architects to design his Geneva base for his family home and office. Passionate about design and architecture, this is the 4th home SAOTA have undertaken for him. The others, in Cape Town and Paris and one in Senegal are manifest statements of exciting, even provocative contemporary architecture.

On either side of the 20 metre wide channel sits the two portions that make this house, the main house and the annex. What link the two buildings are the cinemas, spa, auditorium and garages underneath. The main house is a combination of round edged cubes and triangular masses that form the L-shape of the living spaces. A double volume living area with a curved wall on the façade facing the lake, flows into a dining area and kitchen on the ground floor and bedrooms, a lobby and en-suite's on the top level. The top floors are accessed by a glass cylinder encased lift. The annex houses a guest suite and what unifies the two are their materiality and spacial relationship to each other, making the gap in between read rather like a pause in time as opposed to an empty space.

Mark Rielly from ANTONI ASSOCIATES spent a number of days in Geneva with the client to source and specify all interior decor and furniture. The contemporary architectural spaces defined the design direction, which resulted in a modern approach to the interior. The living room is divided into two zones, the formal area and the informal arrangement centred around the feature fireplace – a black suspended flue and fire dish mounted on the floor. Continuity of these two zones was achieved by specifying the same modular sofa, the curved Arne sofa from B&B Italia, but in different configurations. Custom-sized organic-shaped patchwork Nguni rugs were designed for both of these areas.

Clearly, it's a building that revels in counterpoint: sculptural versus man-made versus natural versus high tech. Its poetry is a result of its parts and as the owners remarked, the African-ness of the design is enhanced by the subtlety and elegance of the light – so much more apparent in Europe – that moves through the house, changing from season to season and changing too the character of the spaces.

业主要求SAOTA（Stefan Antoni Olmesdahl Truen Architects）建筑事务所为其设计位于日内瓦的居所和办公室。这已经是SAOTA为他设计的第四所住宅了。其他三所分别位于开普敦、巴黎和塞内加尔，展示的都是令人兴奋、极具影响力的当代建筑。

住宅包括主房和副房两部分，分别坐落在20m宽的通道两侧。这两个建筑通过下方的电影院、水疗中心、礼堂和车库相连接。主房是圆滑立方体和三角锥的混合体，形成“L”形的居住空间。具有弧形外墙的两层式生活区朝向湖面，生活区包括用餐区、底层厨房、卧室、大厅和顶层的套房。业主通过一个玻璃柱面电梯可以到达顶层。副房内设置了一间客人套房。建筑的材质使主、副房有了联系。

来自安东尼建筑事务所的马克·雷利在日内瓦花了几天时间，陪业主选定所有的室内装饰和家具。住宅的内部设计展现了现代化的建筑设计理念。客厅分为两个区域，包括待客场所和起居室，其中后者围绕一个特色壁炉而设置，壁炉由黑色烟道和地面上的炉火盘构成。为取得两个区域的连续性，设计采用了同样的组合式沙发——来自B&B Italia的弧形Arne沙发，不过采用了不同的布局方式。两个区域都有定制尺寸的地毯。

很显然，这栋建筑展现了人造、自然与高科技的多重对比。各部分的组合和谐共生，正如业主的评价：“微妙和优雅的光线使带有非洲风情的设计充满了个性。”光线照射于房间之内，随四季变化而变化，为房间带来不同的光影效果。

Location / 地点:
Geneva, Switzerland
Date of Completion / 竣工时间:
2010
Area / 占地面积:
7,612 m²
Architecture / 建筑设计:
SAOTA - Stefan Antóni Olmesdahl Truen Architects
Interior Design / 室内设计:
ANTONI ASSOCIATES
Landscape / 景观设计:
N/A
Photography / 摄影:
SAOTA

Materials: Glass, Stainless Steel, Stone, Timber, Concrete, Marble.

SOUTH EAST ELEVATION

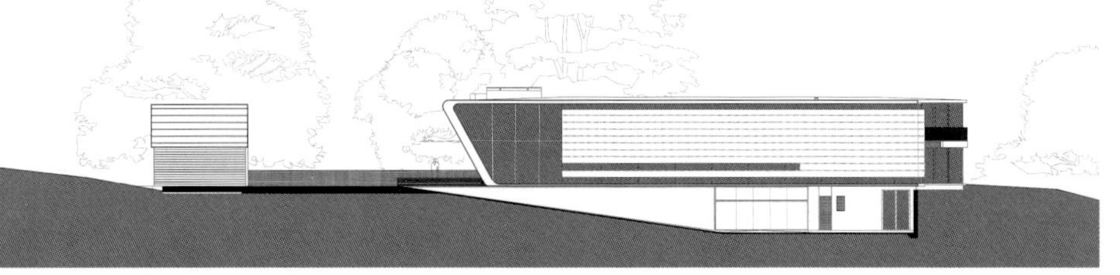

NORTH WEST ELEVATION

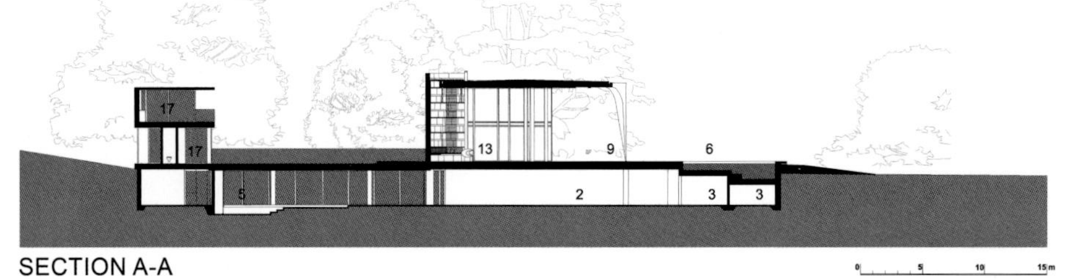

SECTION A-A

LEGEND

1. DRIVEWAY
2. GARAGE
3. STORE
4. OFFICE
5. THEATER
6. POOL
7. STEAM ROOM
8. SAUNA
9. TERRACE
10. STAFF ROOMS
11. ENTRANCE GATE
12. POND
13. LOUNGE
14. DINING
15. KITCHEN
16. SCULLERY
17. BEDROOM
18. DOUBLE VOLUME

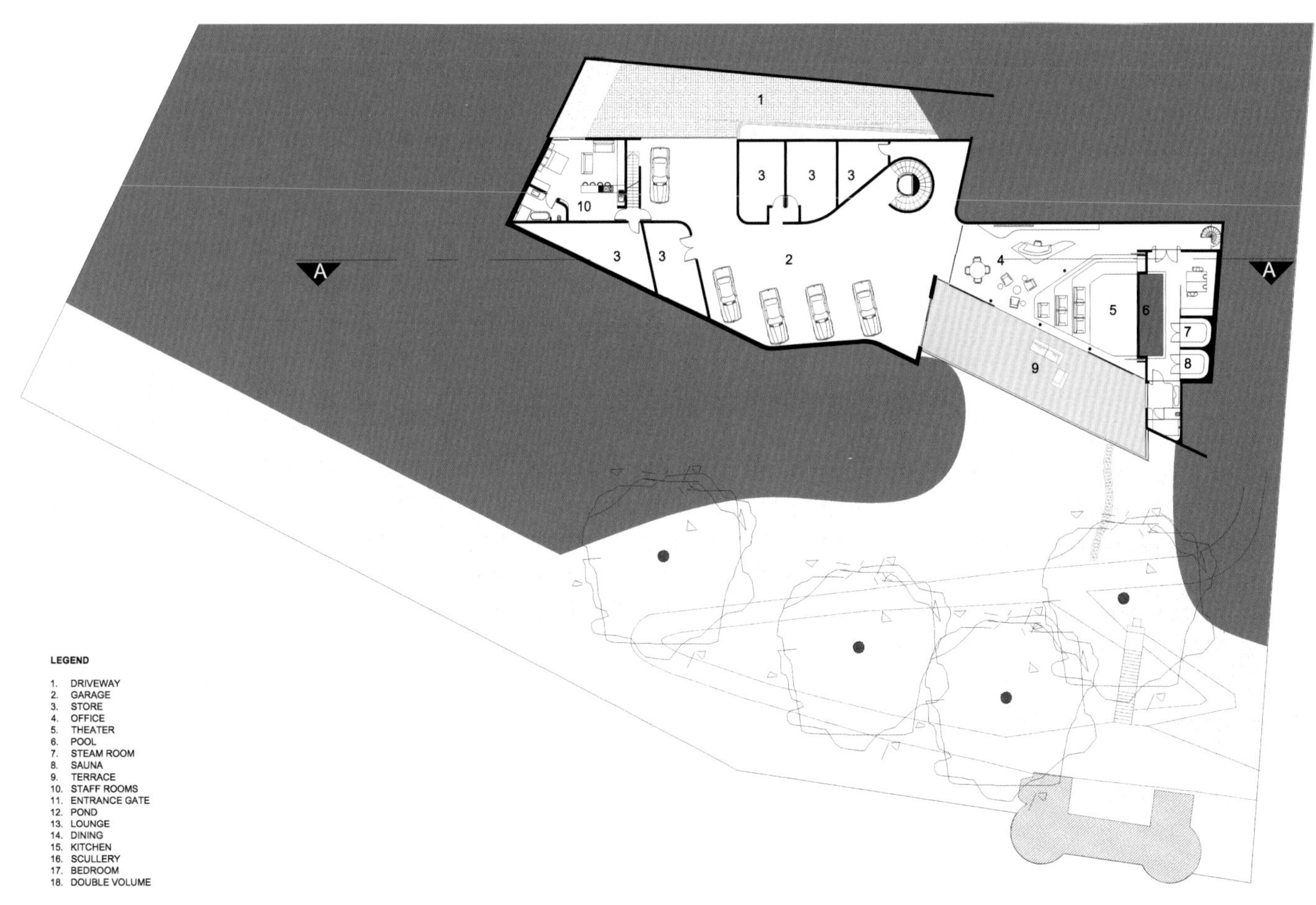
LEGEND
1. DRIVEWAY
2. GARAGE
3. STORE
4. OFFICE
5. THEATER
6. POOL
7. STEAM ROOM
8. SAUNA
9. TERRACE
10. STAFF ROOMS
11. ENTRANCE GATE
12. POND
13. LOUNGE
14. DINING
15. KITCHEN
16. SCULLERY
17. BEDROOM
18. DOUBLE VOLUME
A
A

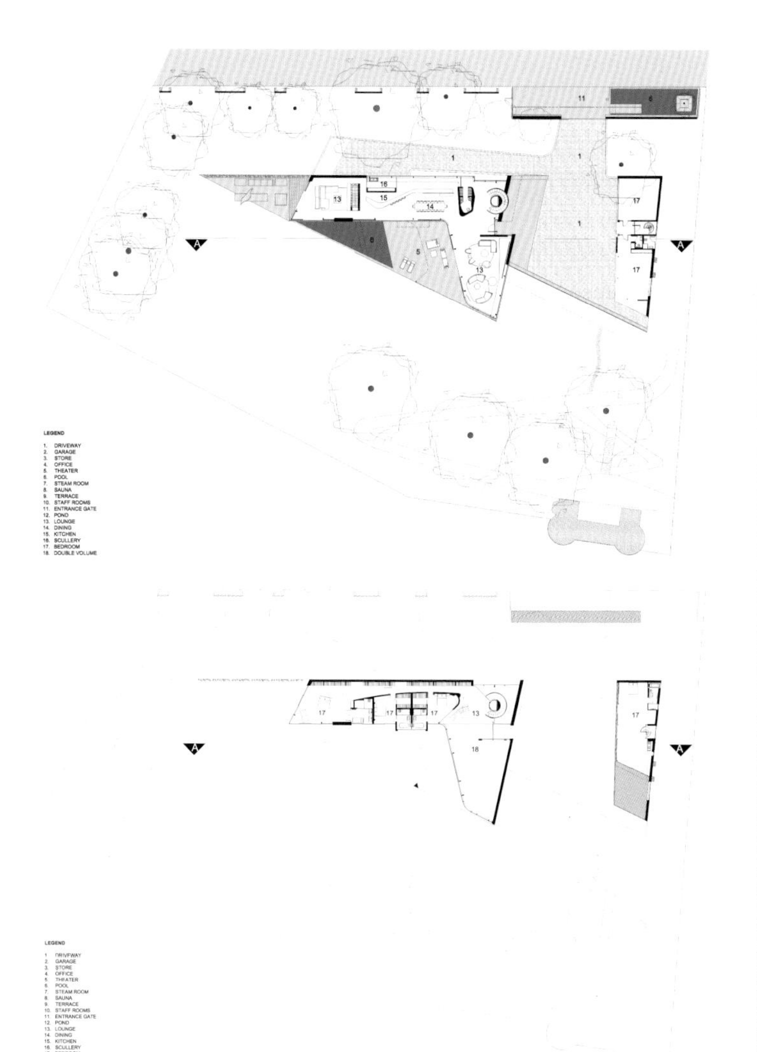
LEGEND
1. DRIVEWAY
2. GARAGE
3. STORE
4. OFFICE
5. THEATER
6. POOL
7. STEAM ROOM
8. SAUNA
9. TERRACE
10. STAFF ROOMS
11. ENTRANCE GATE
12. POND
13. LOUNGE
14. DINING
15. KITCHEN
16. SCULLERY
17. BEDROOM
18. DOUBLE VOLUME
LEGEND
1. DRIVEWAY
2. GARAGE
3. STORE
4. OFFICE
5. THEATER
6. POOL
7. STEAM ROOM
8. SAUNA
9. TERRACE
10. STAFF ROOMS
11. ENTRANCE GATE
12. POND
13. LOUNGE
14. DINING
15. KITCHEN
16. SCULLERY
17. BEDROOM
18. DOUBLE VOLUME

Topographic House - Single House in Llavaneres

地形楼——Llavaneres独栋楼

The house lies along its longitudinal axis: the view on the landscape, towards the sea, is – again – one of the most important values for the project, which unfolds itself and gets mixed with topography. The building is horizontal, elongated, sinuous, catching the landscape, moving delicately in order to obtain the best views, without shutting spaces.

The project begins with the topographical understanding of the place. By re-drawing the contour lines, possible spaces to occupy appear, from the underground parking to the hanging planes of the studio. These curves, rising from the concrete retaining walls, lose weight, but still get visible and construct spaces through long overlaid metal beams.

When beams are coincident, they build the horizontal structure, and through them, light allows having a complete topographical perception. Curves flow all along the site, and get blurred in its limits. They get blended with topography. This house will be like a great window to frame the place, the sea and the village: a thin panoramic space where a more built architecture can be seen above, the studio. It has the most privileged views and it occupies the upper zone of the house.

This big 'skeleton' is coated of zinc, a very fabric-like material that can be used in such zigzag shapes, on the stone areas and on walls which surround other spaces of the house.

Location / 地点:
Barcelona, Spain
Date of Completion / 竣工时间:
2008
Area / 占地面积:
700 m²
Architecture / 建筑设计:
Josep Miàs
Interior Design / 室内设计:
Josep Miàs – Blanca Tey, Victoria Ayesta
Landscape / 景观设计:
Josep Miàs
Photography / 摄影:
Adrià Goula, Jordi Anguera

这个住宅沿着场地的纵向轴线展开设计。场地面朝大海，拥有美好的自然景观，为该项目带来极大的优势。该建筑为横向结构，体量细长、弯曲，深植于景观之中，以精致的建筑结构捕获最佳的视野。同时，该建筑没有产生封闭的空间。

该项目从研究场地的地形入手。建筑师重新勾勒出场地的轮廓线，从地下停车场到工作室顶部的潜在空间都一一被展现出来。

金属梁的结合形成了建筑的横向结构。在光线的作用下，通过这些横向结构可以看到完整的地貌。这些曲线沿着场地流动，在场地尽头逐渐消失，与地形融为一体。整座住宅就是一个大型窗户，它形成一个全景空间，身处其中的人们可以看到大海和村庄美丽的风景。工作室位于建筑的顶层，是这座建筑中视野最好的地方。

建筑“骨架”的石材区域和墙体部位镀上了一层锌，这种材料质地如织物一般，十分独特。

Materials: Concrete, Glass, Steel.

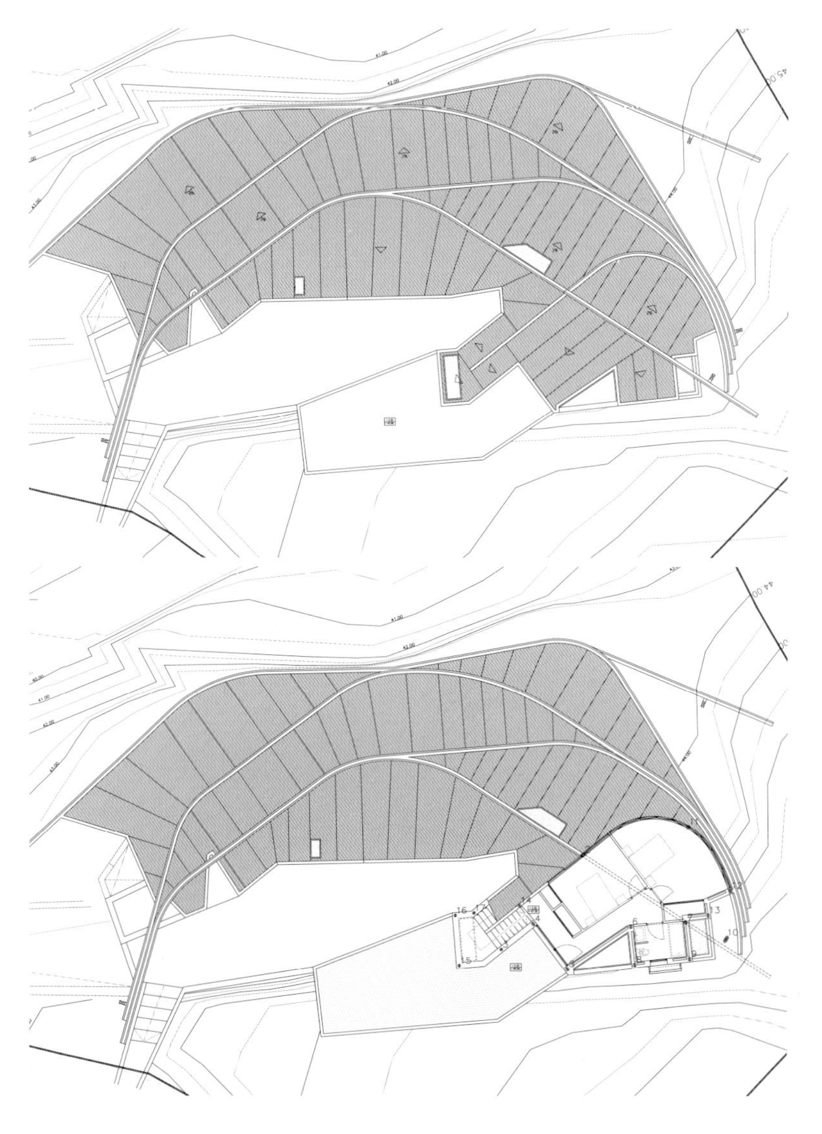

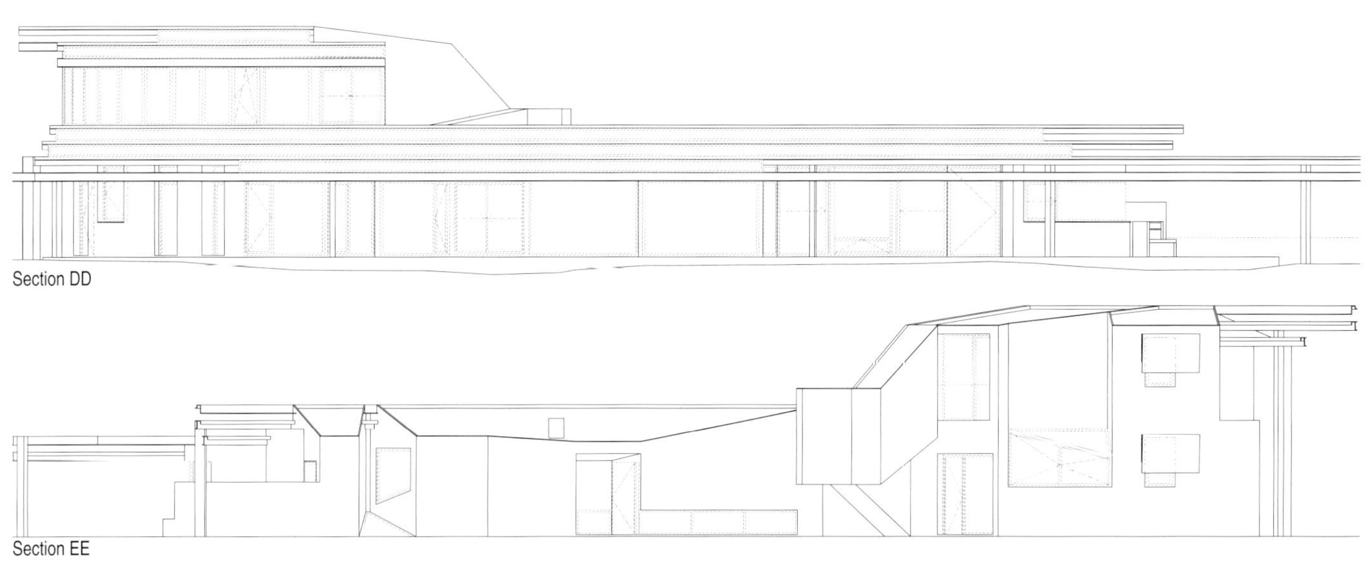

Section DD

Section EE

Plan Underground Level

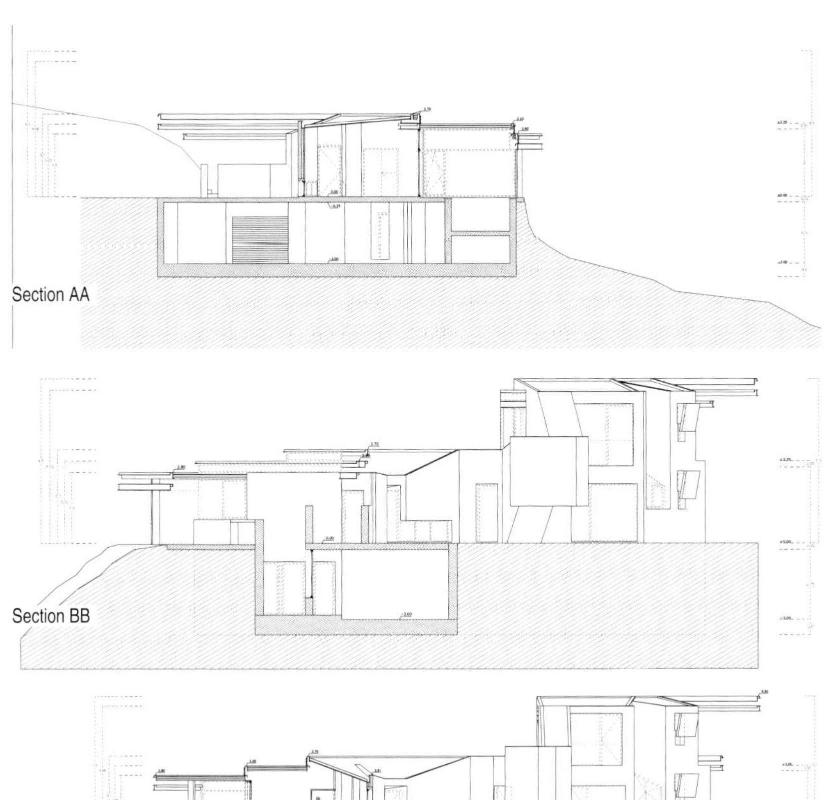

Section AA

Section BB

Section CC

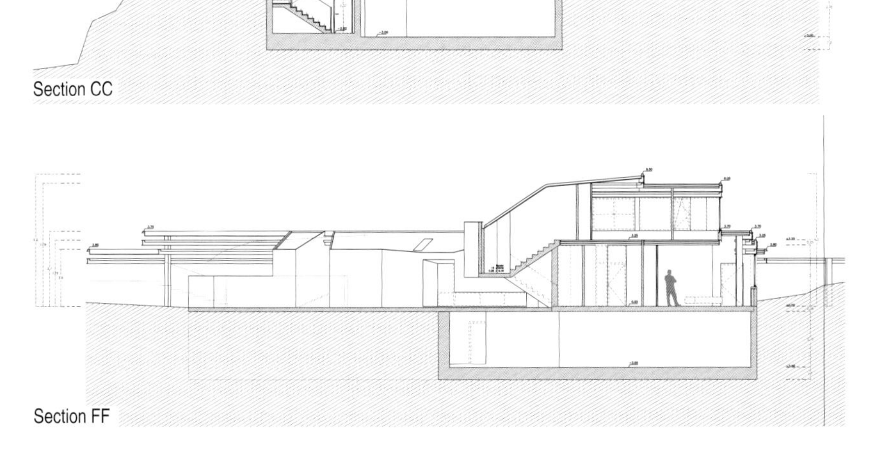

Section FF

Urban Oasis in Indonesia

印尼都市绿洲

A luxury home is able to accommodate the entire needs of rooms and for all the family activity. The resort-style home with 1,000 m^2 building area above the 650 m^2 land area are using the best choice for the natural material, like the display on the secondary skin that uses a wood or marble which asserted the natural element in this house. This house equipped with different rooms that could accommodate the lifestyle of the residents, like a home theatre or fitness room. On the first floor there is a large space for open living area with fishpond, patio area, and swimming pool which became the centre of the family activity.

A strong different feeling is deeply felt in this house. Starting from the entrance area to the foyer area and then to the pantry area and the dining area are designed to converge. The connectivity between the first floor and the second floor is presented by the design of void.

A modern resort house dimension has an attractive value when a few vertical gardens presented as the expression of the green design. This house is representative as the successful of managing the tropical climate became a gift to create the resort house in the middle of the heat in the city. By placing the vertical garden in the basement area near the study room, in the stairs area and also in the near fishpond area that represent a high pedestal with some plants above which can be enjoyed from the living area, stairs area and the upper floor. Design of the height of swimming pool placed in position between first floor height and basement. The water of swimming pool is designed like waterfall from the surface and makes a sound like water fall. Placement of trees around the house has been calculated in the middle of construction thus providing an opportunity for them to grow. And the placement of vines in the back of wall that giving the privacy aspect.

Some of the design of this residence is a function expression which is then poured in interesting shape. This house is more than just a residence that tries to combine modern concept with all the natural elements of exploration integrity, but also have more value for priority to a dynamic modern lifestyle, without forgetting the concept of feng shui, and still give priority to the meaning of being together with family.

这个豪华的居所能够满足业主对房间和家庭活动的所有需求。这栋度假风格的住宅建于650m^2的土地上，建筑空间达1000m^2。建筑材料尽可能选用天然材料，例如二层建筑外墙采用木材，其他区域的外墙则选用天然大理石。住宅中配有不同的功能空间，如家庭影院和健身房，为居住者提供了多样的生活方式。一楼拥有一个大型的开放式生活区，设有鱼塘、露台和泳池，是家庭的活动中心。

房屋内部的设计则给人带来一种截然不同的感受。入口、门厅、厨房和就餐区的起点交汇于一点。一楼和二楼之间通过挑出的结构相连接。

几处垂直绿化体现了绿色设计理念，使这个现代化的度假屋更具吸引力。这栋房屋在炎热的城市环境中脱颖而出，是驾驭热带气候的代表性项目。垂直绿化被设置在书房附近的地下室区域、楼梯区域以及鱼塘附近的区域，形成葱郁的竖向景观，无论从生活区、楼梯处还是上层空间，都能欣赏到这个垂直绿化。泳池位于一楼和地下室之间。泳池的池边可溢出水，形成一个小瀑布。树木的布局是在建造过程中精确计算形成的，从而为树木提供了自由的生长环境。墙体后面种植了藤本植物，营造了私密的氛围。

房屋的一些功能性设计以有趣的形态展现出来。该住宅不仅展现了现代化设计手法和自然元素的结合，还打造了一个充满活力的现代生活方式，同时考虑风水问题，阐述了家庭生活的内涵。

Location / 地点:
Jakarta, Indonesia
Date of Completion / 竣工时间:
2010
Area / 占地面积:
1,000 m^2
Architecture / 建筑设计:
Wahana Cipta Selaras
Interior Design / 室内设计:
Ivone Xue
Landscape / 景观设计:
Elegant Flora
Photography / 摄影:
Fernando Gomulya (Tectography)

Floor: Marble, Parquet, Wooden Deck;
Walking Path: Granite, Pebbles;
Illumination: Floor Washlights, Fluorescent Light, Indirect Light, Spotlight;
Furniture: Duco Finish, High Pressure Laminate Finish, Colour Mirror;
Plants: Plumeria, Spathodea, Mandevilla, Thunbergia.

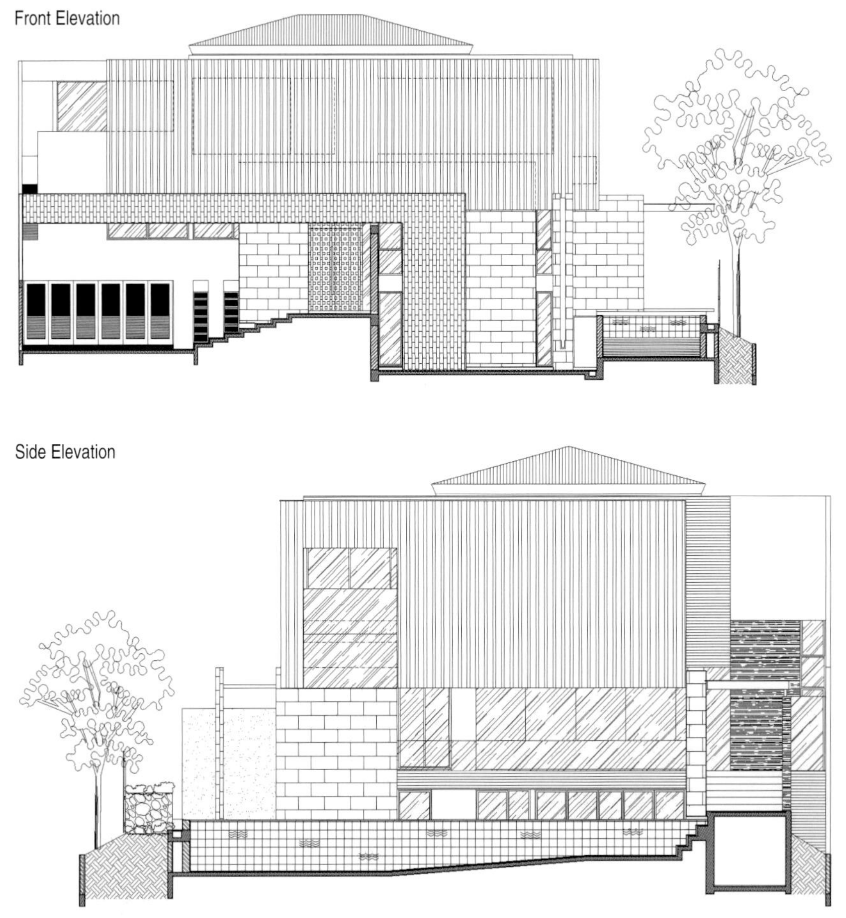
Front Elevation

Side Elevation

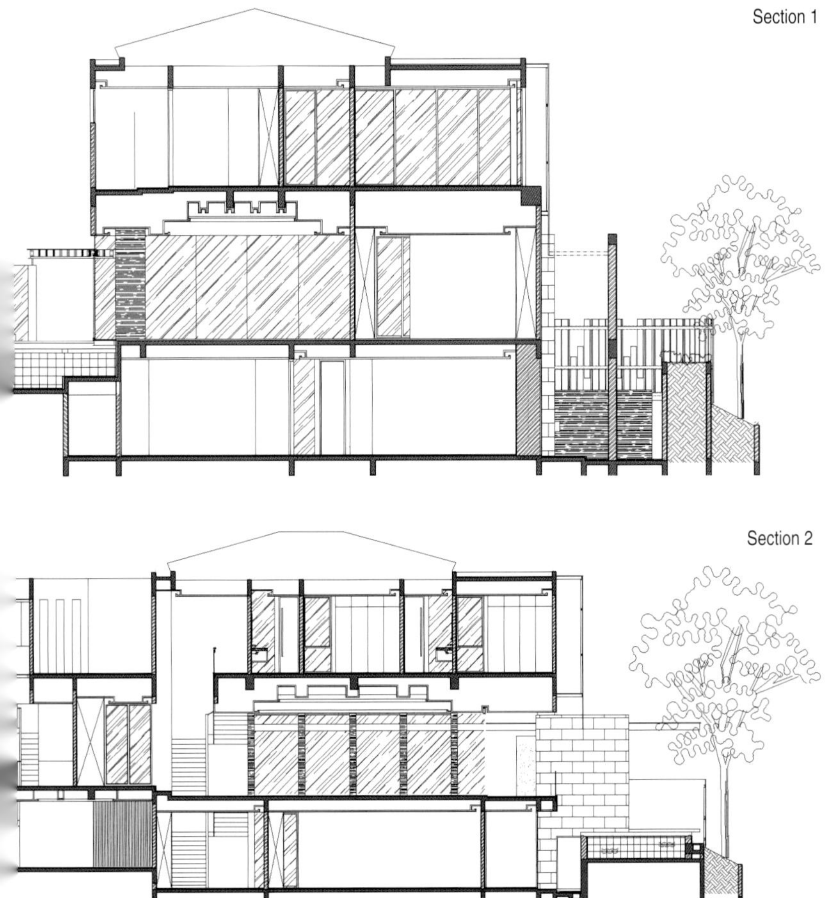
Section 1
Section 2

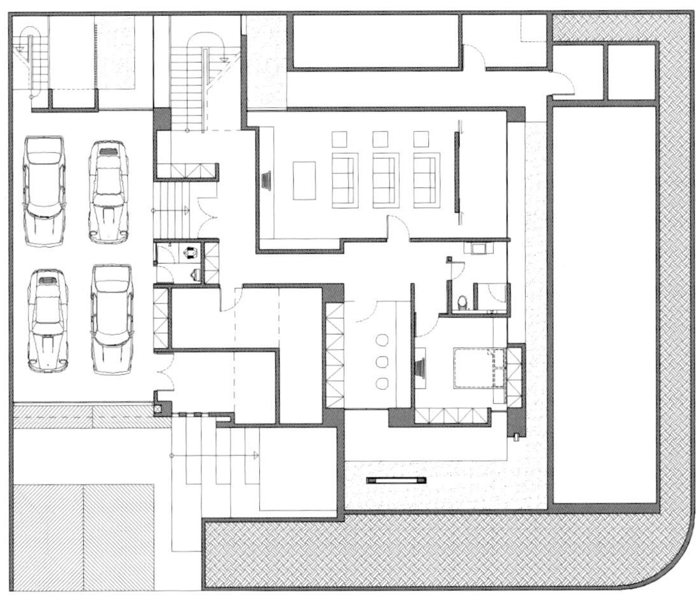

Plan Level Basement

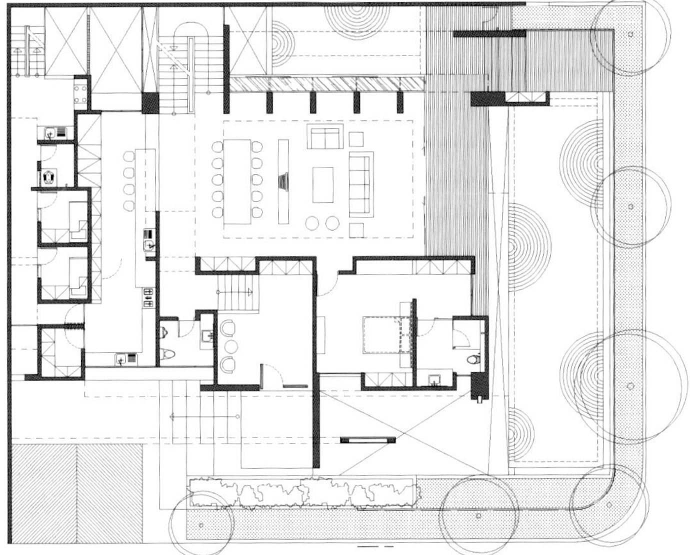

Plan Level Ground

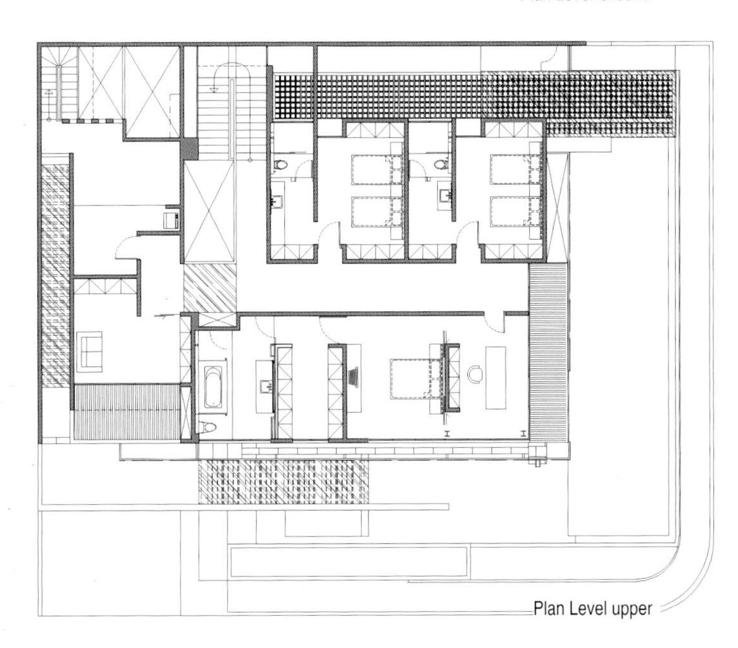

Plan Level upper

Villa Castela Residence

卡斯蒂利亚别墅

Built on a sloped site (30 degrees) in the city of Nova Lima, Brazil, the house uses dramatic cantilevers to emphasize the extremity of its position. The architects have chosen this concept not only for aesthetic reasons, but above all to reduce the interference of the building mass in the topography, keeping the site as natural as possible. This project was based on four main concepts: little interference in the site, better use of natural resources, integration to the surroundings, and a generous urban presence.

As the architects placed the ground level of the residence 7 metres below the street , they were able to preserve the pedestrian view of the forest, at the same time that keeping the privacy of the owners, because the main apertures and the glass walls are oriented to the east, on the opposite side of the street. The urban impact of the residence is minimized, in benefit of the beautiful view of the woods. As the climate of this region is very good, the right orientation of the doors and windows prevents the use of artificial climate. Solar voltaic cells are placed on the roof.

The constructed area is 650 m^2, divided into three floors: the basement, where the owners can enjoy leisure facilities such as sauna, Jacuzzi, and a wine cellar; the ground floor, where the living area and the kitchen are located, integrated to the outdoor swimming pool and the external terrace, making this the centre of the house, and the first floor, where the occupants can obtain privacy in the bedrooms.

The form is generated by the engineering of the concrete structure, which is robust and sculptural and, at the same time, light and contextualized with the surroundings, not a self-referred structure. The concrete maintains its texture, and the masonry is painted terracotta, for low maintenance reasons. (the ground has a red dust, iron ore dust).

Architects used as few columns as possible in order to preserve the existing site. Unsurprisingly, given the exquisite surroundings, the largest proportions of the building face outwards down the hill with views of the forest. Glazed elevations make the most of these views and also of the sunrise to the east.

这栋住宅位于巴西的新利马市，建在一个坡角为30度的斜坡上。住宅采用大跨度的悬挑结构来彰显它极端的建筑环境。设计师采用这种设计不仅是出于美感的营造，更多的是为了减少建筑对地形造成的影响，从而尽可能地使场地环境保持自然。该项目的设计理念主要有四个：对场地的影响尽可能小、更好地利用自然资源，和周围环境相融合，以及大方的建筑外观。

住宅的第一层比建筑地平低7m，从而保留了行人能观赏到的森林景观。房屋的主要开口和玻璃幕墙是朝东的，并未紧邻街道的一侧，可以保证户主的私密性。都市因素对住宅的影响被降至最低，从而让业主可以更好地享受森林的美丽景色。当地的气候十分宜人，于是设计师在门和窗口的朝向设计中充分利用了当地的气候特色，使房屋不需要利用设备来调节温度。屋顶设置了太阳能光伏板。

住宅的建筑面积为650m^2，一共分为三层。在地下室，屋主可以尽情享受悠闲的时光，这里配有桑拿房、大浴缸以及酒窖；一层则是整个建筑的主要空间，生活区、厨房都分布在这个区域，并且与户外游泳池和露台一体化集成；二层则是较为私密的卧室。

建筑的外部是混凝土结构，结实而有雕塑感，同时也显得很轻盈，与环境相融合，而不是自成一派。混凝土原本的外观被保留下来，砖石部分则因考虑到维护成本而被涂为赤褐色（场地有红色的铁矿石灰尘）。

为保持现有的场地条件，建筑师尽可能少地使用支柱。优美的的环境无疑使建筑的大部分都能朝向自然幽静的森林景观。建筑的玻璃立面更是将这些景观和日出的美景展现到极致。

Location / 地点:
Belo Horizonte, Brazil
Date of Completion / 竣工时间:
2009
Area / 占地面积:
650 m^2
Architecture / 建筑设计:
Anastasia Arquitetos
Interior Design / 室内设计:
Anastasia Arquitetos
Landscape / 景观设计:
Anastasia Arquitetos
Photography / 摄影:
Jomar Bragança

Floors: Travertine Marble, Wood;
Walking Paths: Stone Miracema.

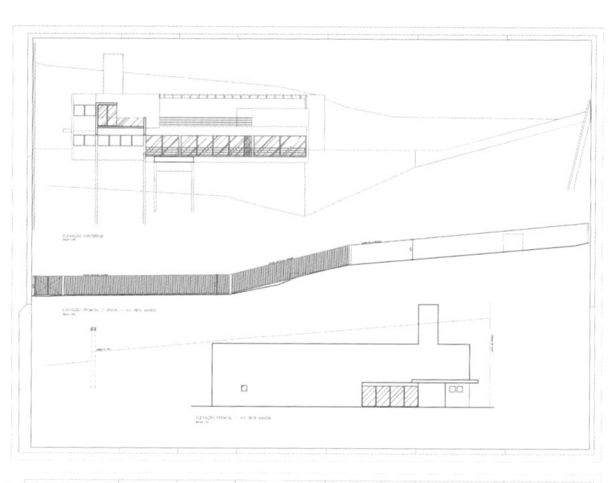

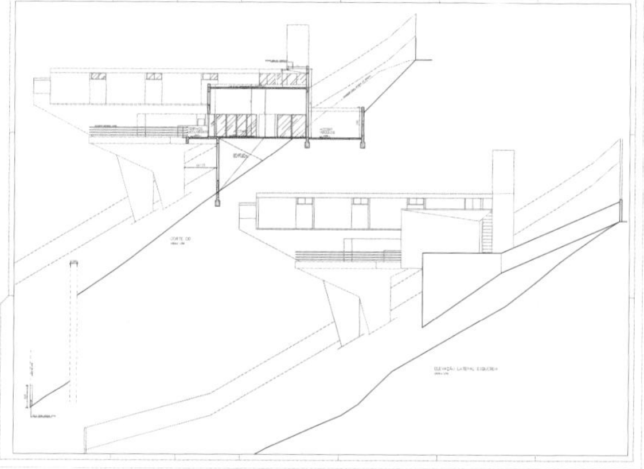

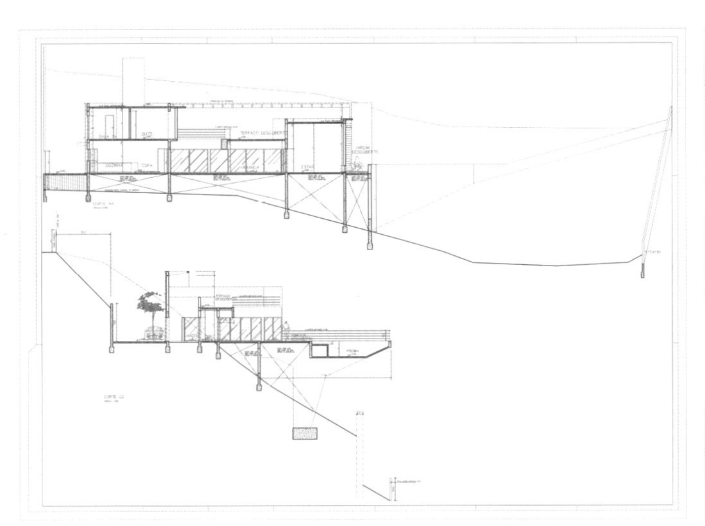

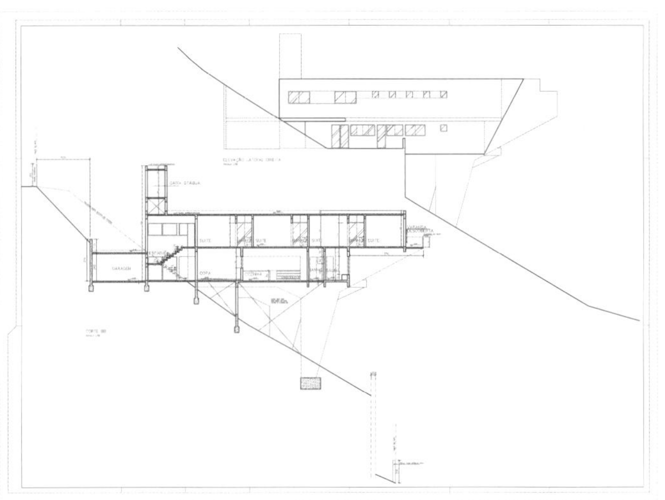

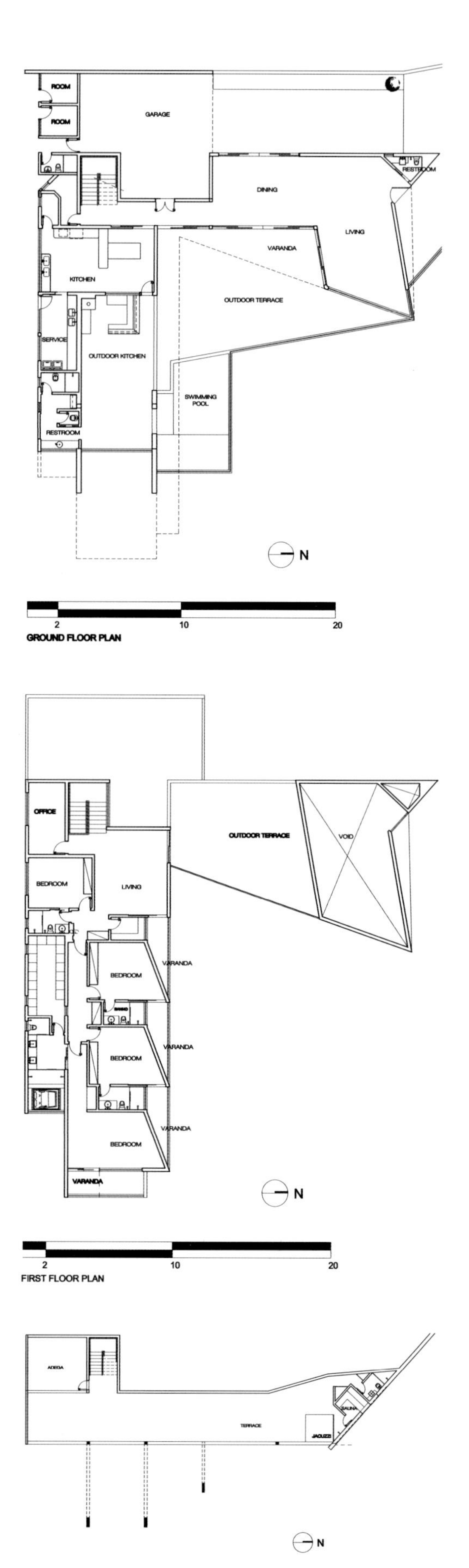

GROUND FLOOR PLAN

FIRST FLOOR PLAN

BASEMENT PLAN

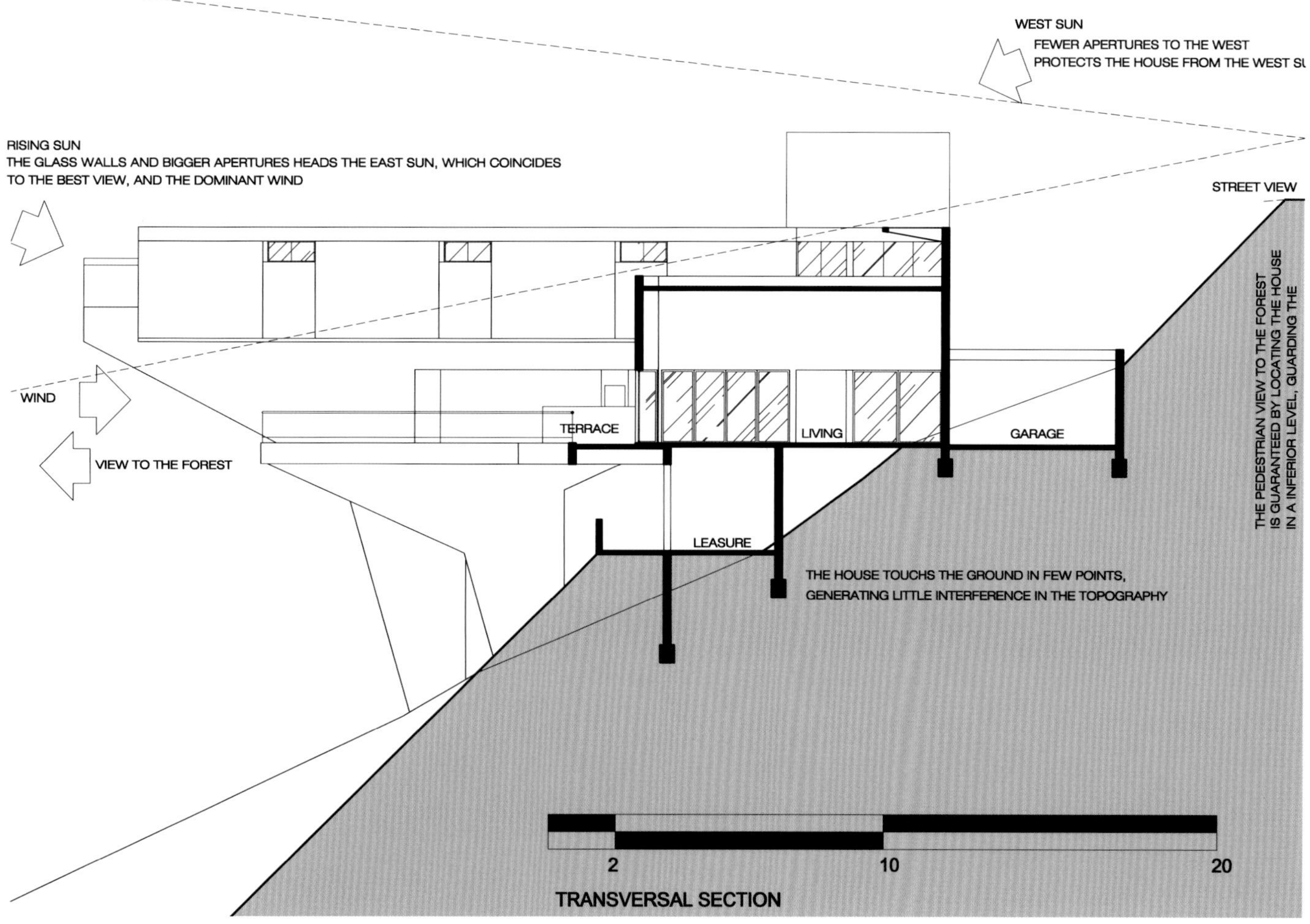
WEST SUN
FEWER APERTURES TO THE WEST
PROTECTS THE HOUSE FROM THE WEST SU
RISING SUN
THE GLASS WALLS AND BIGGER APERTURES HEADS THE EAST SUN, WHICH COINCIDES
TO THE BEST VIEW, AND THE DOMINANT WIND
STREET VIEW
WIND
VIEW TO THE FOREST
TERRACE
LIVING
GARAGE
LEASURE
THE HOUSE TOUCHS THE GROUND IN FEW POINTS,
GENERATING LITTLE INTERFERENCE IN THE TOPOGRAPHY
THE PEDESTRIAN VIEW TO THE FOREST
IS GUARANTEED BY LOCATING THE HOUSE
IN A INFERIOR LEVEL, GUARDING THE
2
10
20
TRANSVERSAL SECTION

Villa Veth in the Netherlands

荷兰Veth别墅

Location / 地点:
Hattem, the Netherlands
Date of Completion / 竣工时间:
2011
Area / 占地面积:
6,475 m^2
Architecture / 建筑设计:
123DV
Interior Design / 室内设计:
123DV
Landscape / 景观设计:
123DV
Photography / 摄影:
Christiaan de Bruijne

Villa Veth is a modern, customized villa, a private residence for a family of four. It is situated on a large parcel of land by a forest near the idyllic town of Hattem in the eastern part of the Netherlands.

The driveway on the entrance side leads to a carport situated below the house. This places the cars out of sight and gives the impression of the house partially hovering. It is a simple architectural approach with a great visual effect. The house looks sleek and abstract on the outside, but has a warm and cozy interior.

The ground floor and principal living area of the two-storey residence is divided into two. On one side are the master bedroom and two kids' bedrooms plus two small studios. The other half of the floor plan is taken up by an open-concept living area that includes the kitchen, dining and living spaces. The furniture in the living room is all custom designed. The kitchen, storage space, fireplace, piano and audio equipment form an integral part of the wall unit.

The living area of this bungalow with woodland is orientated towards the south. The curved glass wall enclosing the living area towards the spacious terrace is designed to visually minimize the boundary between inside and outside. From the inside of the house this provides a maximum experience of its surroundings. Supported with a single column clad with reflective metal, the low-profile roof extends to cover an outdoor patio. The large canopy and floor heating allow inhabitants to enjoy the terrace during cool autumn days.

Veth别墅是一栋根据现场环境而量身定制的现代别墅，供一个四口之家居住。它坐落在一片被森林围绕的宽阔的场地上，靠近荷兰东部的田园小镇哈特姆。

入口的车道直接通向房子下方的车库，让汽车被隐藏在视野之外，同时，悬挑的建筑也给人以轻盈空灵之感。通过简单的建筑手法获得显著的视觉效果和功能满足。房屋的外观豪华而抽象，但室内却温馨而舒适。

这栋住宅为两层结构，底部为车库，上部作为主要生活区被划分为两部分。一部分是主人卧室、两个孩子的卧室外加两个小型工作室。另一部分采用开放理念的生活区布局，包括厨房、用餐区和生活空间。客厅的家具都是定制设计的。厨房、储存室、壁炉、钢琴和音响设备构成墙壁单元整体的一部分。

这栋位于林地边上的平房将生活区朝南而设。弧形玻璃幕墙围绕着生活区，面向宽敞的露台，旨在从视觉上将室内外的边界弱化至最低。这个构造让业主可以在室内拥有极佳的室外环境体验。建筑悬挑的部分由一个单独的柱子支撑，柱尖面镀有反射金属层，低矮的屋顶向外延伸，成为露台的顶棚。由于有宽敞的屋顶与地热采暖系统，即使是寒冷的秋日，闲坐于露台上也同样可以享受周边优美的自然风景。

Material: Dark Gray Stucco, Frameless and Curved Glass, Power Floated Concrete Floor.

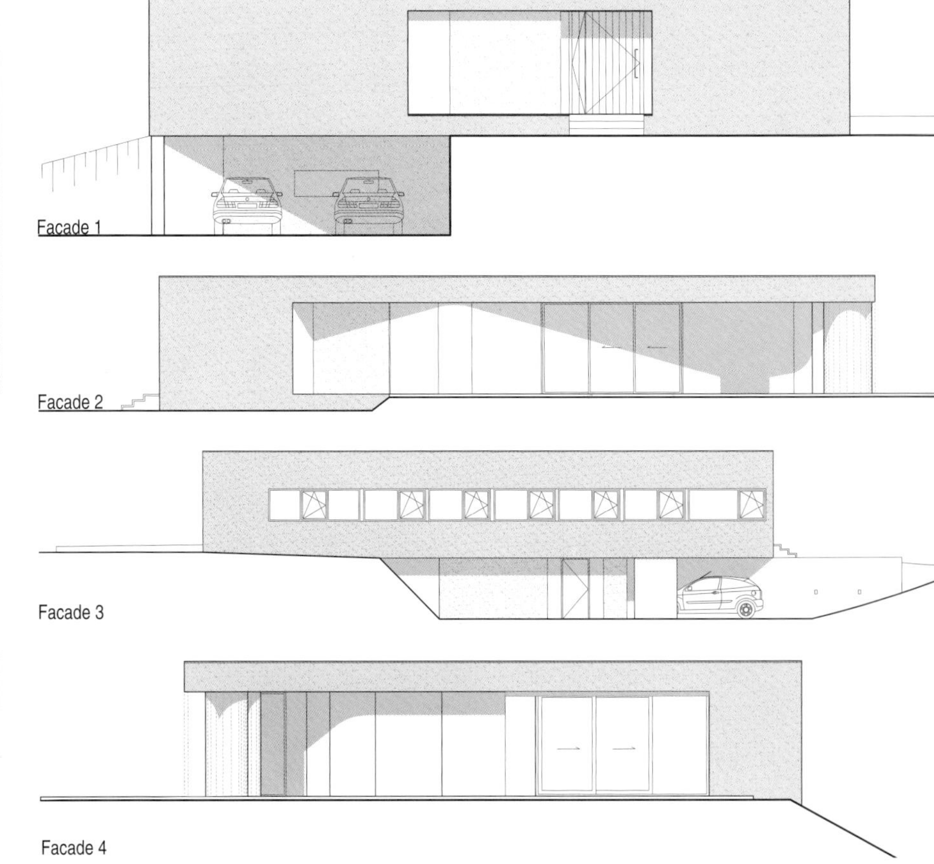
Facade 1
Facade 2
Facade 3
Facade 4

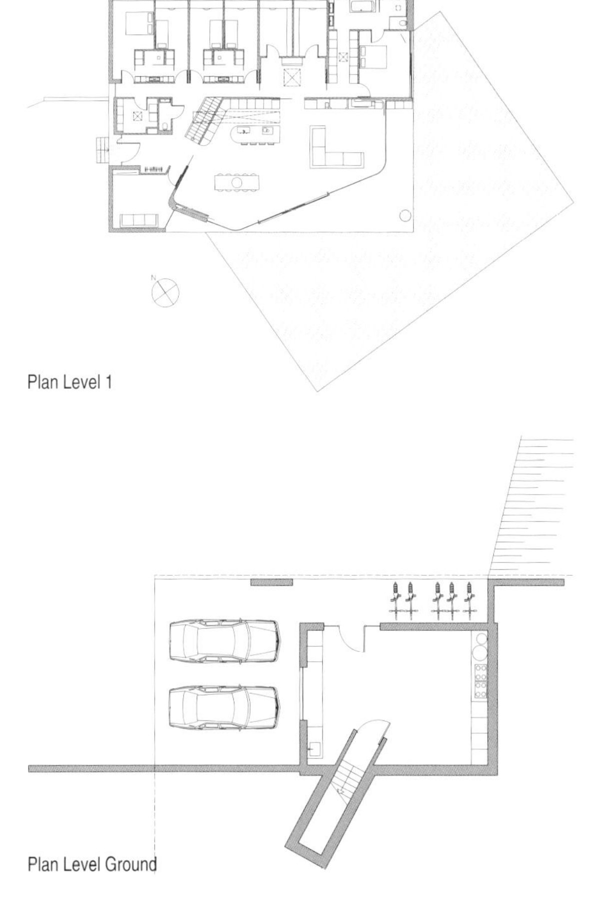

Plan Level 1

Plan Level Ground

Villa Amanzi in Thailand

泰国Amanzi别墅

Nestled in a cascading, west facing ravine with a dramatic slab of rock defining the northern edge and a stunning outlook over the azure blue of the Andaman Sea to the south, the commission; to do this demanding but ultimately spectacular site justice, was both daunting and exciting.

The defining elements are the rock and the view. They dominate at every juncture. They resonate on first approach, through the migration from public to private space, in the living and in the family areas, in the gardens, in the bedrooms; and they continue to command respect down the tropical jungle steps that arrive at a secluded rock platform, flanked by the same seam that welcomed you 60 m above. Constant references to these elements instill a feeling of solidity that contrasts with the openness of the house, reinforcing the dynamism and vibrancy that pays homage to the magic of the location.

The home grows out from the rock; the bedroom element rests between it and the wing that strikes the perpendicular, rising vertically from the slope. This composition defines the open living and dining space that is simply a transition between two garden areas. It is intimate but open and the uninterrupted clear span creates a bridge under which the conventions defining indoor space disappear. The goal of the design is to make the home harmonize with its location and the surroundings. This is achieved by carefully analyzing the survey and topographical information and designing the building to tuck into the site, capitalizing the drama of the rock that runs through the home and defines it; from first approachs all the way down to the rock pools at the ocean front. Cantilevered over a massage sala, the swimming pool completes the composition. It is the focal point that draws the eye to the view and instills a calmness that provides balance with the energy of the architecture.

Sustainability has been a real driving force for the design of Villa Amanzi. Phuket enjoys a monsoon climate where the wind direction is predominantly east to west in the dry monsoon and opposite in the wet season. This has enabled them to harness the monsoons cooling potential by enabling the west and east sides of the house to be opened up. This works so well that the owner has commented that they have never actually used the air conditioning in the living and dining space.

这栋别墅隐映在陡峭、朝西的峡谷中，峡谷中有一块显著的巨石清晰地勾勒出场地北部的边界，同时面向蔚蓝的安达曼海。这块场地景色壮观，但在此施工却很艰难，这让建筑师既兴奋又紧张。

本项目最大的特色是将建筑建造于石山之间，拥有壮丽的自然景观。奇景佳色自然地渗透到房屋的每一个角落：一方面，无论是从公共空间踏入私人空间，还是在生活区、家庭区、花园、卧室，处处都能欣赏到美景；另一方面，沿着热带丛林中的台阶而下，来到房屋一侧下方60m处一块岩石平台上，这里同样能欣赏到美不胜收的景色。目不暇接的石山和美景使建筑深深地扎根于此，并与房屋开敞的结构形成对比，建筑显得活力十足、生机盎然。

建筑从岩石中脱颖而出，卧室位于与石山形成直角的侧翼部分。开放式客厅及用餐区，既私密又开放。同时，连续的跨越结构如同一座桥，使下方对室内规划的约束减少。从一开始，设计的目标就是使建筑与特殊的场地和周围的景观有机地结合在一起。建筑师仔细地分析了地形特征，将建筑巧妙地嵌入山体，并对岩石地貌加以有效利用。游泳池悬挑于按摩室的上方，为整个设计画上完美的句号。它作为景观的焦点，烘托了宁静的氛围，与建筑完美结合。

可持续发展是Amanzi别墅设计的基本原则。普吉岛在旱季，风主要由东吹向西，在雨季则相反。于是，建筑师将建筑的东、西侧敞开，利用风来降低室内的温度。这种手法的运用效果非常显著，业主表示他们在起居空间和用餐空间都未使用过空调设备。

Location / 地点:
Phuket, Thailand
Date of Completion / 竣工时间:
2008
Area / 占地面积:
2,664 m²
Architecture / 建筑设计:
Original Vision Ltd.
Interior Design / 室内设计:
Original Vision Ltd.
Landscape / 景观设计:
Original Vision Ltd.
Photography / 摄影:
Helicam Asia Aerial Photography, Marc Gerritsen

Materials: Tempered Safety Glass, Stainless Steel, Terrazzo, Washed Terrazzo, Bamboo Flooring, Teak Decking, Sandwash Outdoor Stairways, ICI Paint, Cement Rendered Walls.

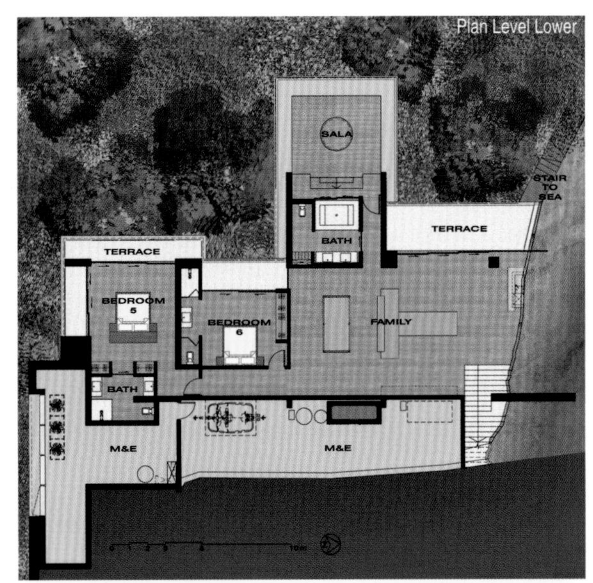
Plan Level Lower
SALA
STAIR TO SEA
TERRACE
BATH
TERRACE
BEDROOM 5
BEDROOM 6
FAMILY
BATH
M&E
M&E

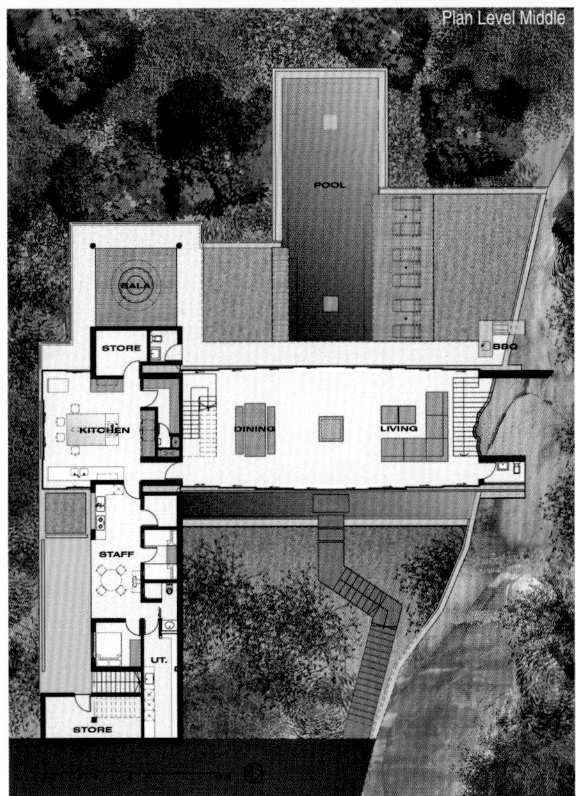
Plan Level Middle
POOL
SALA
STORE
BBQ
KITCHEN
DINING
LIVING
STAFF
UT.
STORE

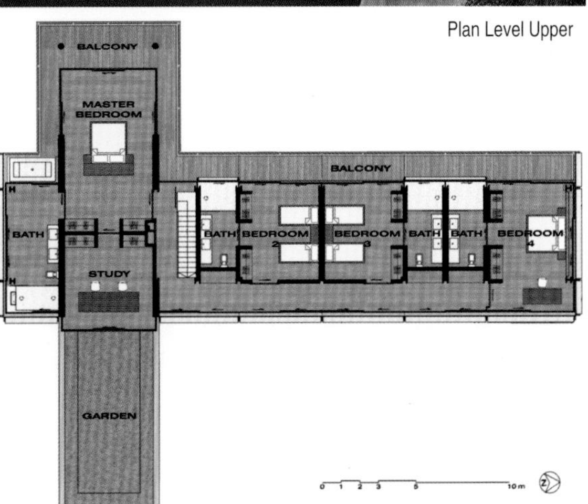
Plan Level Upper
BALCONY
MASTER BEDROOM
BALCONY
BATH
BATH
BEDROOM
BEDROOM
BATH
BATH
BEDROOM 4
STUDY
GARDEN

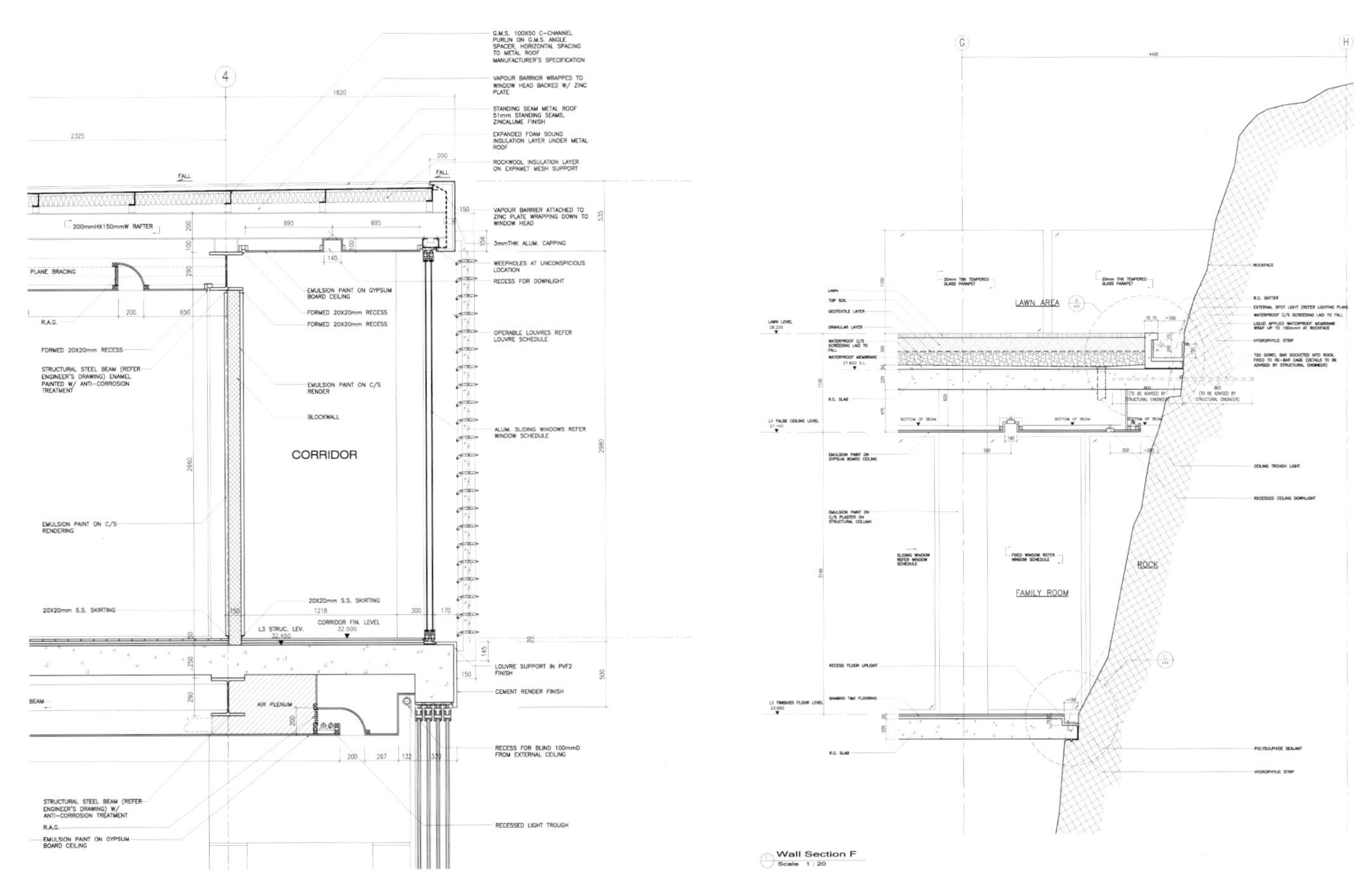
G.M.S. 100X50 C-CHANNEL PURLIN ON G.M.S. ANGLE SPACER, HORIZONTAL SPACING TO METAL ROOF MANUFACTURER'S SPECIFICATION
VAPOUR BARRIOR WRAPPED TO WINDOW HEAD BACKED W/ ZINC PLATE
STANDING SEAM METAL ROOF 51mm STANDING SEAMS, ZINCALUME FINISH
EXPANDED FOAM SOUND INSULATION LAYER UNDER METAL ROOF
ROCKWOOL INSULATION LAYER ON EXPAMET MESH SUPPORT
VAPOUR BARRIER ATTACHED TO ZINC PLATE WRAPPING DOWN TO WINDOW HEAD
3mmTHK ALUM. CAPPING
WEEPHOLES AT UNCONSPICIOUS LOCATION
RECESS FOR DOWNLIGHT
OPERABLE LOUVRES REFER LOUVRE SCHEDULE
ALUM. SLIDING WINDOWS REFER WINDOW SCHEDULE
LOUVRE SUPPORT IN PVF2 FINISH
CEMENT RENDER FINISH
RECESS FOR BLIND 100mmD FROM EXTERNAL CEILING
RECESSED LIGHT TROUGH
200mmHX150mmW RAFTER
PLANE BRACING
R.A.G.
FORMED 20X20mm RECESS
STRUCTURAL STEEL BEAM (REFER ENGINEER'S DRAWING) ENAMEL PAINTED W/ ANTI-CORROSION TREATMENT
EMULSION PAINT ON C/S RENDERING
20X20mm S.S. SKIRTING
BEAM
STRUCTURAL STEEL BEAM (REFER ENGINEER'S DRAWING) W/ ANTI-CORROSION TREATMENT
R.A.G.
EMULSION PAINT ON GYPSUM BOARD CEILING
EMULSION PAINT ON GYPSUM BOARD CEILING
FORMED 20X20mm RECESS
FORMED 20X20mm RECESS
EMULSION PAINT ON C/S RENDER
BLOCKWALL
CORRIDOR
20X20mm S.S. SKIRTING
CORRIDOR FIN. LEVEL 32.500
L3 STRUC. LEV. 32.450
AIR PLENUM
FALL
LAWN AREA
FAMILY ROOM
ROCK
Wall Section F
Scale 1 : 20

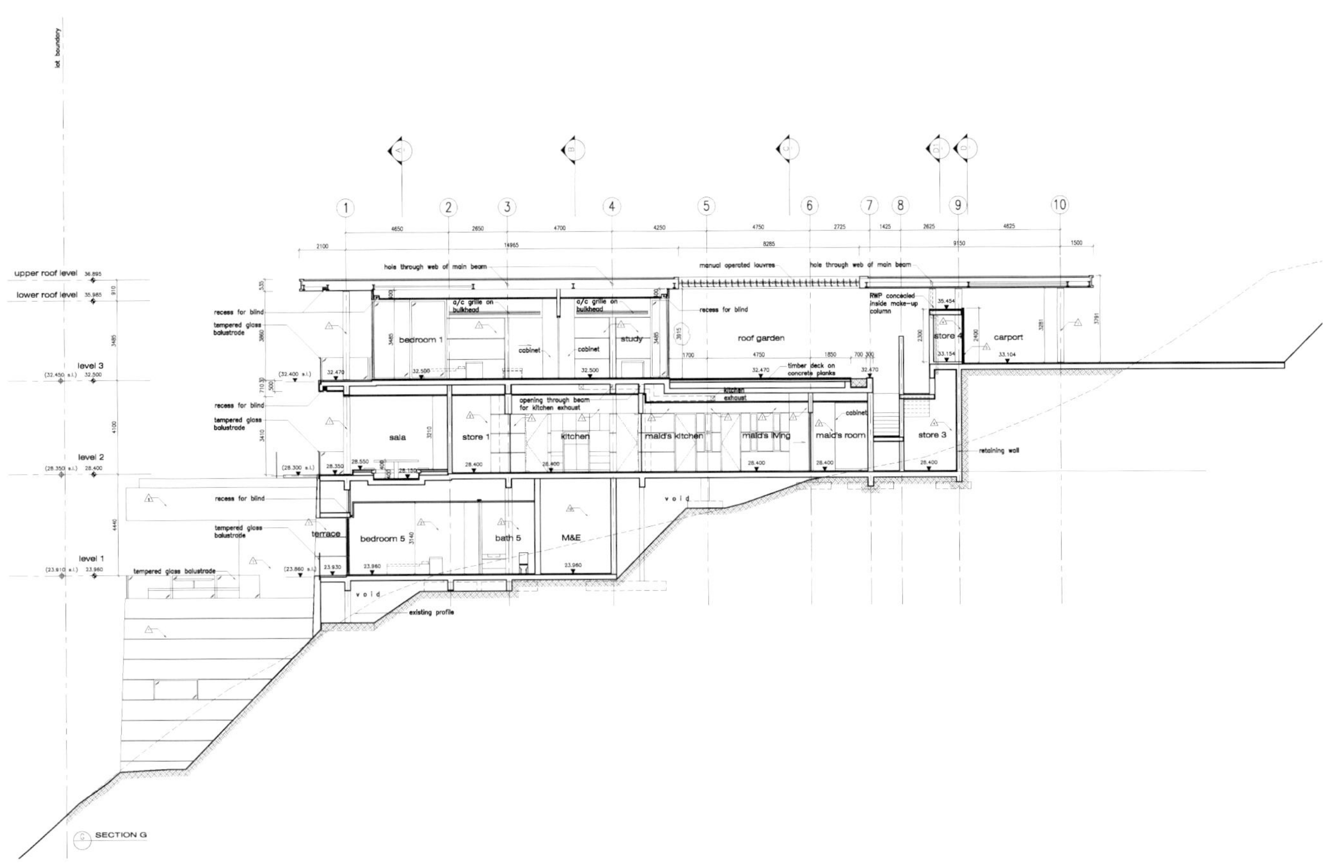
lot boundary
upper roof level 36.895
lower roof level 35.985
level 3
(32.450 s.l.) 32.500
level 2
(28.350 s.l.) 28.400
level 1
(23.910 s.l.) 23.960
hole through web of main beam
manual operated louvres
recess for blind
tempered glass balustrode
a/c grille on bulkhead
bedroom 1
cabinet
study
roof garden
timber deck on concrete planks
RWP concealed inside make-up column
store 4
carport
opening through beam for kitchen exhaust
kitchen exhaust
sala
store 1
kitchen
maid's kitchen
maid's living
maid's room
store 3
retaining wall
terrace
bedroom 5
bath 5
M&E
void
existing profile
SECTION G

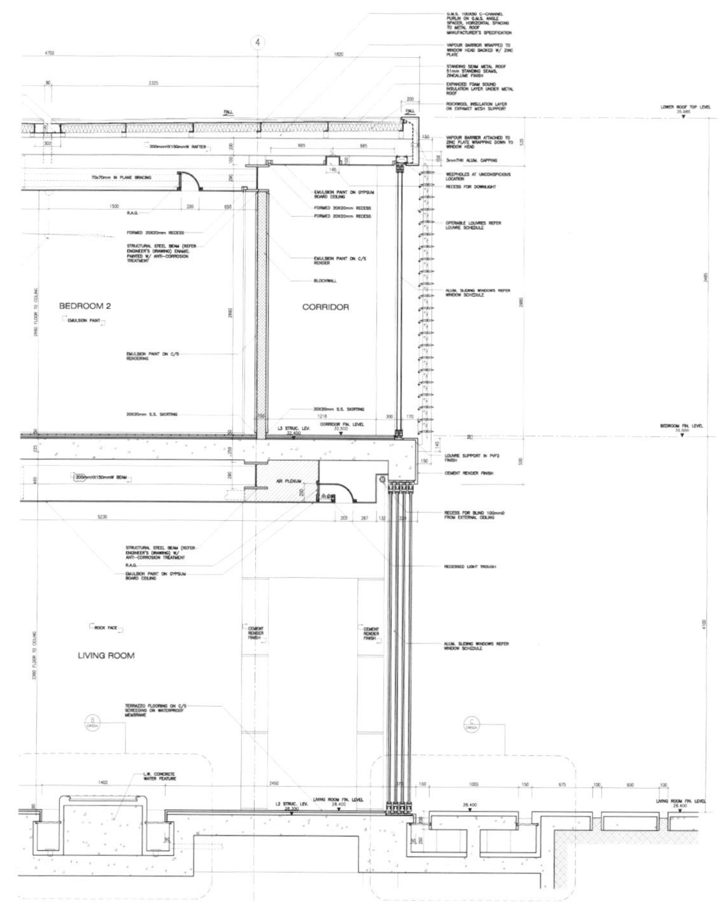
BEDROOM 2
CORRIDOR
LIVING ROOM

lot boundary
A
B
C
D
E
F
G
H
2625
5000
4400
1300
rock profile in front
roof gutter
manual operated alum. louvre
alum. cladding
upper roof level
lower roof level
alum. louvre
motorized alum. louvre
roof garden
timber deck on concrete planks
level 3
kitchen exhaust
maid's living
bath
timber fence
level 2
service path
sunken planter behind
existing profile
400mmW drainage channel

Villa GM in Italy

意大利GM别墅

This villa, like a garden pavilion hung with a spectacular view of the sea, is part of a complex of houses located in Marina di Ragusa, the seafaring village of Ragusa, on a plot of land with beautiful views overlooking the Mediterranean and a stretch of coast in the direction of the island of Malta, distant about sixty miles off, and that in a bright day you can see clearly. The design of the villa derives from the influence exercised by the program of the Case Study Houses (CSH) implemented in the '50s by John Entenza and the magazine he founded "Art & Architecture". The house that is more representative of the program is certainly the case study houses of Pierre Koenig's Stahl House, masterfully photographed by Julius Shulman, became an icon of American lifestyle in the famous photo of the living room of the house with the background on the amazing night view of Los Angeles. The position of the batch of the project and the cultural similarities with that program became the essence of contemporary absolutely present after more than fifty years, has determined the main choices that affect the architecture of the house.

The villa has an L-shaped ground plan shape, is set around a large swimming pool with a sun terrace paved with planks of larch treated with a white primer. The continuity of the interior of the living room and is secured by a glass wall that continues to spread around the perimeter of the house facing the sea view. Compared to the garden, the house is almost suspended, because of a continuous and smooth edge, detached from the ground, surrounds the house, determine the line of coverage, the line connects with the base, is defined by vertical sidewalls.

Two walls demarcating the inlet side and the opposite border of the pool are independent of the structure and connected with it through a high window and continuous thought of as individual plates that slide, too detached from the line of soil and structure. Architecture is dry and clear, made so well by the economy of the materials used, steel and wood frame, glass for the sidewalls and cement floors for both internal and external. The whole house is a tribute to its architect Pierre Koeing, perhaps the most brilliant architects with Craig Ellwood Americans who have given it a great contemporary American and world architecture.

这栋别墅是一个复合住宅区的一部分，位于拉古萨的一个航海村。别墅犹如一个花园展馆，坐拥壮观的海景。别墅的设计深受CSH项目的影响。CSH，即住宅佳作分析，由约翰·伊斯坦和他创办的杂志《艺术与建筑》于20世纪50年代发起。CHS包括一项案例分析，展现的是尤里乌斯·舒尔曼拍摄的由建筑师皮埃尔·凯尼格所设计的施塔尔别墅的一系列照片，其中有一张客厅照展现了洛杉矶令人称奇的夜景。这个项目成为代表美国生活方式的象征性作品。这项案例深刻地影响了同时期的建筑，而50年后这些案例的影响力依旧，案例中建筑场地和文化的相似性对本项目的建筑方案设计起到了决定性的作用。

别墅的地面规划呈现“L”形，设计围绕一个大型游泳池展开，别墅中还设置了白漆饰面的落叶松木板铺就的阳光露台。室内起居空间的连续性是通过一个个玻璃墙实现的，在景观衬托下，建筑几乎是悬浮着的。

两堵墙划定了入口和泳池对面的边界，独立于房屋结构，并通过高大、连续的窗体相连，窗体由单独可滑动的板块结构组成，与建筑和地面脱离。建筑的框架采用钢材和木材，侧面采用玻璃材料，建筑在内、外部都铺设了水泥地板，因而显得清爽而敞亮。这栋别墅是对建筑师皮埃尔·凯尼格的一次致敬，他同克雷格·埃尔伍德一起为美国乃至世界的建筑艺术作出了巨大的贡献。

Location / 地点:
Ragusa, Italy
Date of Completion / 竣工时间:
2010
Area / 占地面积:
1,250 m²
Architecture / 建筑设计:
Architrend Architecture
Interior Design / 室内设计:
Architrend Architecture
Landscape / 景观设计:
Architrend Architecture
Photography / 摄影:
Moreno Maggi

Materials: Steel, Glass, Wood.

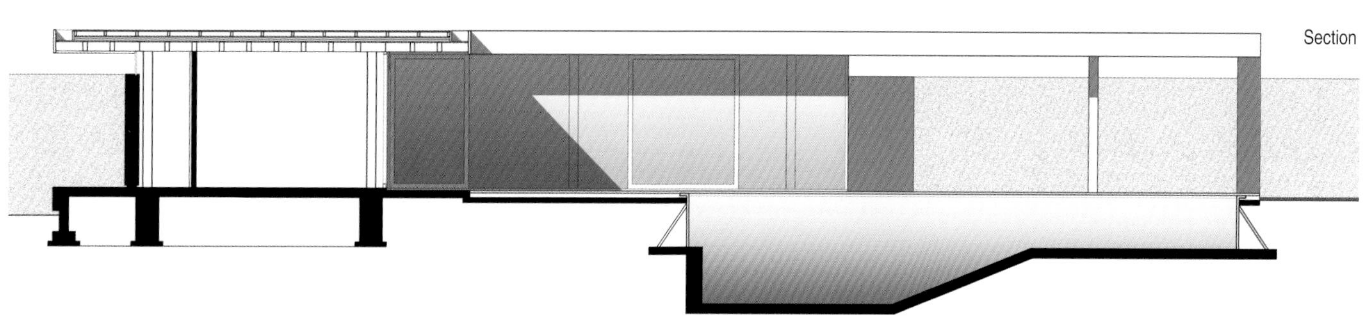

Section

Elevation

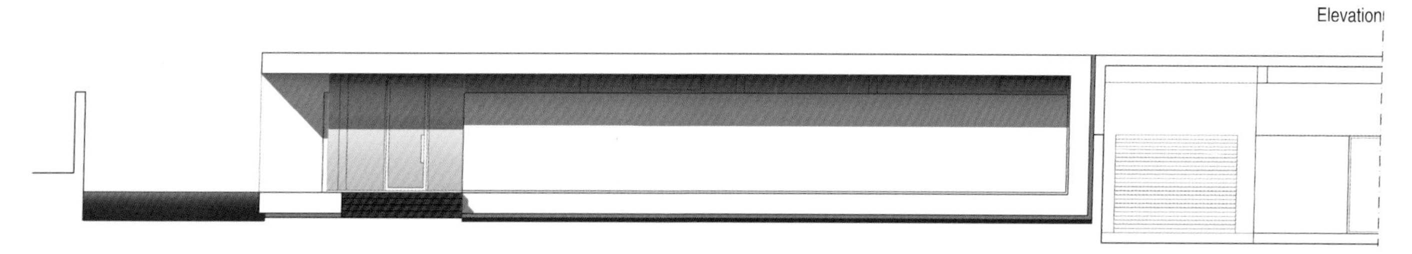

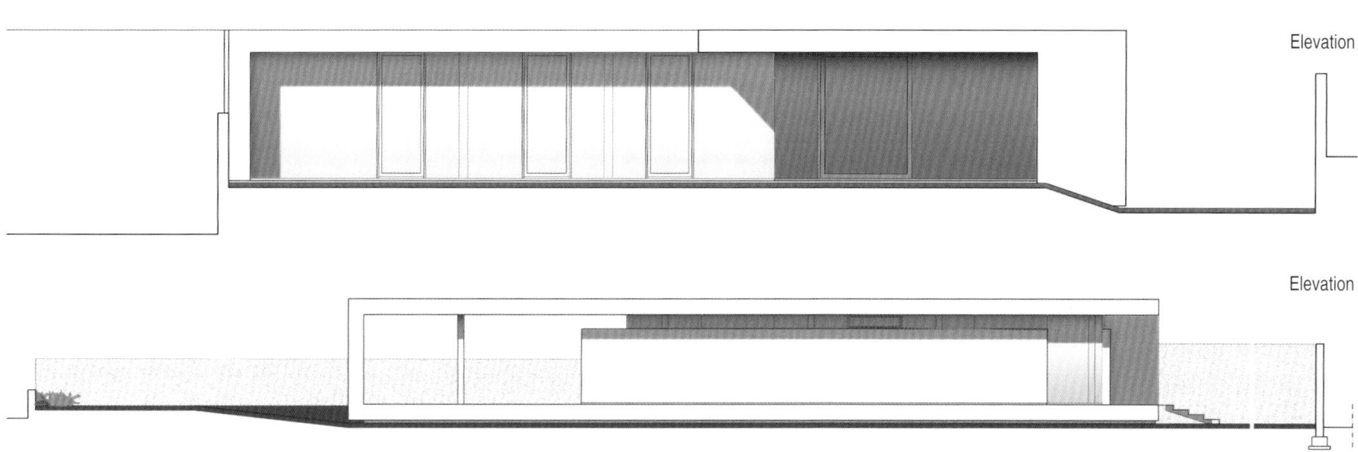

Elevation

Elevation

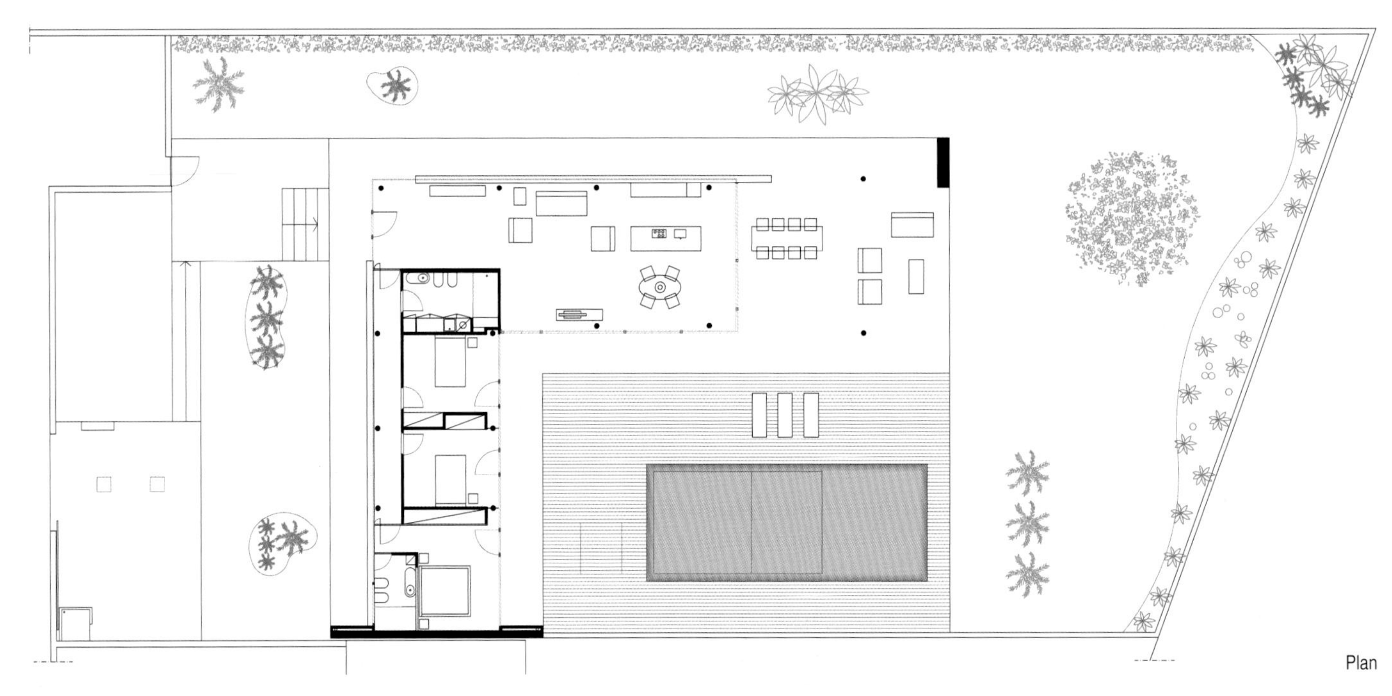

Plan

House in Alcobaça

阿尔科巴萨居所

In the elaboration of the project for this holiday house, located in Alcobaça, due to the client's fascination by the landscape's beauty, the architects tried to produce an architecture that minimized disturbances to the environment. The architects opted for a house with a mono-oriented plan, which was rationalized and simplified to be inscribed in the plot's width, and a panoramic view to the west above the level of the pool, extended over its entire length in a large terrace protected from the sun by a concrete overhang.

On the existing terrain level, the roof is planted to allow a better integration of the house in the landscape, limiting heat loss in winter and providing natural cooling in the summer. The latter is reinforced by transverse ventilation integrated into the ceiling and adjustable horizontal openings in the wall of corten steel on the eastern side.

Entering the sloped garden or the grass stairs torn on the roof garden, the far landscape slowly disappears to be replaced by the atmosphere of a planted courtyard dominated by a single olive tree. At this level, covered in corten steel to integrate gates, doors and vents, the house is closed to preserve its privacy.

Entering the house, people find from the living room a new perspective, this time framed by the wooded hills, and animated by the water plan of the pool.

Instead of providing the environment for the object "house", the land was shaped to blend with the architecture and integrate it as if it were a work of Land Art. Over time, the gardens were designed to allow the local flora to return to the place that belongs to it, to strengthen the feeling of being able to inhabit the landscape.

这栋住宅位于阿尔科巴萨（Alcobaça），鉴于业主对周围景观的迷恋，建筑师决定将项目对环境的干扰降至最低。建筑被设计为单一朝向，结构合理而简洁，很好地利用了场地的特征，同时拥有西侧池塘景观的全景视野。房身沿其形体的长度伸展开来，形成大型的露台，并通过混凝土屋檐遮挡阳光。

房顶与场地的地面处于同一水平面，不仅使房屋与场地更好地融合，更能够在冬季减少房屋热量的散失，在夏季提供天然的降温。同时，为了强化夏季制冷的功能，设计师在建筑吊顶处集成了横向通风装置，并在东侧的耐候钢墙体上设计了横向展开的通风口。

进入倾斜的花园或屋顶花园上的草地，远处的风景慢慢消失，映入眼帘的是一片以一株橄榄树为主导景观的庭院。房屋为封闭结构，保护了业主一家的隐私，其外部由耐候钢覆盖。

走进房屋，人们可以看到树木繁茂的的山地，这一景观在池塘水景的映衬下更显生机。

建筑师并非简单地将房屋放在环境之中，而是使建筑与场地环境相协调，使这栋住宅成为一件“大地艺术品”。随着时间的推移，花园将重现属于本地的植物群落，增强建筑的本土化气息。

Location / 地点:
Alcobaça, Portugal
Date of Completion / 竣工时间:
2011
Area / 占地面积:
270 m²
Architecture / 建筑设计:
Topos Atelier de Arquitectura, Lda
Jean Pierre porcher
Margarida Oliveira
Albino Freitas
Interior Design / 室内设计:
Topos Atelier de Arquitectura, Lda
Jean Pierre porcher
Margarida Oliveira
Albino Freitas
Landscape / 景观设计:
Topos Atelier de Arquitectura, Lda
Jean Pierre porcher
Margarida Oliveira
Albino Freitas
Photography / 摄影:
Xavier Antunes

Façades: Concrete, Corten Steel, Glass;
Floors: Creme Marfil Stone;
Walls: White Painted Wood, White Painted Wall Plaster;
Walking Paths: Concrete;
Illumination: Floor Washlights, Recessed Ceiling Illumination;
Plants: Olive Tree, Cork Oak, Heather, Thyme, Azaleas, Rosemary, Japanese Ivy;
Others: Grass, White Gravel, Evergreen.

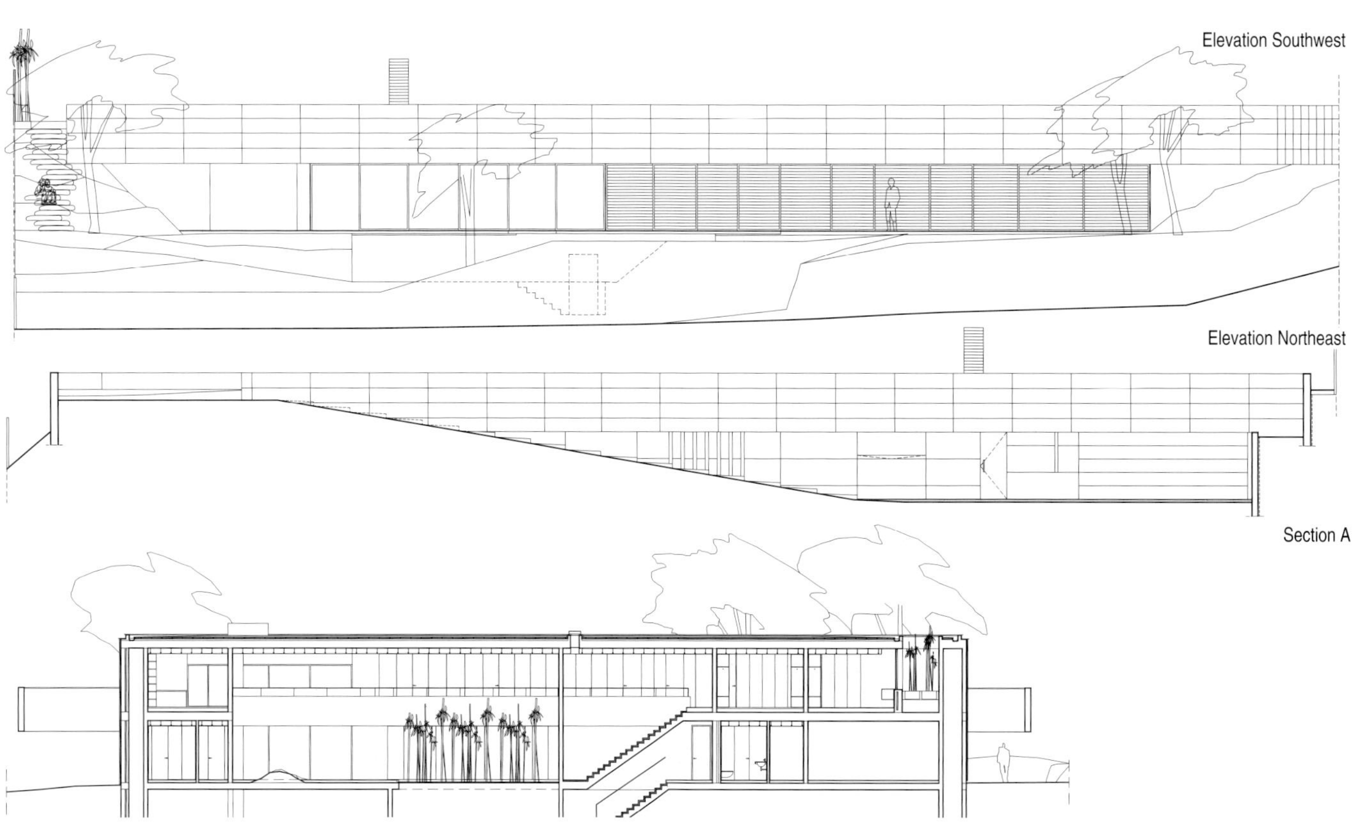
Elevation Southwest
Elevation Northeast
Section A

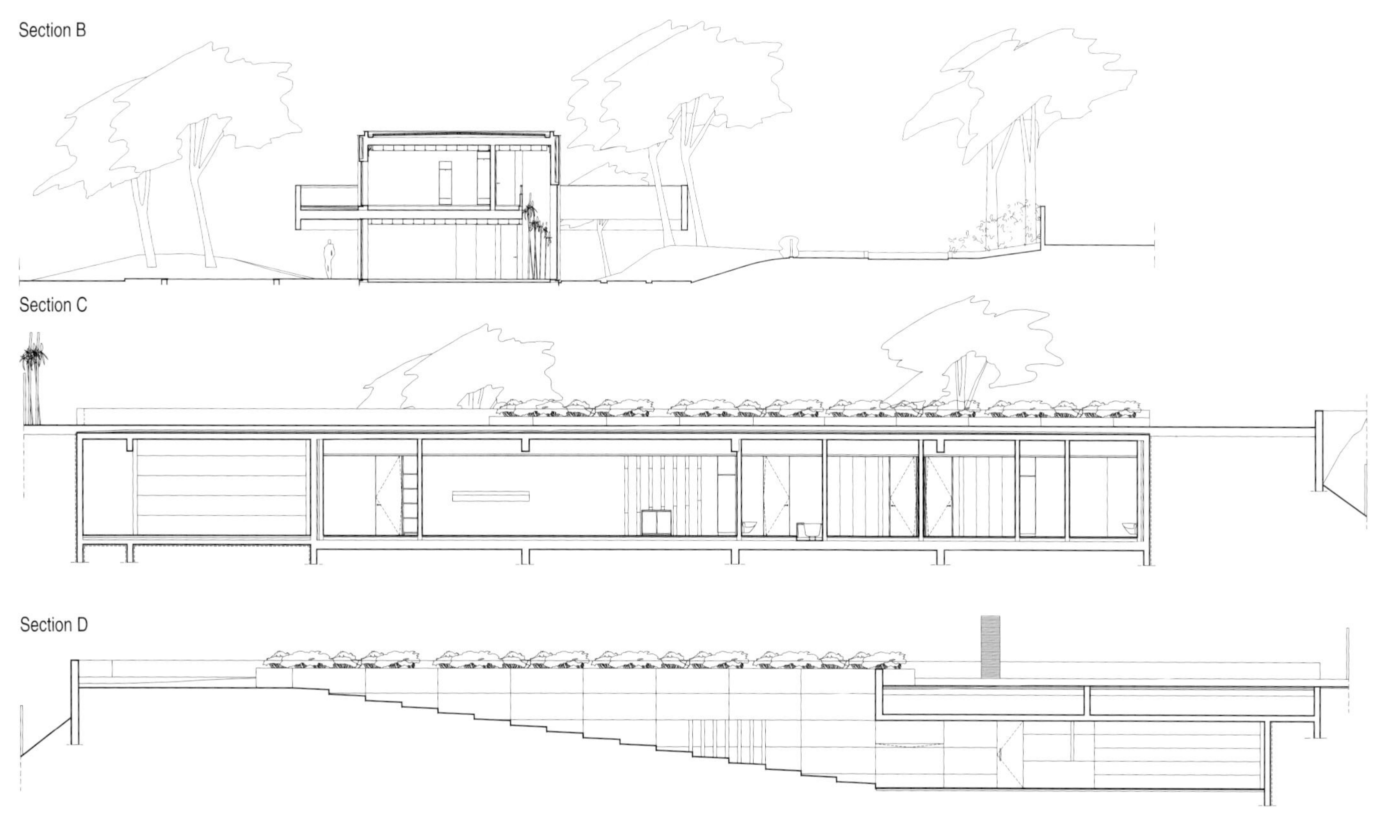
Section B
Section C
Section D

Plan General

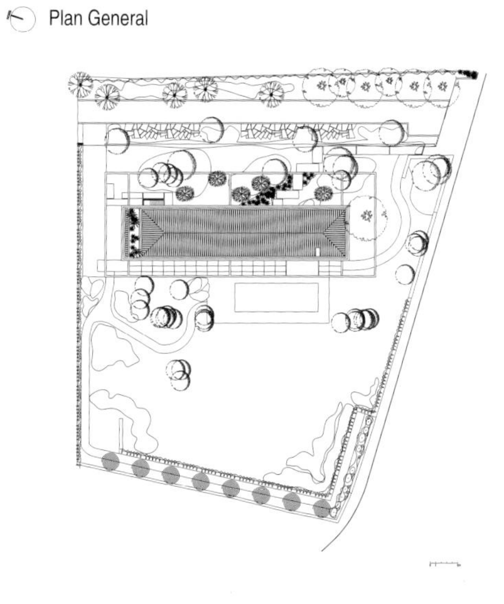

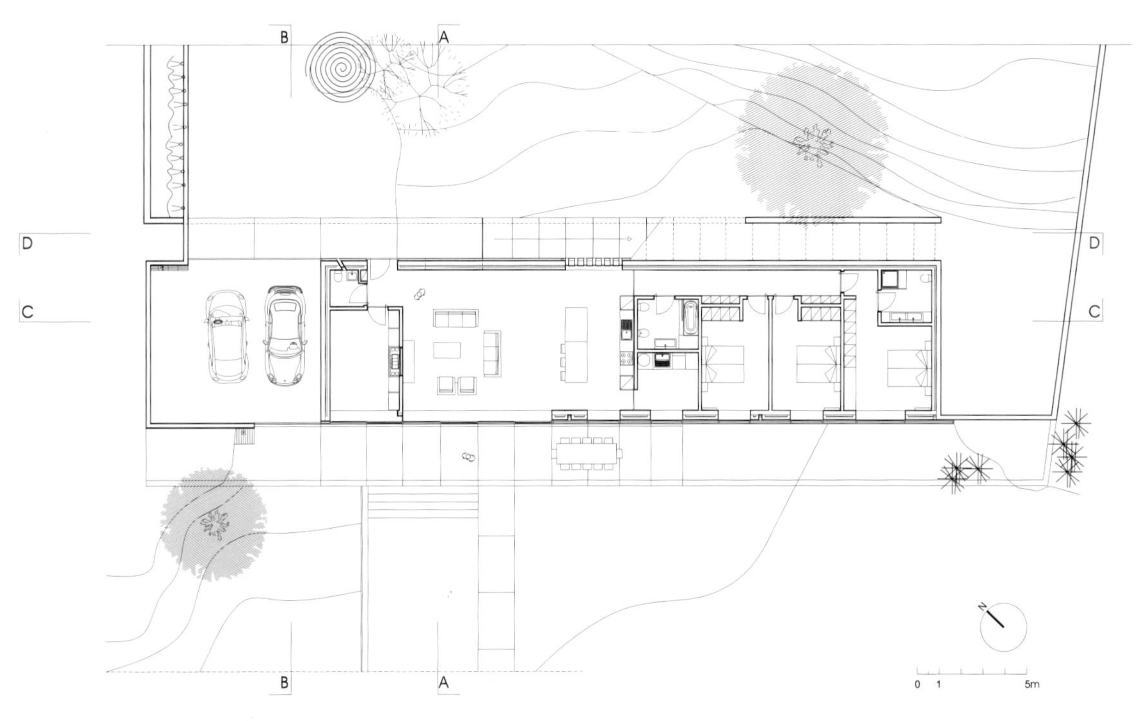

Floor Plan

Shell in Japan

日本贝壳建筑

A large shell shaped structure finds itself in the middle of the woods. It is hard to determine what exactly the structure is, and unlike the surrounding caves and rocks, it clearly is not a part of nature – nor is it a ruin. A frame, a shape, made at a completely different place for a completely different purpose. Within this shell shaped structure one could find floors constructed, wall separating spaces, and rooms furnished. The scenery conjures a SF film-like image, in which locals inhabit over an abandoned spacecraft. With time, trees start to grow encircling the spacecraft, harmonizing it into the landscape.

Desiring a place that would be occupied frequently over many years and yet at the same time be in sync with nature, the architects came up with the aforementioned scenery of a large shell structure floating above ground. Being in sync with nature isn't about yielding to nature – it's about coexistence. The existence of the structure depends on its power to endure nature. By isolating living space from the wilderness, and upgrading its quality as a shelter, the house would be protected from nature and provide a comfortable environment. With this, the house would be taken care of and used frequently and continuously. Specifically in case of villas, frequent use is what leads it to blend in with its surroundings. Despite the general avoidance of concrete material in the region, its usage and the lifting structure have helped the villa protect itself from the humidity.

Leaving the boundary between human life and ambiguous nature is a Japanese virtue. Yet, this ideal can only be achieved through meticulous attention and care of the wilderness on a daily basis. This might be attainable at the home. It goes without saying that villas should not only be functional spaces for the weekend. Their greatest goal is to provide people with good rest, leisure, and picturesque views that never become dull – all in the vicinity of nature. In the style of many modern sculptures, the architects aimed to enhance the surrounding nature by incorporating it within the spatial structure.

这栋壳形建筑静静地屹立在一片丛林之中。其结构特征很难用语言准确描述，不同于周围的山洞和岩石，它显然不属于自然的一部分，也不是一座废墟。它拥有奇特的框架、怪异的外形，并建于特殊的环境中，它具有独特的用途。壳形结构的内部有地板、墙壁隔开的空间和经过装饰的房间。这幅场景不禁让人联想起科幻电影中的影像——一群土著民以废弃的飞船为家园。随着时间的推移，树木开始生长并包围“飞船”，使其和周围景观相协调。

上述的大型壳状建筑是设计师精心打造的，旨在建立一个可以长期居住的住房，同时与自然融为一体。与自然同步，不等于屈服于自然，而是建筑与自然达到共存。而建筑的存在则取决于它承受自然的能力。设计师将生活空间与旷野相隔离，并强化建筑的防护性能，房屋可以很好地防护外界不利的自然因素，为居住者提供舒适的环境。这样，房子在正常维护下便可以长期使用下去。特别是作为一栋别墅建筑，频繁的使用频率，决定它要与周围环境融为一体。尽管该地区一般避免使用混凝土材料，设计师仍然对混凝土材料加以利用，并设计了向上升起的结构，使得建筑能够很好地应付防潮问题。

保持人与自然含蓄朦胧的关系是日本的一种民族理念。然而，达成这种理念的前提是每天都要对环境进行精心的维护。这在家里是可以实现的。毋庸置疑的是，别墅作为周末的休闲住所，它们最大的目标就是为主人提供良好的休闲空间以及永不乏味的自然美景。建筑师将建筑的外观赋予强烈的雕塑感，并力求将建筑与自然有机结合，从而提升环境体验。

Location / 地点:
Karuizawa, Japan
Date of Completion / 竣工时间:
2008
Area / 占地面积:
1,711 m²
Architecture / 建筑设计:
ARTechnic Architects
Interior Design / 室内设计:
ARTechnic Architects
Landscape / 景观设计:
ARTechnic Architects
Photography / 摄影:
Nacasa & Partners Inc.

Roof and Exterior Walls: Exposed Concrete with Penetrative Sealer Finish;
Openings: Steel and Aluminum Sash;
Terrace flooring: Ulin Flooring with Penetrative Paint Finish;
Interior Walls: Teak Plywood with Oil Finish.

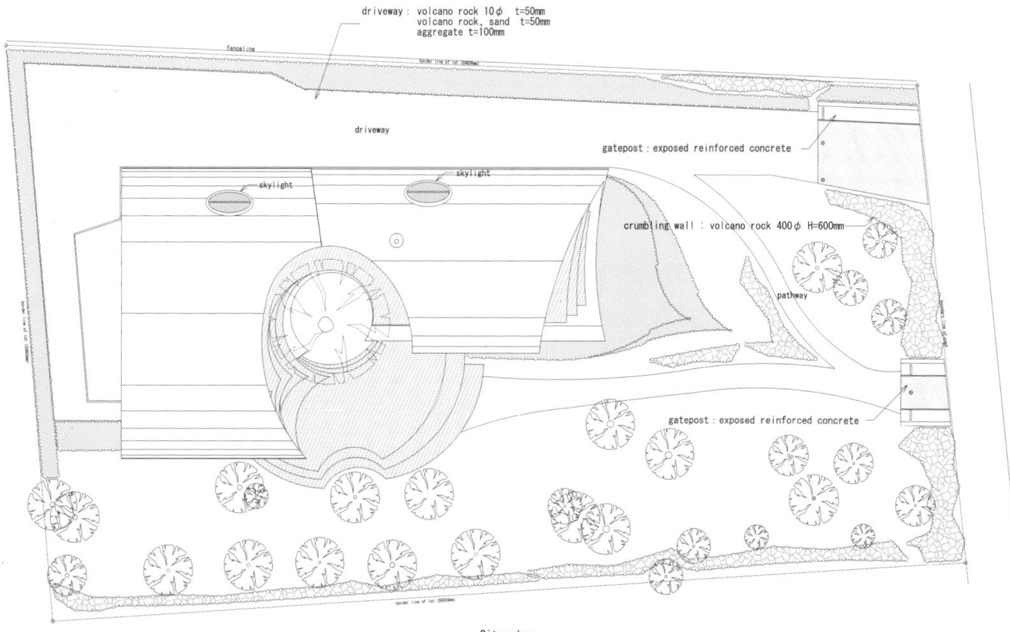

Site plan
S=1/150

Illustration of Air Conditioning System

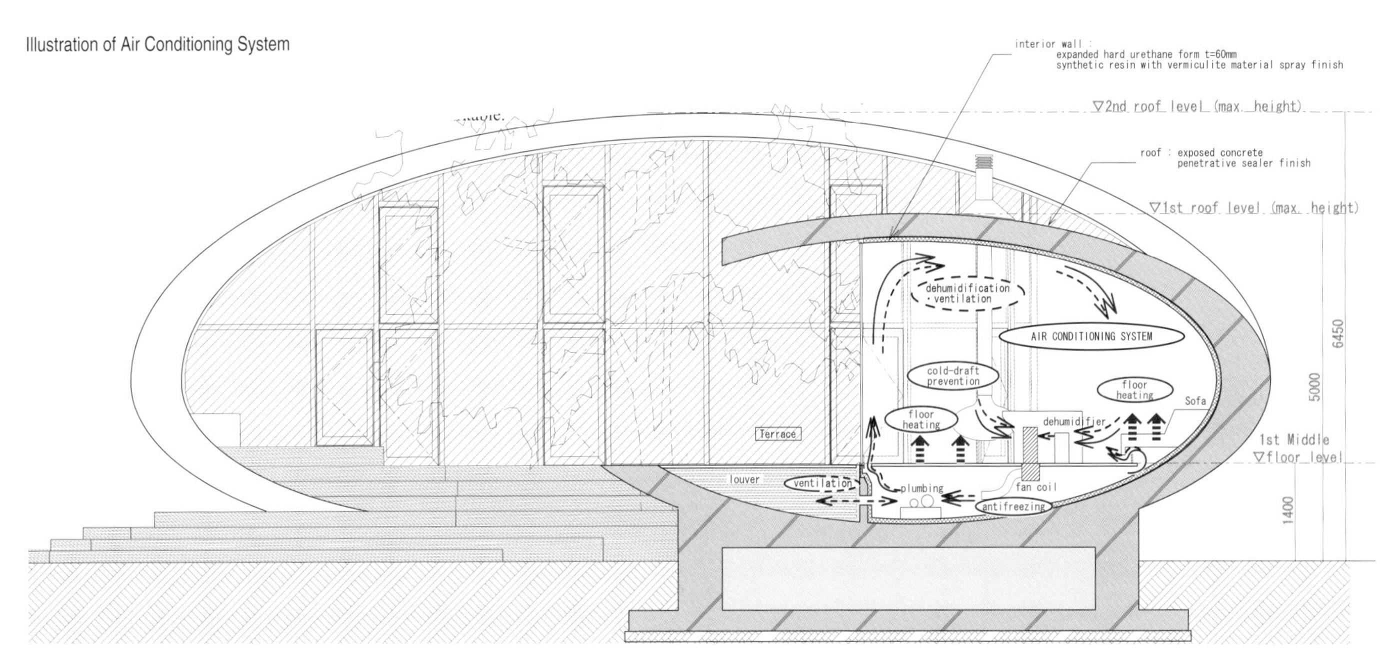

Elevation A

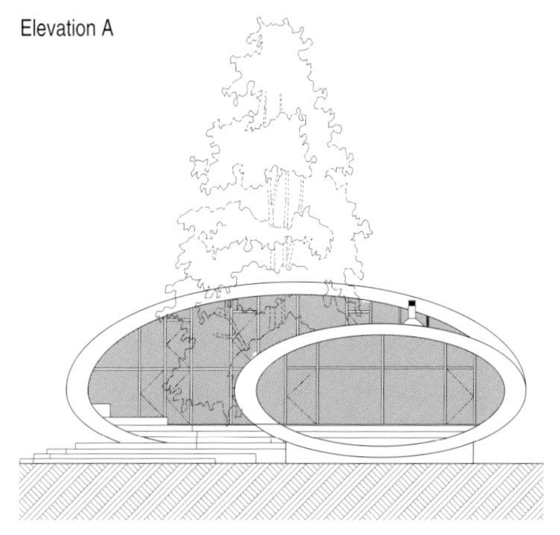

Elevation B

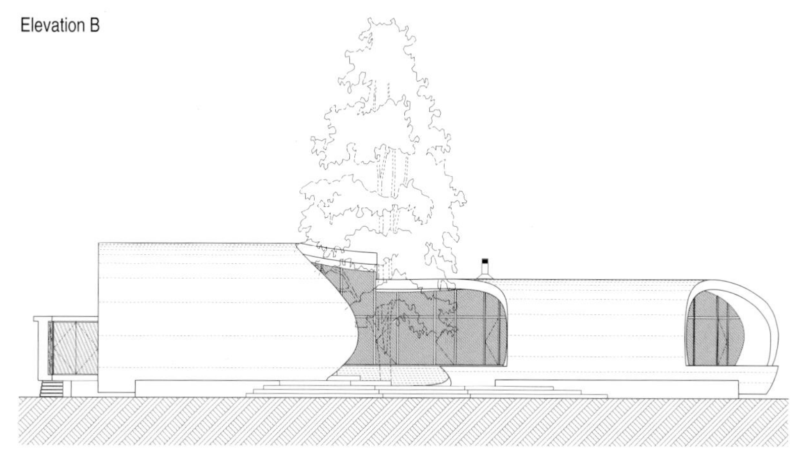

Elevation C

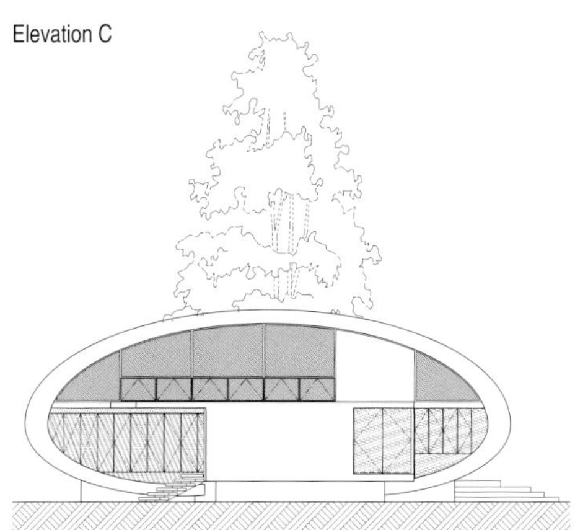

Elevation D

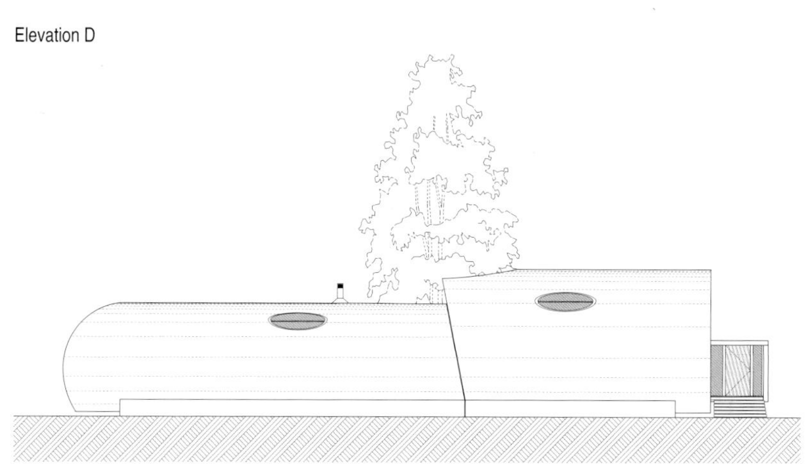

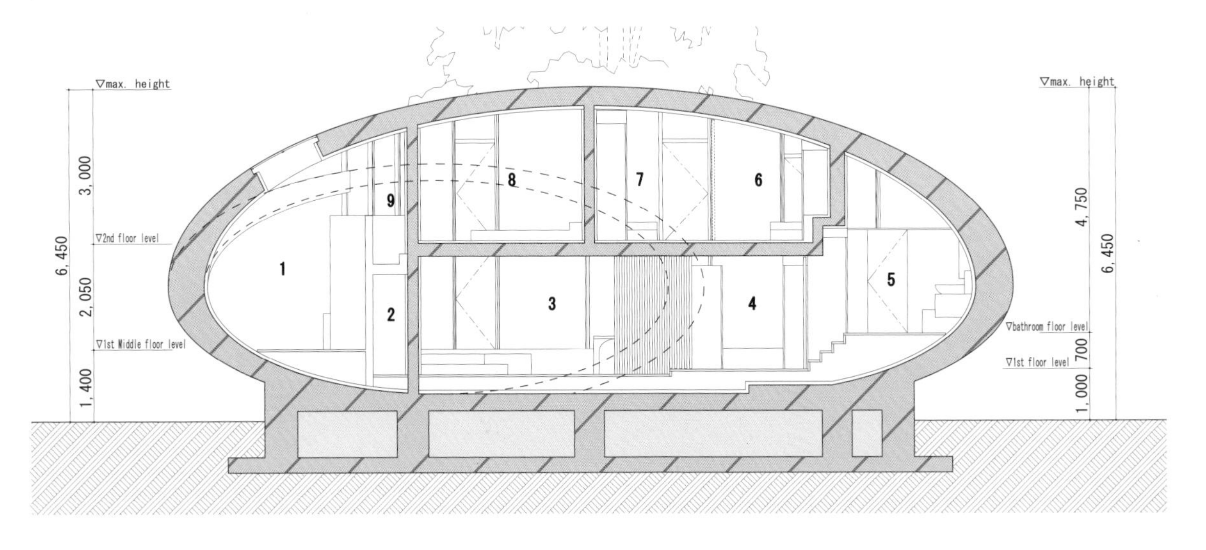

East-West Section
Scale=1/100

1. Hallway
2. Storage
3. Master Bedroom
4. Closet
5. Bathroom
6. Bedroom #3
7. Bedroom #2
8. Bedroom #1
9. Balcony

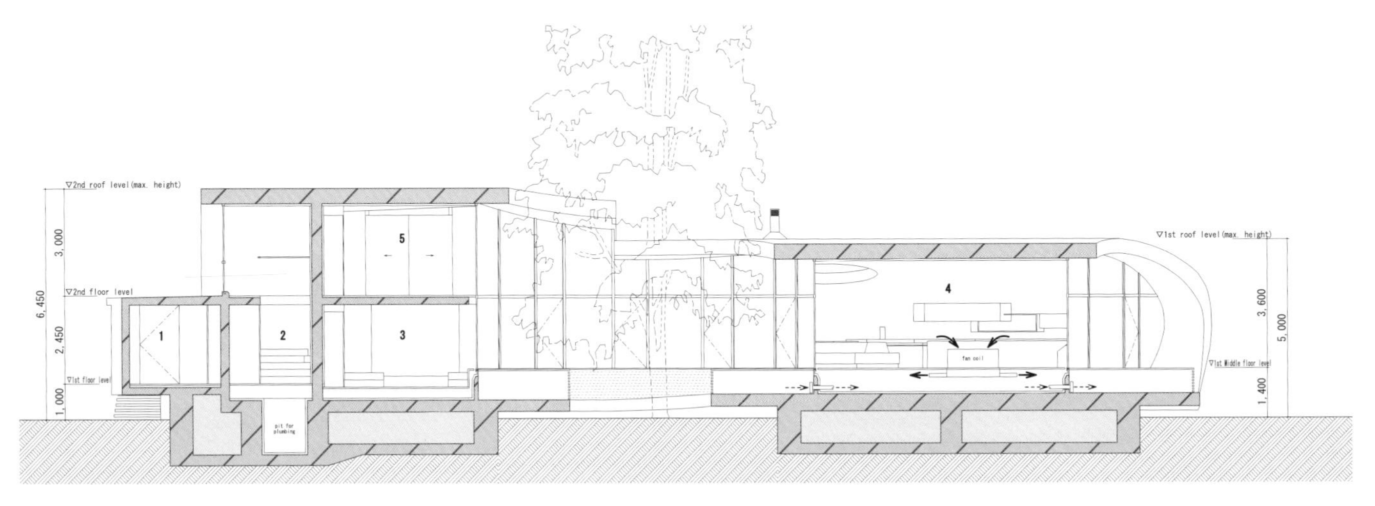

0 1000 5000

North-South Section
Scale=1/100

1. Entrance
2. Hallway
3. Master Bedroom
4. Living
5. Bedroom #1

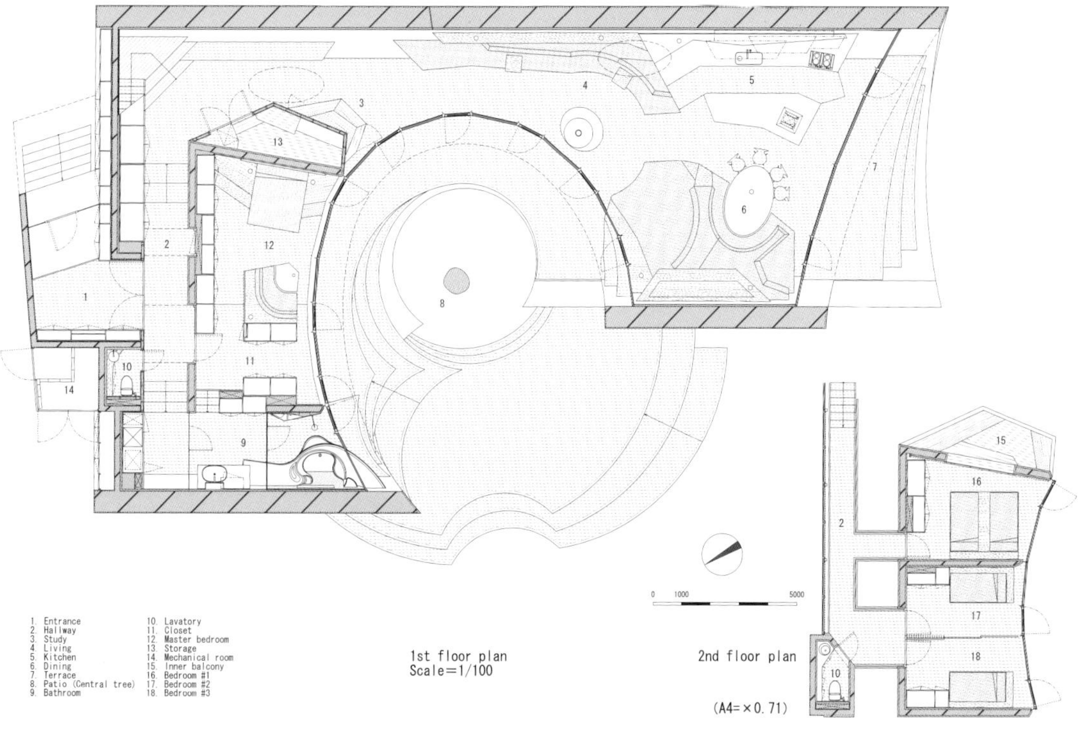

1. Entrance
2. Hallway
3. Study
4. Living
5. Kitchen
6. Dining
7. Terrace
8. Patio (Central tree)
9. Bathroom
10. Lavatory
11. Closet
12. Master bedroom
13. Storage
14. Mechanical room
15. Inner balcony
16. Bedroom #1
17. Bedroom #2
18. Bedroom #3

House in Bom Jesus
邦热苏斯住宅

The building is located on a hillside that develops in terraces above the city of Braga, allowing the view of the sea between the blue hills surrounding the city. The client wanted a black and white house with large terraces overlooking the valley. Successive drawings of the terraces led architects, to maintain the volumetric simplicity and free the overview of the entire valley, to involve the house in a heroically suspended ring, which came with its horizontality to emphasize the horizon and offset the verticality of a grove of oaks. Regulating the relationship of the ring with the ground, and of the ground with the ring to ensure the access and privacy in relation to neighbouring houses, creates a transitional space, space patio and space garden, closely related to the intimacy of dwellings .

This patio, surrounded by a suspended wall, extends the internal space limits beyond the window panes. Landscaped and wooded it controls the density of light that establishes with its daily movement an ongoing dialogue with the whiteness of the walls of the living room and the marble floors. In the morning the patio's vegetation filters the sun's diagonal light and projects diffuse shadows. When the sun exaggerates, the whiteness of the ring reflects the south vertical light tempered by the solid and protective shadows of the trees. At the end of the day the suspended wall of the ring absorbs the golden and horizontal light of the sunset and avoids the glare from the sun. In the two-storey high living room, the sliding glass, with 5.30 m in height, searches in its relation with the sky the clouds movement.

The house has an underground floor with technical areas and storage. The ground-floor reunites the service and social spaces: garage, showers, laundry, kitchen, living room, hall, and restrooms. On the 1st floor are located the private spaces of the bedrooms and the library, and levelled with the top of the oaks are large terraces overlooking the city.The house was to be a wooden box but due to maintenance issues it is now coated with patinaed zinc.

该建筑坐落在布拉加（Braga）高处的一个山坡上，透过建筑四周的群山可以看到美丽的海景。业主希望建立一座黑白色调的房屋，并拥有可以俯瞰山谷的宽敞露台。露台拥有连续的轮廓，这使建筑保持简洁的形体，并拥有观赏整个山谷的视野。建筑师将房屋设置在一个庞大的横向环状结构内，借以强调其水平的外观。控制环状结构与地面的相互关系可以确保出入口的隐私性，并创造出与住房密切联系的过渡空间、庭院空间和花园空间。

庭院被悬接的墙体所围绕。经过美化，植被繁茂的庭院遮挡了一部分光线，这些随时间变化的光线同客厅的白色墙壁和大理石地板形成不间断的“对话”。上午，庭院的植被过滤太阳的光线后，投射出婆娑的阴影。当阳光增强时，环状结构的白色表面会反射南部垂直射下的光线，这些光线通过浓密的枝叶后显得格外缓和。到了午后，环状结构的悬接外墙吸收了金色的夕阳光线。在两层楼高的客厅中，5.30m高的滑动玻璃掩映了蓝天、白云的景观。

建筑设有地下一层，容纳着专门的功能区和存储区。底层综合了服务空间和社交空间：包括车库、淋浴房、洗衣房、厨房、客厅、大厅和洗手间。一层设有私人空间，即卧室和藏书室，还有与橡树顶端基本齐平的大型露台，在此可以俯瞰全城。这栋房屋原计划采用木质的外观，但考虑到日常维护的因素，最终采用了钝化锌的涂层作外墙表面。

Location / 地点:
Braga, Portugal
Date of Completion / 竣工时间:
2009
Area / 占地面积:
700 m^2
Architecture / 建筑设计:
Topos Atelier de Arquitectura, Lda
Jean Pierre porcher
Margarida Oliveira
Albino Freitas
Interior Design / 室内设计:
Topos Atelier de Arquitectura, Lda
Jean Pierre porcher
Margarida Oliveira
Albino Freitas
Landscape / 景观设计:
Topos Atelier de Arquitectura, Lda
Jean Pierre porcher
Margarida Oliveira
Albino Freitas
Photography / 摄影:
Xavier Antunes

Façades: Patinaed Zinc, Glass;
Floors: White Sivec Marble, Panga Panga Wood;
Walls: White Painted Wood, White Sivec Marble, White Painted Wall Plaster;
Walking Paths: Concrete, Yellow Gravel;
Illumination: Floor Washlights, Recessed Ceiling Illumination;
Plants: Oak Tree, Chestnut Tree, Strawberry Tree, Birch Tree, Laurel Tree, Fruit Trees, Heather, Holly, Azaleas, Japonese Ivy, Escallonia;
Others: Grass, Evergreen, Granite.

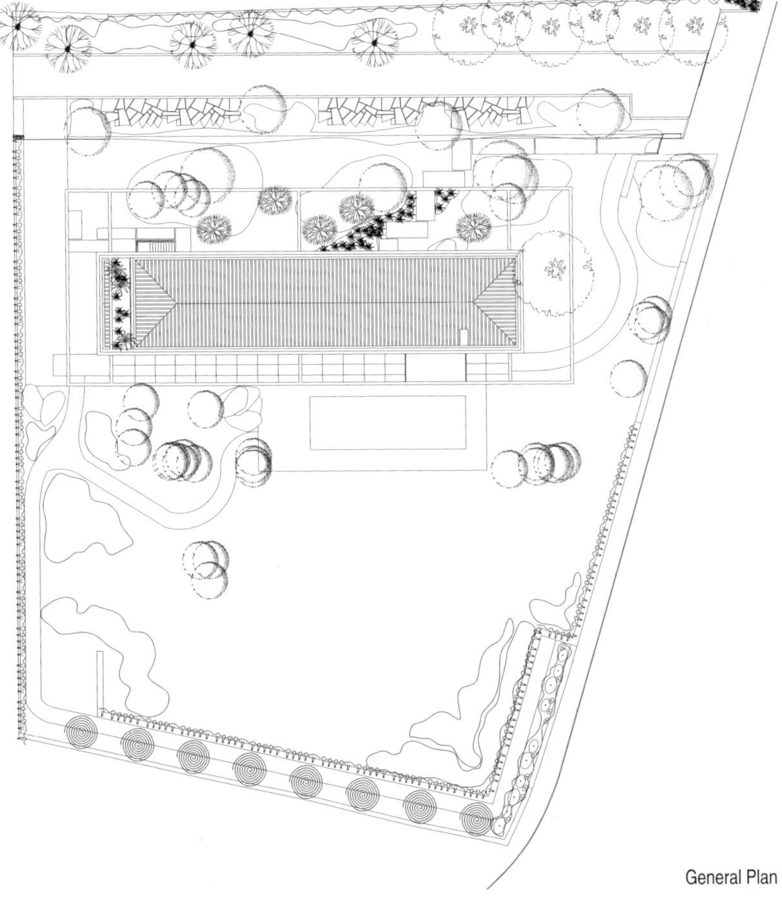

General Plan

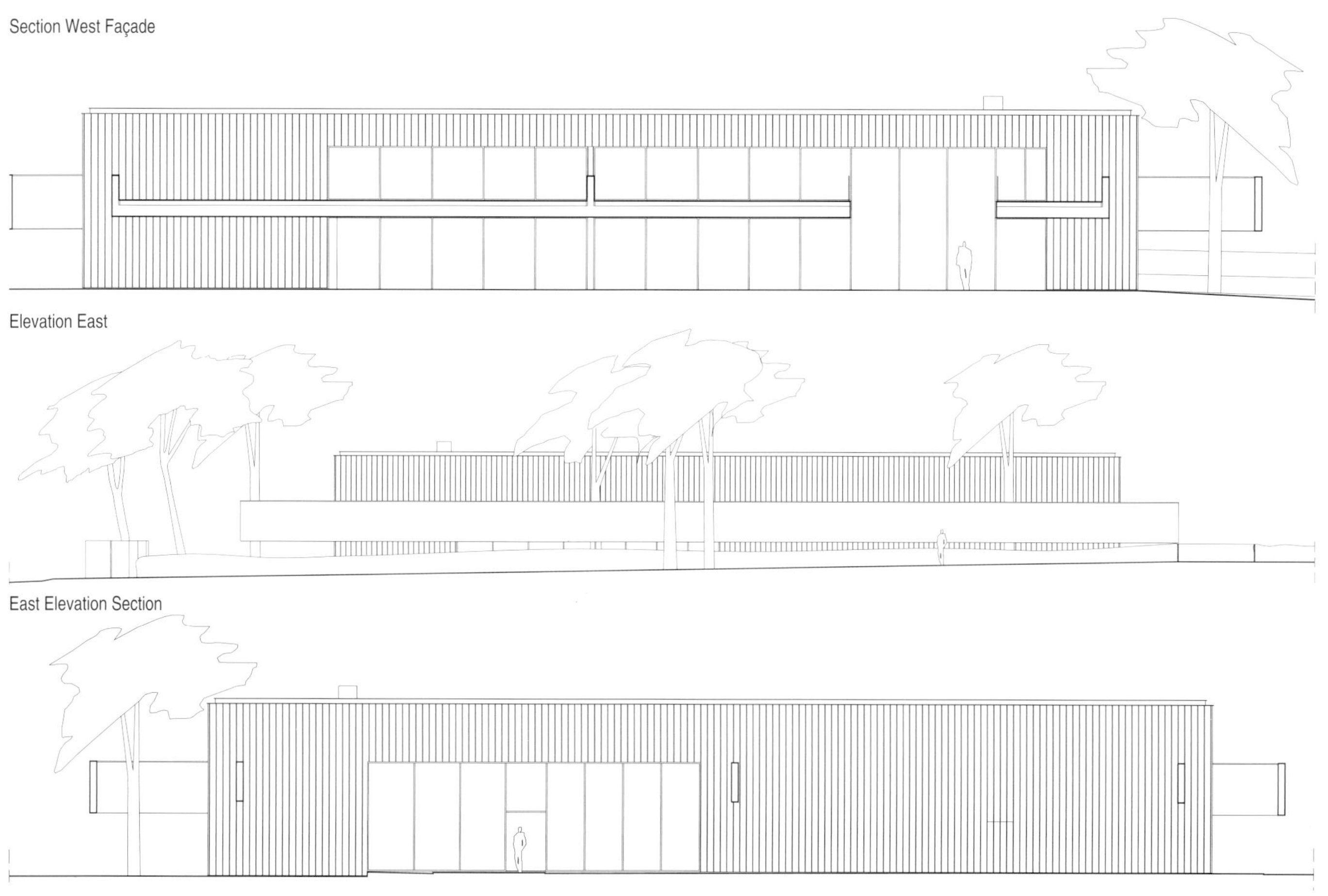

Section West Façade

Elevation East

East Elevation Section

Elevation North

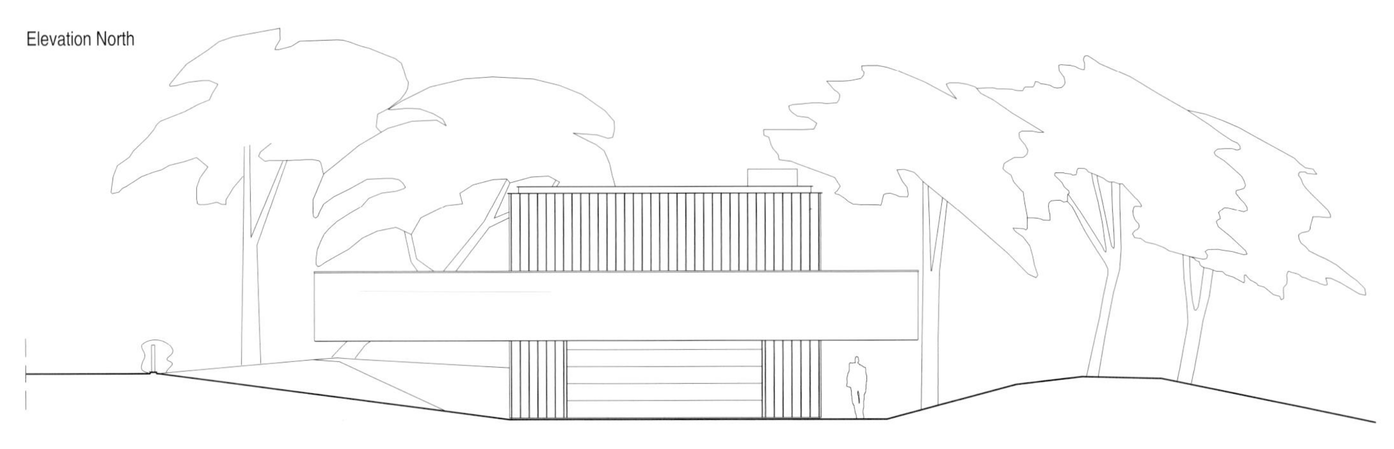

Elevation South

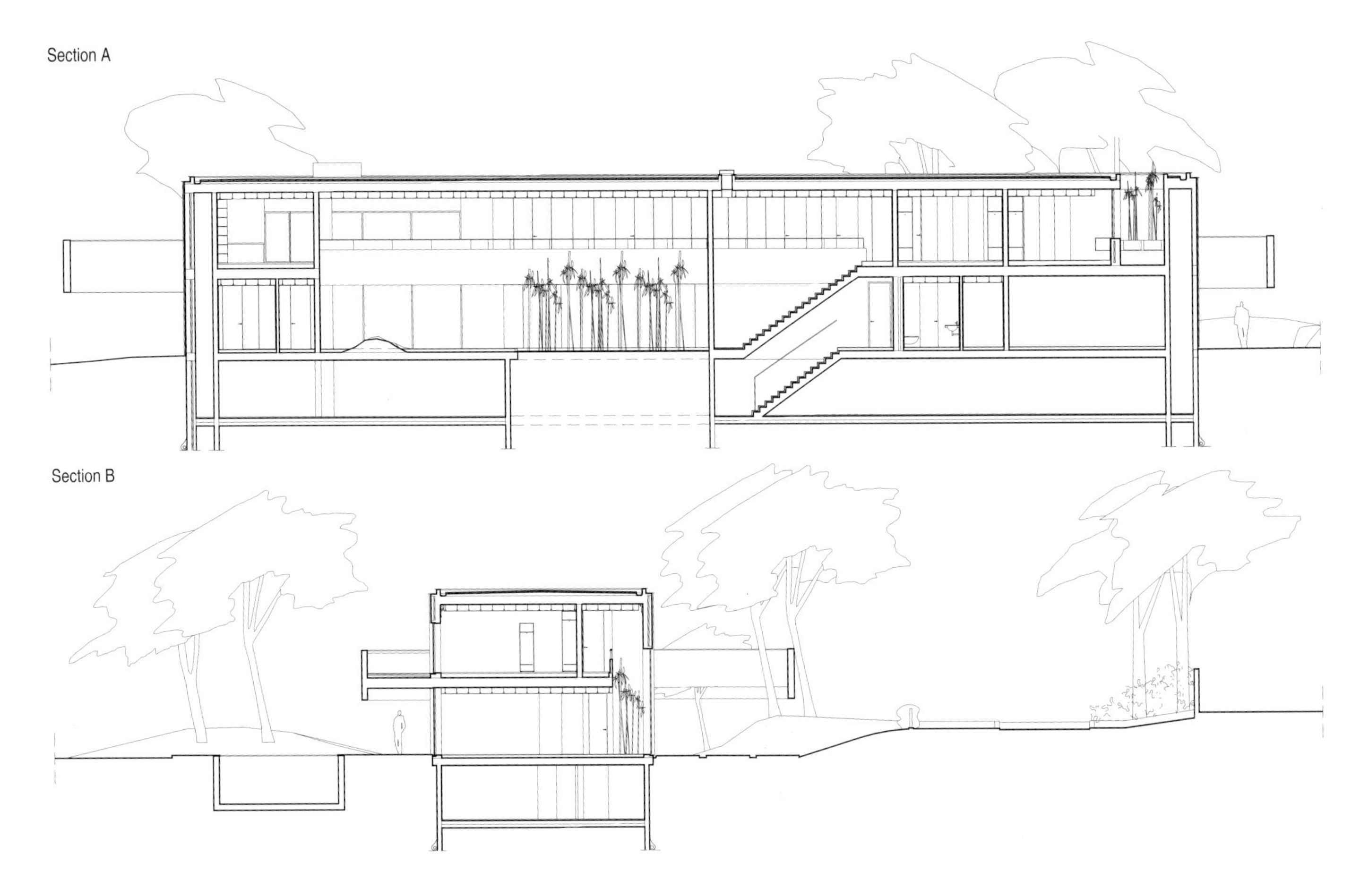
Section A
Section B

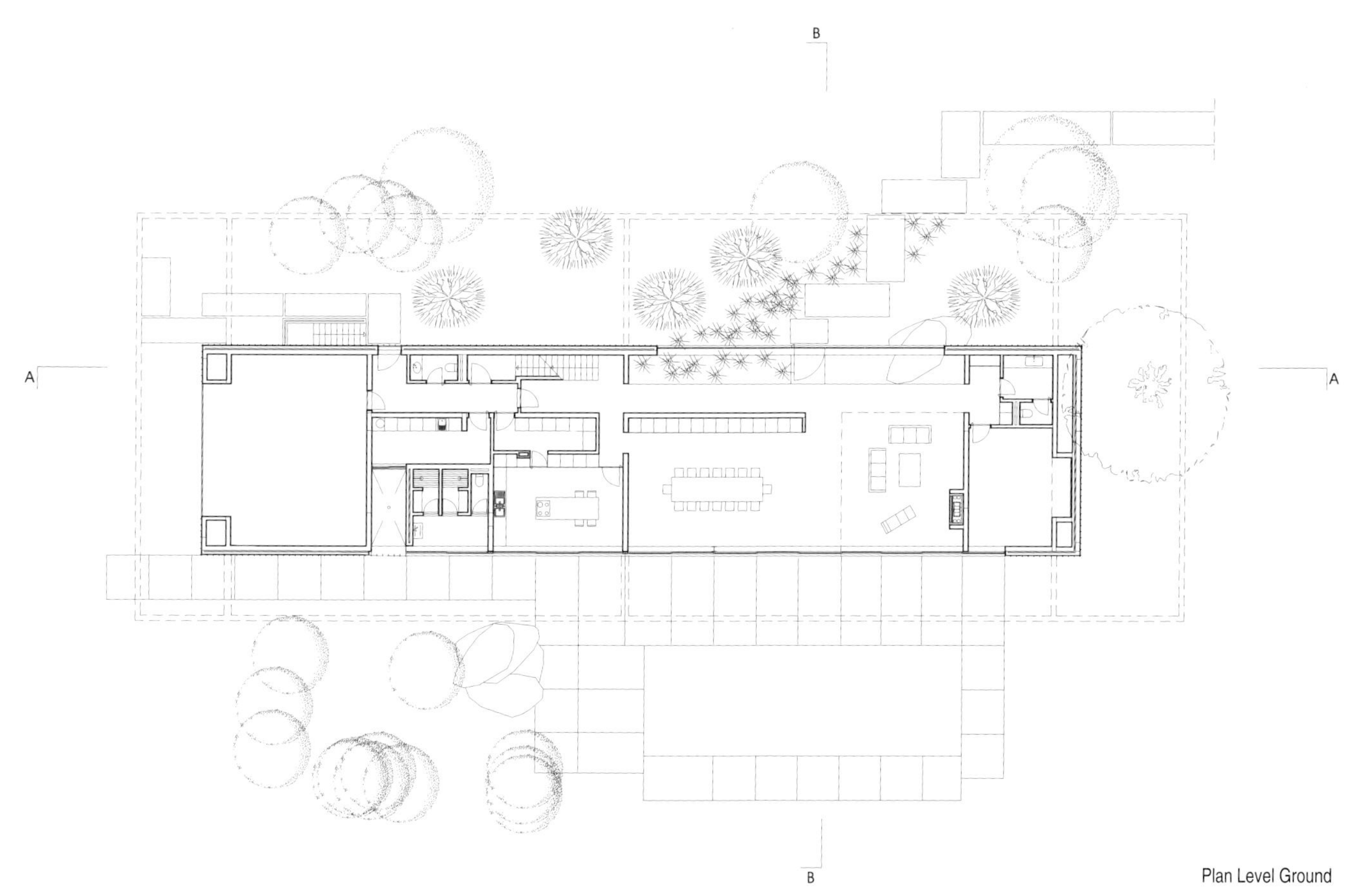

Plan Level Ground

109 Architectes

Add: 19, Street 38, Sector 1, Baabda, Post office box 40-222, Beirut, Lebanon
Tel: +961 5 922 183; Fax: +961 5 922 330
Web: www.109architectes.com

109 Architectes is an international design studio skilled in all aspects of designing and planning. Since 2002, their diverse team of architects has worked collectively to develop visionary solutions that meet each client's individual needs. They prioritize new ideas, functionality, and a superior level of client service. The result is distinctive, practical designs that realize the project intention at the highest international standards.

123DV

Add: Sint-Jobsweg 30, 3024 EJ Rotterdam, the Netherlands
Tel: +31 (0) 10 478 20 64
Web: www.123DV.nl

123DV Modern Villas currently consists of 10 architects, planners and interior designers. Their mission is to improve people's lives by creating a built environment that bring beauty and imagination alive. 123DV designs and realizes modern villas. Their goal is to create a sense of home in a modern villa. They translate the clients story, tailored to their needs, into the modern form, with attention to detail. Of course, they'll always deliver beautiful, striking designs. But their creativity runs deeper. Because the ideas reflect who the client is, what they stand for and where they want to be, they produce powerful, lasting results.

123DV Modern Villas focuses on both architecture and interior. From product design to realization. Their disciplines include architecture and interior design - from kitchens to furniture. A modern villa in their opinion does not only consist of the outside. It is especially the connection between the inner and outer world. Therefore, they often include the design for the interior. Besides interior, they also include design for lighting, sound and security. They don't consider sustainability a trend, but as a matter of course.

1024 Architecture

Add: 27 passage Courtois, 75011 Paris, France
Tel: +33 143 55 1024; Fax : +33 144 93 5011
Web: www.1024architecture.net

1024 Architecture is a company created by Pierre Schneider and François Wunschel, both co-founders of the EXYZT collective. 1024 Architecture focuses on the interaction between body, space, sound, visual, low-tech and hi-tech, art and architecture...
They make audio-visual installations, micro-architecture, urban intervention, performances, exhibitions and others...

3LHD

Add: N. Bozidarevica 13/4, Zagreb, HR-10 000, Croatia
Tel: +385 1 2320200; Fax: +385 1 2320100
Web: www.3lhd.com

3LHD is an architectural practice, focusing on integrating various disciplines – architecture, urban planning, design and art. 3LHD architects constantly explore new possibilities of interaction between architecture, society and individuals. With contemporary approach, the team of architects resolves all projects in cooperation with many experts from various disciplines. Projects, such as Memorial Bridge in Rijeka, Croatian Pavilion in EXPO 2005 in Japan and EXPO 2008 in Zaragoza, Riva Waterfront in Split, Sports Hall Bale in Istria, Centre Zamet in Rijeka, Zagreb Dance Centre in Zagreb and Hotel Lone in Rovinj are some of the important highlights.

They represented Croatia at the Venice Biennale 2010, 12 International exhibition of architecture, and took part in the 2008 exhibition "Mare Nostrum" on the second International Architecture Biennale in Rotterdam, and in a group exhibition in Boston at Harvard University: "New trajectories: Contemporary Architecture in Croatia and Slovenia ."

The work of 3LHD has received important Croatian and international awards, including the award for best building in Sport category on first World Architecture Festival WAF 2008, IOC/IAKS Bronze Medal Award 2009 and IOC/IAKS Silver Medal Award 2011 for best architectural achievement of facilities intended for sports and recreation, AR Emerging Architecture Award (UK), the ID Magazine Award (USA); and Croatian professional awards Drago Galić (2008), Bernando Bernardi (2009; 2005), Viktor Kovačić (2001), and Vladimir Nazor (2009; 1999).

a21studĭo

Add: 2/10 Nguyen Huy Luong, Ward 14, Binh Thanh District, Hochiminh City, Vietnam
Tel: +84 983 146 665
Web: www.a21studio.com.vn

a21studĭo is a newly setup company since 2010. After re-organising the management system, the architects are working as architectural designers. Their service is designing architecture and interior for different types of buildings from residential, hotel, resort, coffee shop, office. One of their members has several architecture awards such as national award, bamboo international design...which he gained before setting up the company.

AART Architects

Add: A/S Aboulevarden 22, 5.Sal-8000 Arhus, Denmark
Tel: +45 8730 3286; Fax: +45 8730 3296
Web: www.aart.dk

AART Architects is a high-performance team of 55 architects, constructing architects and designers who work with the community as a value-creating element. With offices in Denmark and Norway and several first prize projects in international project competitions, they are one of the trend-setting studios in Scandinavia.

As architects, they strive to give meaning and intimacy to everyday life by creating vibrant spaces for social communities and achieving the highest architectural quality. In order to lead the development of a sensuous and innovative architecture, they have been engaged in several research and development projects over the years.

In continuation hereof, they have organized the studio into four research teams: Integrated Design, Green Design, Health Design and Value Design. The four research teams form a common thread in their professional development and give added value to each project by developing new methods within the fields of project engineering, energy optimisation, health-promoting architecture and user experience.

As a research-based studio, they make theur living from the human factor in the form of creativity, innovation and collaboration. Their most important resource is thus dedicated employees who, by combining evidence-based studies with a nuanced empathy, develop beautiful, empathetic and functional solutions.

ACXT Architect

Add: Monasterio del Escorial, 4 Madrid, Spain
Tel: +34 91 444 11 50, Fax: +34 91 447 31 87
Web: www.acxt.net

The architecture service company ACXT began as an association of professionals seeking to provide quality solutions in the field of architecture, and a feasible alternative for the career development of architects interested in quality architecture. The philosophy of this association of professional architects is to provide a framework enabling its members to collaborate and maximise their synergies, all within an understanding of architecture in its fullest technical and cultural dimension.

ACXT was originally part of the IDOM group, an extension into the world of architecture of the principle of professional association that inspired the foundation of IDOM more than forty years ago, the principle subsequently adopted as the basis for the foundation of ACXT. Its internal operations are based on the commitment of associated professionals with a view to improving upon the services available in today's market which ranges from the model of the architecture studio to the service company.

ACXT bases its technical foundation, business philosophy and quality services not only on the solidity and experience of IDOM but also on the professional capacity of the partner architects and their multi-disciplinary approach to the practice of architecture.

The ACXT philosophy is based on respect for traditional values, still in vogue today, generated during the secular exercise of the architecture. ACXT is in favour of the original architecture, understood in terms of respect for the creative values historically provided by the architect throughout the intellectual and constructive process, demonstrating true vocation to meet the needs of the client. ACXT aims to provide a quality response to the immense social prestige which involves the exercise of architecture in both its commercial and cultural dimension.

Anastasia Arquitetos

Add: Rua Orenoco, 137 Loja 01, Belo Horizonte, MG, Brazil
Tel: +33 1 41 72 27 27
Web: www.archi5.fr

The office Anastasia Arquitetos was founded in 2001 by architects Johanna Anastasia Cardoso and Tomas Anastasia Rebelo Horta. Since then the office acts in various types of architectural design, including residential, commercial, offices, buildings and industries.

Johanna Anastasia Cardoso graduated in 1996 from PUC - Pontificia Univesidade Catolica. She attended post graduate degree in design and interior design in ISAD – Higher Institute of Architecture and Design - in Milan, Italy. She worked in the office of architect Paolo Vasino, in Milan, and in the office of architect Marco Antonio Anastasia, in Belo Horizonte, Brasil, before founding Anastasia Arquitetos.

Tomás Anastasia Rebelo Horta graduated in 1996 by Universidade Federal de Minas Gerais. He worked on several offices with projects in many different scales and types, including the office of architect Marco Antonio Anastasia, in Belo Horizonte, Brasil, Oscar Niemeyer, in Rio de Janeiro, GSI Architects, in Cleveland (US), Botti Rubin, in São Paulo, before founding Anastasia Arquitetos.

archi5

Add: 548-50 rue Voltaire 93100 Montreuil, France
Tel: 55 31 3282-1334; Fax: 55 31 3282-1334
Web: www.anastasiaarquitetos.com.br

archi5 was created in 2003, despite the youth of the agency they are working together for more of 14 years. They are involved in each agency's project. They ensure the consistency of the concept, from the first sketch to the completion.archi5 designs, research and designs projects planning, architecture and infrastructure.

They share a sensitivity and ethics: a willingness to build right and print in time the fragility of an image.The agency counts 35 people between Paris [archi5prod] et Warsaw [a5architekci] where Jakub Wroniszewski manages the workshop.archi5 regularly joins with partners and consultants covering all fields of competence of planning, architecture and landscape.

Recently, they built the Marcel Sembat high school in Sotteville lès Rouen or the Lamoricière school in Paris. They are also doing urbanism projects in Argentina and Brazil. Various projects are ongoing such as the Stade Léo Lagrange in Toulon or several housings, cultural or offices projects in France.Recent success in competitions were added to their credentials as the National Archaeological Museum in Rabat, high schools and Louise Michel Louise Aragon Gisors or school group Jules Verne at Châtenay Malabry.

Architrend Architecture

Add: Via Padre G. Tumino 97100, Ragusa, Italia
Tel: +39 0932 652661; Fax: +39 0932 652661
Web: www.architrend.it

Architrend Architecture is an architecture firm that approaches design as an interdisciplinary venture where minimum form takes maximum effect. The entire project is considered starting from client's needs as well as site and environment exploration through all design process to design and build high quality and value added architecture, built as figured it out in the design process.

Starting from 1989 the two partners Gaetano Manganello and Carmelo Tumino design and build public buildings, resorts and single family houses with a contemporary, minimalist and direct approach.

To house their new office, Manganello and Tumino had designed themselves a building for their headquarter in the outskirts of Ragusa. Made of reinforced concrete and covered in glass, this building is a strong character tending to get minimal, with the least formal means, the ultimate functional outcome.

Along with being published in major national and international media such as Casabella, Electa, Interni, Edilizia e Territorio, A10, Archdaily, Architrend Architecture has received many prestigious design awards including the 2009 "G.B. Vaccarini Quadranti di Architettura" award, the 2010 and 2011 "Ance Catania" awards. They have lectured and exhibited at many universities as well as at several national and international design conferences.

Arquitetosassociados

Add: Rua Palmares, 17 . Santa Lúcia .
30360-480 Belo Horizonte . MG . Brasil
Tel: +55 31 3261 7446
Web: www.arquitetosassociados.arq.br

Arquitetosassociados is a collaborative studio dedicated to architecture and urban design based in Belo Horizonte, Brazil. Each project is treated as a unique and specific work, for what a particular organization of the group is set, allowing the emergence of various design groups inside the studio with eventual external collaborators. This dynamic modus operandi improves the answers to the specific demands of each project and blurs some authoral issues, while allowing a permanent change and reinvention of the group, followed by a continuous improvement of its conceptual basis.

The work of each of the five members of the studio's core takes Brazilian modern architecture as a departure point in different points of view, restating and rethinking some of its concepts to reach adequate and innovative answers to local problems in the realm of architecture and its relation to the city.

Developed in parallel to a docent practice, the work deals with a great range of scales and programmes, from individual houses to public buildings and urban design, always committed to rethink programmatic and construction issues beyond its ordinary sense. The permanent research on architecture is allowed by a regular participation in competitions, parallel to the reconsideration of the vernacular construction in some small scale typologies.

A contemporary approach focused on the design of the main infrastructural elements recognizes the virtues of indeterminacy and seeks for an open design that could increase the life of buildings while allowing changes in use, transformation and growth. This also allows some radical approaches on real estate market considering flexibility, mutability and change.

Art'Ur Architectes

Add: 31, Rue Saint-Didier, 75116, Paris, France
Tel: 01 47 27 53 90; Fax: 01 47 27 19 30
Web: www.art-ur.net

Art'Ur Architectes – Philippe Pascal, Eric de Chambure, Muriel Teynier and Julien Mogan - is a private company limited by shares, comprising 20 architects and staff. Their young and dynamic team is divided between two agencies in France: one in the heart of Paris, not far from the Eiffel Tower, and the other by the Garonne river in Bordeaux.

Art'Ur Architectes is regularly selected to participate in architectural competitions for major public projects, high schools, middle schools, primary schools, and in any building project where, as part of a strong environmental drive, wood is predominant.

They always aim at revealing the symbolic, social and cultural dimensions of their projects, in order to set humans at the heart of their processes, at the heart of their projects, and at the heart of life.

ARTechnic Architects

Add: #3, 3-34-1, kamimeguro, Meguro, Tokyo, 153-0051, Japan
Tel: 81-3-5768-8718; Fax: 81-3-5786-8738
Web: www.artechnic.jp

ARTechnic Architects has a strong belief in the ability of architecture to enrich daily life. They aspire to improve the quality of life and the environment through innovation and design excellence. The studio has extensive experience in residential and commercial design and development through to interior and furniture design. They are a creative studio with passion, drive and focus on innovation based on clients' needs, the environment and social requirements.

BATLLE I ROIG ARQUITECTES

Add: Manuel Florentín Pérez 15,
08950 Esplugues de Llobregat - Barcelona
Tel: (34) 93 457 98 84
Web: www.batlleiroig.com

Enric Batlle lectures in Urbanism and Landscape Architecture at the Vallès School of Architecture (ETSAV) and on the Master's degree in Landscape Architecture at the Polytechnic University of Catalonia (UPC). He gained his master's degree in Landscape Architecture and his doctorate in 2002 with the thesis "The Garden of the Metropolis", for which he received the extraordinary doctorate prize from the Doctorate Commission of the UPC. He lectures in Landscape Architecture in the Department of Urbanism and Territorial Development at the ETSAV, and is involved in the Architecture and Project workshops there, and in the master's degree in Landscape Architecture at the UPC.

Joan Roig has taught Architectural Projects at the ETSAB since 1984, and has lectured at the Barcelona School of Agriculture and has been visiting professor at the following schools, among others: the Academie van Boukunst (Rotterdam, Holland); Delft University of Technology (Holland); the ILAUD-International Laboratory of Architecture and Urban Design (Urbino, Italy); the IUAV-Istituto Universitario di Architettura (Venice, Italy); the Faculty of Architecture (Genoa, Italy); the ITU, Faculty of Architecture of Istanbul (Turkey); Washington University (St. Louis, USA); the University of Illinois (Chicago, USA); the École Supérieure du Paysage de Versailles (France); the Technische Universität (Munich, Germany); Universität Stuttgart (Germany), and the School of Architecture of Navarre (Pamplona).

BDP

Add: 16 Brewhouse Yard, London EC1V 4LJ, UK
Tel: +44 [0]20 7812 8028
Web: www.bdp.com

BDP is the foremost interdisciplinary practice of architects, designers, engineers and urbanists in Europe. BDP works closely with users, clients and the community to create special places for living, working, shopping, culture and learning across Europe, Africa, Asia and Australia.

Founded in 1961, they now employ more than 1000 architects, designers, engineers, urbanists, sustainability experts, lighting designers and acoustics specialists in 16 studios across the UK, Ireland, Netherlands, UAE, India, and China. BDP has a leading track record in all major sectors including health, education, workplace, retail, urbanism, heritage, housing, transport and leisure. BDP combines expertise across disciplines, locations, sectors and all major building types to deliver a truly integrated way of working – resulting in high quality, effective and inspiring built spaces.

Bio-architecture formosana

Add: Shilin District, Taipei City 111, Taiwan of China
Tel: 886-2-28313377; Fax: 886-2-28387712
Web: www.bioarch.com.tw

Bioarchitecture Formosana is a collaborative practice renamed in 1999 after the union between Kuo Ying Chao Architects (established in 1995) and Ching-Hwa Chang Architects (established in 1991). The Company also cooperates with environmental engineering consultants (ecoscape formosana) in order to complete the overall design of the environment in each case.

The Company works with a professional team of skilled staffs, including architects, landscaping architects and other elite designers from both Taiwan of China and other parts of the world. Their goal is to provide architecture, landscape planning and several other design services based in Taiwan of China, but also expand their visions of the world. With their understanding of sustainability design, they monitor the process of planning designs as well as the constructions. They uphold the belief that people can live in harmony with the environment through dedication, careful planning, and design. Through the practice of Green Architecture, they hope to create a beautiful place where man and nature can become integrated.

C. F. Møller Architects

Add: Europaplads 2, 11,8000 Aarhus C, Denmark
Tel: +45 8730 5300
Web: www.cfmoller.com

C. F. Møller Architects is one of Scandinavia's oldest and largest architectural practices. Their work involves a wide range of expertise that covers programme analysis, town planning, master planning, all architectural services including landscape architecture, as well as the development and design of building components.

Simplicity, clarity and unpretentiousness, the ideals that have guided their work since the practice was established in 1924, are continually re-interpreted to suit individual projects, always site-specific and based on international trends and regional characteristics. Over the years, the company won a large number of national and international competitions and awards.

Their work has been exhibited locally as well as internationally at places like RIBA in London, the Venice Biennale, the Danish Architecture Centre and the Danish Cultural Institute in Beijing. Today C. F. Møller Architects has app. 320 employees. Their head office is in Aarhus and they have branches in Copenhagen, Aalborg, Oslo, Stockholm and London, as well as a limited company in Iceland.

Carlos Martinez Architekten

Add: Schnabelweg 8 CH-9442 berneck, Switzerland
Tel: +41 71 727 99 55, Fax: +41 71 727 99 44
Web: www.carlosmartinez.ch

Carlos Martinez was born in Widnau in 1967 and both of his parents are Spanish emigrants. Between 1970 and 1973, he spent his primary year in Asturias primary school. During 1988-1992, he studied at Abendtechnikum St.Gallen and got federal diploma as architect FH in 1992. The next year in 1993, he founded office Koeppel&Martinez and then transformed it to CarlosMartinez Architekten in 2003. During 2002-2010, he was an expert of architecture of the federal commission of art.

Christopher Charles Benninger Architects

Add: 'INDIA HOUSE', 53 Sopan Bagh, Balewadi,
Pune 411 045, Maharashtra, India
Tel: +91 20 6510 2331/ 32
Web: www.ccba.in

Founded by Prof. Benninger along with the Managing Director A.Ramprasad Naidu in the year 1995 the firm has grown into an internationally known 'design house'. Christopher Charles Benninger Architects with a team of 40 Architects, creates designs ranging from capital cities and new towns; educational campuses and corporate headquarters; housing estates and complexes; hotels resorts and hospitals; down to the design of chairs and art works. The entire range of materiality plays a role in the studio's search for beauty. In the end it is not the "things" that the studio designs, but the transcendental experience of the people using them, looking

at them, or just being in them which is the essence.
For the studio, the good life exists just a step outside of materiality, in a mystic twilight zone, which is called architecture.

The firm's work has been published in international and national journals and magazines such as Ekistics (Greece), Spazio-e-Societe (Italy), AIArchitect (USA), Cities (UK), Architectural Record (USA), ZOO (UK), Business Week (USA), Architects' Newspaper (USA), Arquitectura Viva (Spain), World Architecture (UK), MIMAR (UK), Habitat International (UK), Architecture+Design, Indian Architect and Builder, Inside-Outside (India) and many others. Mainstream magazine Business Week (USA) named the Mahindra United World College of India as among the ten "Super Structures of the World" in the year 2000.

Christian de Portzamparc

Add : 1, Rue de l'Aude – F-75014 Paris, France
Tel: +33 1 40 64 80 00
Web: www.portzamparc.com

Born in 1944, Christian de Portzamparc studied at the École des Beaux-Arts in Paris / France from 1962 to 1969. Soon after completing his studies, he set his mark on the landscape of the new town of Marne-la-Vallée with his Water Tower, a "poetic monument". In 1975, he designed a neighbourhood in Paris with 210 dwellings: Rue des Hautes-Formes was completed in 1980 and marked a turning point in the history of urban design.

Christian de Portzamparc works all over the world, theorizing on the present and future of the city, fragmentation, a case-by-case approach and the "open block" concept. From an individual building to an entire neighbourhood, the city is fundamental to his work, which is dominated by three overlapping themes.

Christian Kronaus

Add: Schoenbrunnerstrasse 59 1050 Vienna, Austria
Tel: +43 1 5480015
Web: www.kronaus.com

Architecture and Planning are processes that end with results that have to comply with the clients' aesthetic and economic requirements. Their competent employees and partnerships guarantee that they can meet these goals. Office organisation and clear project management structures allow them to meet the schedule, economic and aesthetic goals that a project defines.

Cino Zucchi Architetti

Add: Via Revere, 8 20123 Milano, Italia
Tel: +39 02 48016130; Fax: +39 02 48016137
Web: www.zucchiarchitetti.com

The Cino Zucchi Architetti studio is constantly searching new spatial solutions for contemporary life in the delicate and rapidly changing context of the European landscape. Its goal is to combine innovation and research with professional completeness capable of responding to complex programs on any scale and employing, when needed, a well-established net of specialized consultants.

The studio has designed and realized many commercial, public, industrial and residential buildings, public spaces, renewal plans and master plans for agricultural, industrial and historical areas. It has participated in numerous national and international competitions and is active in the field of interior and exhibition design. The works of the studio have received widespread critical acclaim and have been published in magazines both in Italy and abroad, and have been the subject of numerous exhibitions.

Major recent works of CZA include a large master plan for the Keski Pasila area in Helsinki, residential and office buildings for the former Alfa Romeo-Portello area in Milano, the Group M headquarters in Assago (Milano), the new Lavazza headquarters in Turin, and the extension and renovation of the Turin National Car Museum which was recently awarded the Inarch/Ance 2011 prize. The projects of the studio have been published in books and magazines worldwide.

CM Mimarlık Ltd.

Add: Ihlamur Yolu Çehreli 1 Apartmanı 65/6 Topağacı 34360, Nişantaşı / Şişli Istanbul, Turkey
Tel: +90 533 687 01 99/+90 212 232 62 97; Fax: +90 212 296 66 06
Web: cmmimarlik.com.tr

Led by Cem Sorguç, "CM Design and Architecture" was first founded in 2001, Istanbul, and since 2004 has been operating under the name "CM Architecture Ltd."

CM Mimarlik design and operation intellect seeks the accurate harmony of project components such as clients,construction cost economy, physical conditions, time, endusers and functional criteria, along with the aesthetic and scientific conditions, with no interest in trendy fashions, and creates contemporary, technological, environmental and working products as final outcome.

Coop Himmelb(l)au

Add: Spengergasse 37 1050 Vienna, Austria
Tel: +43 (0) 1546 60-0; Fax: +43 (0) 1546 60-600
Web: www.coop-himmelblau.at

Coop Himmelb(l)au was founded by Wolf D. Prix, Helmut Swiczinsky, and Michael Holzer in Vienna, Austria, in 1968, and is active in architecture, urban planning, design, and art. In 1988, a second studio was opened in Los Angeles, USA. Further project offices are located in Frankfurt, Germany; Paris, France; and Hongkong, China. Coop Himmelb(l)au employs currently 150 team members from nineteen nations.

Over the course of the past four decades, Coop Himmelb(l)au has received numerous international awards. In 2011 the office received the Wallpaper* Design Award 2011 (Category: "Best Building Sites") for the project Dalian International Conference Center, the Dedalo Minosse International Prize for the project BMW Welt as well as the Red dot award: product design (Category: "Architecture") for the Central Los Angeles Area High School #9 of Visual and Performing Arts, to list just a few.

Ctrl G arquitectos

Add: Crr 33 # 5 G 13 Apto 301 Medellin, Colombia
Tel: (57) 3104189532:
Web: www.ctrlgarquitectos.com

Ctrl G arquitectos current members are Catalina Patiño and Viviana Peña. The office received the first place award from the International Competition of Sustainable Gardening at the Art and Landscaping Biennale of Canaries and Tenerife, Spain, 2008; and, a year later in association with 51-1 Studio, the first price in the competition for Modern Art Museum complementary facilities. In 2009, Ctrl G arquitectos built the scenery for the Christmas Show organized by Noel Cookies Co. and along with Camilo Restrepo's Office the dance academy Playdance, in Medellín, Colombia.

Ctrl G arquitectos + Federico Mesa built the San Antonio de Prado and Pajarito La Aurora kindergartens in Medellín city, commissions assigned as a result of the public competition of 2009, and developing the projects for two other kindergartens, Santo Domingo Savio y Carpinelo.

Damilano Studio Architects

Add: Via Vecchia di Cuneo 128 12011 Borgo S. Dalmazzo (CN), Italia
Tel: +39 0171 262924
Web: www.damilanostudio.com

Born in Cuneo in 1961 Arch. Duilio Damilano graduated in architecture in 1988 from the Politecnico di Torino, supervisor R. Gabetti. Since 1989 he began his professional career as an associate. In 1990 he opened Damilanostudioarchitects developed through collaborations with artists and designers, new paths in the field of architecture, both in Italy and abroad. Damilano Studio Architects develop a language with a contemporary approach for continuous research in the modeling of architectural forms.

Particular attention is paid to light, transparent and reflective surfaces in projects that take a force of expression that sometimes cancels or enhances the built volumes.

ECDM Architectes

Add: 7 passage Turquetil, 75011, Paris, France
Tel: +33 1 44 93 20 60; Fax: +33 9 58 72 33 21
Web: http://ecdm.eu

Founded after being the recipient of the Albums de la Jeune Architecture award in 1993 and the Villa Médicis Hors les Murs – scholarship in 1996, Emmanuel Combarel and Dominique Marrec Architects (ECDM) is leading for 18 years a work focused working on defining a living environment through an architectural project. The architecture that materializes the firm's approach is underpinned by the evolutions and mutations of the society.

The dynamics of the project emerge from a confrontation with the context, and a hierarchization of the problematics induced by the program and the site. Environmental quality, landscape, uses, ways of life, and technical choices are all structuring elements of the office's projects. The architecture that materializes the firm's approach is underpinned by the evolutions and mutations of their society. It tends to be a simple, sober architecture following a rigorous logic, with no preconceptions, nostalgia or stylistic preoccupations.

Emilio Faroldi Associati

Add: Borgo Lalatta 10, 43121 Parma, Italy
Tel: 0039 02 89056202
Web: www.emiliofaroldiassociati.it

With offices in Parma and Milan, the studio addresses issues related to cities, their various hinterlands, the architecture of which they are made up and the technologies required for their construction, explored in terms of the functions and activities that they accommodate in tandem with the combination of their form and composition.

By bringing together theoretical activities and design practices, it seeks to define a cultural reference scenario, through a professional research process that is viewed as a setting for testing and further investigating architectural and construction-related issues, perceived through a multiscalar and interdisciplinary interpretation of the architect's role, divided between humanistic and scientific cultures.

The studio's professional activity is carried out in the form of design applied at multiple levels: urban planning, architecture and the building fabric itself throughout its entire life-cycle. Landscaping, sustainable development and technological innovation, as applied to the complexities of location, space and contemporary circumstances, are the primary elements addressed in the design method.

The organisation's goal is to provide integrated architectural and technological design services that encompass the planning, construction and architectural management phases. Architects and specialists in the sector collaborate on work that is customised to each specific professional assignment based on the services required. A flexible approach enables them to collaborate with other professional organisations in order to develop joint designs both in Italy and abroad.

Erhard An-He Kinzelbach

Add: Zhidafangbo 8 310003 Hangzhou, Zhejiang, China
Tel: +86 134 56768141
Web: www.knowspace.eu

The studio is concerned with spatial and programmatic organisations that combine local differentiation with overall coherence. Its objective is to identify synergies between practice, design research and teaching and to proliferate them as sources of knowledge, The organization of this knowledge serves as the foundation for the conceptual engagement with spatial and programmatic complexity, and the making of space, material and form – both at the scale of the building as well as the city.

The studio understands its role as that of a moderator and organizer of the complex interactions between cultural, political, economic and material forces and the multiple agents that determine a project. In this sense, they act rather as competent trans-disciplinary mediators than as narrowly specialized experts.

EXP Architectes

- Antoine Chassagnol, Nicolas Moskwa, Maxime Vicens

Add : 23, rue du Buisson Saint-Louis, 75010 Paris, France
Tel : +33 (0)1 42 38 04 04
Web : www.exp-architectes.com

EXP Architectes is an architectural office set up in early 2008 by Antoine Chassagnol, Nicolas Moskwa and Maxime Vicens, architects and urban planners. After starting to work in agencies before having their own private practice, they have later decided to federate their knowledge, skills, and experience. Their constant care is to respond to, and even to go beyond, the needs and desires of people, with respect to the specificity of the place, paying great attention to the historical, social, and ecological qualities of the site. Their work, the result of collective interpretations and decisions, is a personal collaboration with the clients, whom they always closely associate to their reflexions.

EXP Architectes has great expertise in master planning, collective housing, city renovation, urban and peri-urban space and concentrate on environmental issues: sustainable development, renewable energy sources and energy sufficient planning. They have realized eco districts, new urban districts, and several mixed use districts.

They engage a thorough reflexion on the buildings, the city, dwelling and transport, but also design and art, being constantly at the crossroads between the built and natural development. Their objective is to create not only buildings and cities that function well, but also an inspiring architecture, liable to enhance the quality of life of the individuals and the group, as well as the quality of environment.

Feilden Clegg Bradley Studios

Add: Bath Brewery, Toll Bridge Road, Bath, BA1 7DE, UK
Tel: 01225 852 545; Fax: 01225 852 528
Web: www.fcbstudios.com

Feilden Clegg Bradley Studios was first established in 1978 and with offices in Bath and London they have grown steadily over the years to their present strength of 21 partners and 109 staff.

They have a reputation for sustainable design and innovation and a strong track record in education and community buildings. The practice is at the forefront of innovative housing design, from inner city social housing to new suburban neighbourhoods and private developments, and is working on a series of substantial urban regeneration projects throughout the UK.

They believe the best buildings emerge from a clear concept that then finds its way into the DNA of the details. They also come from a strong working relationship with committed clients and creative consultants who understand the transformational power of architecture. Designing buildings is a process about which they are passionate, leading to a product they hope others will enjoy. It is a way of thinking – a way of life – leading to the creation of form and space that is memorable and inspirational.

FGMF Arquitetos

Add: Rua Mourato Coelho, 923, Vila Madalena, St.Paulo, Brazil
Tel: +55 11 3032-2826
Web: www.fgmf.com.br

Created in 1999, FGMF Arquitetos produces contemporary architecture, without any restraints regarding the use of material and building techniques, seeking to explore the connection between architecture and its environment.

In these few years of existence, they've had the opportunity to deal with a wide range of architectural scales, which enhances their belief that, just as life itself, architecture ought to be plural, heterogeneous and dynamic. Urban realm, cultural facilities, residences, sports complexes, hospitals, schools, commercial buildings and many others are part of the same urban landscape and of the daily life: Knowing how to deal with all these programs is a way to enrich their design, in contrast to a specialized architecture. Based on the professional and academic experience of its associates, FGMF has an innovative and inventive approach. There are no pre-conceived formulae: at every challenge they start from scratch, using design as a research tool.

Their dedication and hard work led the firm to the satisfaction of receiving relevant national and international awards, among which some from the Instituto de Arquitetos do Brasil (IAB), Living Steel, Editora Abril and Dedalo Minosse. In 2010, they were the most awarded practice in Brazil.

FoRM Associates

Add: 154 Narrow Street, London E14 8BP, UK
Tel: 0044 (0) 207 5373 654
Web: www.formassociates.eu

FoRM Associates is a London based design practice that looks holistically at the 'livability' of 21st century cities. Established by Peter Fink, Igor Marko and Rick Rowbotham to follow their shared vision of green urbanism, FoRM brings together their experiences and skills in urban design, architecture and landscape architecture. Their passion about working as an interdisciplinary team brings fresh solutions and thinking into the design process, helping cities to reclaim much of their grey, abandoned and overlooked urban areas to make them 'green', both literally in an environmental sense; and metaphorically as places of new growth and positive change. The practice has completed a wide range of projects from intimate sanctuaries to complex masterplans in Europe and North America.

They are always looking at different ways of delivering sustainable solutions, challenging the traditional technological approach to green urbanism. Truly sustainable developments are such that will grow, adapt and transform cities for decades to come. In the case of the award winning Northala Field Park, it was the recycling of inert waste material that generated £6 million funding to deliver the largest new park in London. Through experience they have developed an understanding of what makes a 'round the clock' public space operable. They see place making as an integral part of the process of instigating new initiatives for economic, social and cultural change as well as transport and regeneration opportunities. By inviting people to join the process of design the practice develops new spaces that inject vitality into cities and embody the values of their inhabitants.

G Architecture

Add: 05, Rue Gambetta, 91300 Massy, France
Tel: 00 33 1 69 20 32 78
Web: www.garchitecture.fr

The architectural agency G Architecture was founded in 2000 by Jean Pierre Gautier, after 20 years of working as an architect in independent status.

The agency production is part of an exploratory approach to introduce a new architectural language in the urban landscape. This approach has greatly expanded, throughout the projects, from individuals houses to housing units, privates and publics facilities, to more specifics projects, encompassing, in overall design phases, the functionals solutions and the socials, environmentals and urban problematics, currently at the center of societal questions.

The diversity of their projects allows them to develop a true culture of architecture at all scales, to adapt to an audience of all kinds, to understand and appreciate the specific requirements of each Client. The Architecture owners are Jean Pierre Gautier, Yann Gautier and Loic Gautier

Gracia Studio

Add: 4492 Camino de la Plaza, San Ysidro, 92173, Canada
Tel: (619) 795 7864; Fax: (619) 269 3103
Web: www.graciastudio.com

Jorge Gracia, main architect of Gracia Studio establishes a culture of design and challenge to the conventions on residential architecture in the Tijuana-San Diego area. His studio is fostering a demand for inexpensive and stylish architecture by focusing on pre-fabrication.

Gracia Studio is based on the technical aspects of building systems that lead to worth living architecture for the users. Due to the need of cost-effective building, their challenge always starts from the construction process for each of their designs. In 2005 Jonathan Bell, architecture editor for Wallpaper Magazine, took special interest in Gracia Studio, and published it as part of the Architects Directory article.

Gracia Studio's design philosophy is using materials that meet the needs of the project completely in a structural form; yet aesthetical and modern at the same time, as project Casa Becerril is shown in Arquine book, The Best of the XXI Century, Mexican Architecture 2007-2008. Completing numerous lectures in different universities and architectural conferences, such as the Dwell on Design event in 2010, and Virginia Tech Architecture School, Gracia Studio shares its innovative design and building with students and general public throughout the United States and Mexico.

In 2012 Gracia Studio launches the first multidisciplinary School of Practical Architecture in the Tijuana-San Diego area, with workshops where students interact directly with the craftsmen, gathering experience while working in conjunction with local artisans such as blacksmiths, carpenters, upholsterers, and industrial painters, form workers and landscapers, just to mention a few. Part of the studio's beliefs is not to compromise design with external factors such as economic concerns, rather take them as an opportunity to exploit and develop creativity and inventiveness in new ways.

Henning Larsen Architects

Add: Vesterbrogade 76, Dk-1620 Copenhagen V, Denmark
Tel: +45 8233 3000; Fax: +45 8233 3099
Web: www.henninglarsen.com

Henning Larsen Architects is an international architecture company with strong Scandinavian roots. Their goal is to create vibrant, sustainable buildings that reach beyond themselves and become of durable value to the users and to the society and culture that they are built into.

Since the company was founded in 1959, they have acquired a comprehensive knowledge of the many aspects of building - from sketch proposals to detailed design, building owner consulting and construction management. Their competences are continuously developed through their close collaboration with global partners, experts and specialists.

Henning Larsen Architects attaches great importance to designing environmentally friendly and integrated, energy-efficient solutions. Their projects are characterized by a high degree of social responsibility – not only in relation to materials and production but also as regards good, social and community-creating spaces.

Hyunjoon Yoo Architects

Add: Mapogu Seokyodong 362-11 4th floor, Seoul, Korea
Tel: 82-2-334-8508; Fax: 82-2-334-8518
Web: 82-2-334-8518

Hyunjoon Yoo (A.I.A.) is a chair of department at Hongik University School of Architecture, principal architect of Hyunjoon Yoo Architects in Seoul, Korea. He graduated from Yonsei University in Seoul with a bachelor's degree in architectural engineering, followed by a Master of Architecture degree from MIT, and received his Master of Architecture degree with Distinction from Harvard. He worked for Richard Meier & Partners, and for the MIT Architecture Representation & Computation Lab. He was a visiting scholar at MIT in 2010.

He has won five awards in international design competitions, including an award and honorable mention at the BSA Unbuilt Architecture Design Awards in the same year. He was awarded as the Korean Young Architects Award in 2009. His work has been selected as one of the Seoul's 15 Architectural Wonders by CNN.

He has been a consultant for many projects, including the remodeling project for the Blue House, South Korea's presidential residence. He has also served as a judge in competitions such as the Korea National Architecture Competition. Books he has authored include "52 9 12", "The Flow of Contemporary Architecture" and "Modernism: Cultural Hybrid between East and West."

iArc Architects

Add:1,2,4 Fl. 264-3 Yangjae-dong Seocho-gu
Seoul Republic of Korea (137-894)
Tel: 822 571 4894
Web: www.iarc.net

Founded in 1986, iArc Architects has been acknowledged both in America and far East for achieving innovative design solution by receiving many awards from both sides of pacific. iArc Architects pursue VALUE DESIGN / SUSTAINABLE DESIGN / DESIGN TECHNOLOGY. iArc Architects pursues intelligent pragmatism as its philosophy to practice architecture. By this reason, the letter "i" of iArc signifies intelligent. They believe the role of architect is imbuing organizational intelligence to material. They choose systematic methodology to achieve this as a design process. iArc Architects also believe architecture is a medium to integrate urban and social fragmentation. By providing relationship to diverse different components, architecture could truly befit to its people. iArc is striving for the open society through open space.

IROJE KHM Architects

Add: 1805 Gardentower Bldg, Unni-dong, Jongro-gu,
Seoul,110-795, Korea
Tel: +82 2 766 1928; Fax: +82 2 766 1929
Web: www.irojekhm.com

As principal of IROJE KHM Architects, Architect HyoMan Kim was born in Seoul in 1955. He received B.A in architecture from Dankook University. He is a concurrent professor of graduate school of architecture Gyeonggi University and an editorial adviser of CONCEPT Magazine.

He achieved many awards in Seoul, Korea. He has exhibited his works in New York, USA "The Good Life", in Torino, Italy "2008 UIA congress, ARCASIA TODAY", in Osaka, Japan by JIA "The Modern Architecture to make History Alive".

JBMC Arquitetura e Urbanismo

Add: Rua Pe. João Manoel, 199/41 01411-001
Tel: 5511 3086 3607
Web: www.jbmc.com.br

Being a byproduct of João Batista Martinez Corrêa´s professional experience, which was accumulated in 30 years as an architect at Promon Engenharia, JBMC works with various scales and architectural programs – from urban plans to furniture and object design.

Examples of projects they have been working with include public transportation and infrastructure, industrial masterplans, urban design, public buildings, institutions and private housing, scenography, object design and graphic design. They usually work in partnership with other companies or multidisciplinary teams in order to augment their field of action and the consistency of their projects.

It has always been part of JBMC agenda to produce an architecture that inspires rational energy use and utilizes local resources, in a continuous search for the best possible relation between buildings and the geographical specificities of each location they are inserted in. At an urban scale, this commitment is translated in a permanent debate about the urban problems. As a result, proposals that gather opportunity, creativity and low-cost solutions are produced, such as the monorail link between Sao Judas subway station and Congonhas Airport, in Sao Paulo – a low impact solution that could benefit not only the airport passengers but also the local infrastructure as a whole – and the proposal to increase the area of Ibirapuera Park, where alterations in the street system would allow for a better integration between the park and the public institutions on its surroundings, increasing the green areas and the leisure opportunities in one of the most pleasant places in Sao Paulo.

JOSEP MIAS

Add: Mateu 19 bxs, 08012 Barcelona, Spain
Tel: (+34) 932 388 208; Fax: (+34) 932 388 209
Web: www.miasarchitects.com

JOSEP MIÀS, Architect, COAC 24797-9. Graduated in 1992 in ETSAB_UPC, Escola Tècnica Superior d'Arquitectura de Barcelona_Universitat Politècnica de Catalunya. From 1990 till 2000 he worked as an Associate Architect with Enric Miralles, developing the following projects Círculo de Lectores in Madrid, Sports Hall in Huesca, Centro de Alto Rendimiento in Alicante, Library in Palafolls, Parc dels Colors in Mollet del Vallès, Cemetery in Igualada, Scotish Parliament in Edimburg, Diagonal Mar Park, Santa Caterina Market and Gas Natural Building en Barcelona. In 2000, Josep Miàs founded his own studio, working in landscape and urbanism projects, architecture and interiorism and industrial design, such as the refurbishment of the old city of Banyoles, Barceloneta Market, Golf Fontanals de Cerdanya Clubhouse, Funicular Station and restoration of the Old Historical Building in Tibidabo Amusement Park, Annexa-Puigbert Primary School and iGuzzini Illuminazione España Headquarters. Other projects which are now being finished are the Plug-in 22@ building and a Housing building in Torrebaró, both in BCN and a Pedestrian Bridge in Palafolls.

Among other prizes those which should be underlined are AJAC Best Young Architect Award 2004, Special Mention in Girona Architecture Award 2005, First Prize in Girona Architecture Award 2007, City of Barcelona Architecture and Urbanism Award 2007, Catalonia Construction Award 2009 and Architecture Aplus Award 2011 for the Best Educational Building in Spain. Recently he has received the BUILDING OF THE YEAR 2011 Award, given by the most prestigious architecture website of the world Archdaily, for the iGuzzini Barcelona Corporate Building.

From 1999, he has developed teaching activities in Architecture Schools in Europe, America and USA: Design Department Professor at School of Architecture ETSAB, Design Department Professor at School of Architecture La Salle BCN, Director of the Design Department and Assistant Director at School of Architecture ESARQ_UIC and Professor in UCL London The Bartlett School of Architecture, UNISS Facoltà d'Architettura Alghero, Frankfurt Stadelschule and Harvard GSD.

He has been invited to give lectures and head up workshops in a large number of universities in Spain, Italy, Germany, Denmark, Sweden, Finland, Belgium, Switzerland, United States, Mexico, Uruguay, and Panama among others.

His work, as well as his writings, have been published in international magazines and books, and his work has been exhibited in many architecture exhibition halls as COAC_Barcelona, Nuevos Ministerios_Madrid, Cité de l'Architecture_París, Stadelschule_Frankfurt, Architektur Forum_Zurich and TU University_Berlín.

Currently, he is a Professor at the ETSAB Barcelona, UNISS Alghero and Bartlett_UCL London.

JSA

Add: Sinsenveien 4D, 0572 Oslo, Norway
Tel: +47 22 99 48 99
Web: www.jsa.no

The practice was established in 1995 by Jan Olav Jensen and Børre Skodvin. Starting with 4 architects, the staff has grown to 9 architects in 2009. During these 14 years the office has completed a variety of projects, for public as well as private clients. Projects range from furniture to urban planning, but with weight on building commissions and landscape interventions. The projects are located in a variety of sites and situations. The practice received many awards, The Eric Schelling Award for Architecture in 2008, Forum AID Award in 2007, to list just a few.

Jun Mitsui & Associates Inc.Architects

Add: 5-2-4 Nishi-Gotanda, Shinagawa-ku Tokyo Japan
Tel: +81-3-3491-0419; Fax:+81-3-3491-0418
http://www.jma.co.jp

Jun Mitsui & Associates Inc. Architects and Pelli Clarke Pelli Architects have had more than ten years of successful design collaboration together. In most cases, the two companies work together through the schematic design phase. For projects in Japan, Jun Mitsui & Associates Inc. Architects then completes Construction Documents and construction supervision. Both companies are connected with high-speed digital communication tools. Jun Mitsui & Associates Inc. Architects has a wide international network, collaborating with several foreign designers and architects. Through international collaboration, Jun Mitsui & Associates Inc. Architects is able to perform building design, urban design and interior design worldwide.

KNOWSPACE architecture + cities

Add: KNOWSPACE architecture + cities, Zhidafangbo 8, Donghe Jiayuan 2-1-702, Hangzhou 310003, P.R. China
T el: +86 134 56768141
Web: www.knowspace.eu

KNOWSPACE architecture + cities was founded in New York and has since functioned as a studio for practice and research on architecture and cities.

KNOWSPACE architecture + cities is concerned with spatial and programmatic organizations that combine local differentiation with overall coherence. Its objectives is to identify synergies between practice, design research and teaching and to proliferate them as sources of knowledge. The organization of this knowledge serves as the foundation for the conceptual engagement with spatial and programmatic complexity, and the making of space, material and form.

Many of the studio's projects view façades not as mere interfaces between inside and outside, between building and city, but use the spatial depth of façades as a key means to organize and zone space, define programs and orchestrate atmosphere and affect. Operative and performative façades play a key role in dealing with the dichotomies between tabula rasa and contextual integration, between global and local, monolithic and differentiated, singular and multiple.

KNOWSPACE architecture + cities understands its role as that of a moderator and organizer of the complex interactions between cultural, political, economic and material forces and the multiple agents that determine a project. In this sense, they act rather as competent trans-disciplinary generalists than as narrowly specialized experts.

Line and Space, LLC

Add: Line and Space, LLC, 627 East Speedway,
Tucson, Arizona 85705
Tel: 520 623-1313; Fax: 520 623-1303
Web: www.lineandspace.com

Line and Space, LLC was founded in 1978 in Tucson, Arizona by Les Wallach, FAIA, to facilitate the design and building of environmentally sensitive architecture that respects and responds to existing site conditions. Their work is known internationally and has been recognized by their peers through the receipt of over 85 architectural and environmental design awards, including the American Institute of Architects (AIA) Committee on the Environment (2-time recipient), the AIA Western Mountain Region Firm of the Year, and the AIA Arizona Sustainable Firm of the Year.

Mack Scogin Merrill Elam Architects

Add: 111 John Wesley Dobbs Avenue, NE Atlanta, Georgia 30303, USA
Tel: 404 525 6869; Fax: 404 525 7061
Web: msmearch.com

Mack Scogin and Merrill Elam, the two principals of Mack Scogin Merrill Elam Architects, Inc. have worked together in architecture for over forty years. The firm, founded in 1984 as Parker and Scogin, later as Scogin Elam and Bray, was formed in order to take full advantage of the complementary skills and talents of the two principals. Mack Scogin and Merrill Elam have made the commitment to organize all of the work in a manner that ensures the involvement in the day-to-day development of each project. This keeps the work personal and directed, and brings the best of the firm's collective knowledge and experience to each client.

Magén Arquitectos

Add: Pº Sagasta 54, 7º C, 50006, Zaragoza, Spain
Tel: 0034 976 385 110; Fax: 0034 976 371 495
Web: www.magenarquitectos.com

Magén Arquitectos is an architectural practice founded by Jaime Magén (Zaragoza, 1974) and Francisco J. Magén (Zaragoza, 1980) based in Zaragoza since 2002.

Magén Arquitectos is interested in Architecture as a both technical and cultural fact, deeply focused on unique investigation and development of each project. This method of work allows people to manage different projects, in terms of size, scale and complexity.

The works by Magén Arquitectos have won more than 30 awards, like Giancarlo Ius Gold Medal in 2011, Finalist and Mention Young Project in the XI Spanish Architecture and Urbanism Biennial in 2011, the 1st Prize Bauwelt Preis (Munich, 2007), the 1st Prize SAIE Selection (Bologna, 2010), to list just a few.

The works by Magén Arquitectos have been published in many technical magazines. They have also been exhibited several times in Spain, Germany, Italy and France.

Mario Botta Architetto

Add: Via Beroldingen 26, CH 6850 Mendrisio, Switzerland
Tel: +41 91 972 86 25; Fax: +41 91 970 14 54
Web: www.botta.ch

Architect Mario Botta was born in Mendrisio, Ticino on April 1, 1943. After an apprenticeship in Lugano, he first attends the Art College in Milan and then studies at the University Institute of Architecture in Venice. Directed by Carlo Scarpa and Giuseppe Mazzariol he receives his professional degree in 1969. During his time in Venice, he has the opportunity to meet and work for Le Corbusier and Louis I. Kahn. His professional activity begins in 1970 in Lugano and, from the first single-family houses in Ticino, his work has approached all building typologies: schools, banks, administrative buildings, libraries, museums and sacred buildings.

He has always committed himself to an intense architectural research and in 1996 he was among the prime movers and founders of the Academy of Architecture in Mendrisio, where he teaches and holds the directorship for the academic years 2011-2013. His work has been recognized with important awards and has been presented in many exhibitions.

Markus Scherer Architekt

Add: Sommerpromenade 10, 39012 Meran, Italy
Tel: 0039 0473 490466; Fax: 0039 0473 490467
Web: www.architektscherer.it

Markus Scherer Architekt office's philosophy stresses the importance of the different construction tasks, of the continuing dialogue between old and new, of the co-existence between the new building and its surroundings, and of the constant evolution of a process - be that the change of materials, or the change of an idea during the whole length of the process.

Born in 1962 in Vienna, Architect Markus Scherer studied architecture at the TU Vienna and the IUAV of Venice, where he graduated in 1990 with Vittorio Gregotti and Bernardo Secchi. Founded the A5 Architectural Studio in 1992 with W. Angonese (until June 2001), from June 2001 he had his own studio in Bolzano and in 2003 the studio was transferred to Merano.

Meyers + Associates Architecture

Add: 232 North Third Street, Suite 300, Columbus, OH 43215-2513, USA
Tel: 614 221 9433; Fax: 614 221 9441
Web: www.meyersarchitects.com

Meyers + Associates Architecture is a full-service design firm dedicated to providing exceptional client service and design excellence. The firm understands that design and architecture are highly collaborative endeavors; they seek success through teamwork, open communication and a consistent and accountable process.

Founded in 1999 by Christopher Meyers, AIA, LEED AP, the firm consists of 15 individuals, including 7 licensed professionals with a vast array of project experience. Their diverse and talented staff are actively involved in many local boards and professional organizations including the American Institute of Architects, the Construction Specifications Institute, the International Interior Design Association, and the United States Green Building Council. The staff are fully engaged in sustainable design practice with numerous members being LEED accredited design professionals.

As advocates for design and service excellence, the firm promotes creativity, professional development and innovative service. They have a demonstrated success in creative environmentally smart projects, including their own offices where they have been recognized as the first architecture firm in The State of Ohio to receive LEED Silver for their own office environment.

Ministry of Design

Add: 20 Cross Street #03-01 048422, Singapore
Tel: +65 6222 5780
Web: www.modonline.com

Ministry of Design was created by Colin Seah to Question, Disturb & Redefine the spaces, forms & experiences that surround people and give meaning to the world. An integrated experiential-design practice, MOD's explorations are created amidst a democratic "studio-like" atmosphere and progress seamlessly between form, site, object and space. They love to question where the inherent potential in contemporary design lies, and then to disturb the ways they are created or perceived – redefining the world around them in relevant and innovative ways, one project at a time!

This, they declare, is a real change, not a change for the sake of novelty. Fortified with these aspirations, they begin each distinct project anew by seeking to do 2 things – to draw deeply from the context surrounding each project, but also to dream freely so that they might transcend mere reality and convention. Each MOD project endeavours to be delightfully surprising but yet relevant, distinctly local but still globally appealing.

The response to their ethos has been overwhelming and they've received critical acclaim with a multitude of international award wins and key media coverage – these include being crowned Designer of the Year by the International Design Awards, USA 2010, Rising Star of Architecture by the Monocle Singapore Survey 2010, Gold Key Award Grand Prize Winner 2008, and President's Design Award 2006 & 2008 as well as feature appearances in Wallpaper, Frame and Surface. True to their inter-disciplinary profile, they've also won the Grand Prize in Saporiti Italia's design competition, and Luxury Tower was manufactured for display at the prestigious Milan Design Week 2010.

Nitsche Arquitetos Associados

Add: Rua General Jardim 645 conjunto 12 São Paulo, Brazil
Tel: 55 11 2892 6004; Fax: 55 11 2892 6004
Web: www.nitsche.com.br

Nitsche Arquitetos Associados was founded by Lua Nitsche and Pedro Nitsche in 2000. Born in 25th November, 1972, in Sao Paulo, Brazil, Lua Nitsche graduated from Faculdade de Arquitetura e Urbanismo – FAU (School of Architecture and Urban Planning) of the Universidade de São Paulo – USP (University of São Paulo) in 1996.

Born in 25th July, 1975, in Sao Paulo, Brazil, Pedro Nitsche graduated from Faculdade de Arquitetura e Urbanismo – FAU (School of Architecture and Urban Planning) of the Universidade de São Paulo – USP (University of São Paulo) in 2000.

Nitsche Arquitetos Associados specializes in architecture, urbanism, visual programming and Art. The firm achieved many awards, o melhor da arquitetura award in 2011, jovens arquitetos award in 2009, to list just a few.

OMA

Add: Heer Bokelweg 149, 3032 AD Rotterdam, Netherlands
Tel: +31 10 243 82 00; Fax: +31 10 243 82 02
Web: http://oma.eu

OMA is a leading international partnership practicing architecture, urbanism, and cultural analysis. OMA's buildings and masterplans around the world insist on intelligent forms while inventing new possibilities for content and everyday use. OMA is led by seven partners - Rem Koolhaas, Ellen van Loon, Reinier de Graaf, Shohei Shigematsu, Iyad Alsaka, David Gianotten and Managing Partner, Victor van der Chijs - and sustains an international practice with offices in Rotterdam, New York, Beijing and Hong Kong.

The counterpart to OMA's architectural practice is AMO, a research studio based in Rotterdam. While OMA remains dedicated to the realization of buildings and masterplans, AMO operates in areas beyond the traditional boundaries of architecture, including media, politics, sociology, renewable energy, technology, fashion, curating, publishing, and graphic design.

OPPENHEIM ARCHITECTURE+DESIGN

Add: 245 NE 37th Street, Suite 102, Miami, Florida 33137 USA
Tel: +1 305 576 8404 x612 Fax: +1 305 576 8433
Web: www.oppenoffice.com

Oppenheim Architecture+Design LLP (OAD) is a full service architecture, interior architecture, interior design and urban planning firm located in Miami, Florida. The firm founded by Chad Oppenheim specializes in creating powerful and pragmatic solutions to complex project challenges and has extensive experience in world class hospitality, residential and mixed-use design.

The Oppenheim Architecture+Design LLP design strategy is to extract the essence form each context and relative program – creating an experience that is dramatic and powerful, yet simultaneously sensual and comfortable. The firm's approach begins by a deep understanding an analysis of a clients vision in relation to the project's typology, context, zoning parameters, and financial realities.

Original Vision Ltd.

Add: 22/F, 88 Gloucester Road, Wanchai, Hong Kong, China
Tel: +852 2810 9797; Fax: +852 2810 9790
Web: www.original-vision.com

Over the past two decades, Original Vision Ltd. has been pioneering new concepts in leisure architecture throughout South-East Asia. With projects in Hong Kong of China, Thailand, Indonesia, India, Vietnam and Europe, the company is dedicated to innovation, creativity and quality in design. Their goal is to create environments that are wonderful places and spaces that are breathtaking and inspirational and above all surroundings where people just love to be.

The manipulation of space, light, texture and colour is the core of their philosophy. They bring these elements together to harmonize with function and desire. Architecture is a true art, where the end product is the built form, a living sculpture in and around which people live. They care passionately how their architecture responds to the soul but they are equally concerned about how it performs functionally, economically and environmentally. It is the mastering of these complex and often conflicting criteria that brings them deep satisfaction when their work is successful.

They create through teamwork where their architects and designers are exposed to challenging but rewarding projects in which they can capitalize on their skills and grow with the company towards their ultimate goal, the realization of dreams.

Park Associati

Add: Via Carlo Goldoni, 1-20129-9 Milano, Italia
Tel: +39 02 79 84 52; Fax: +39 02 76 39 06 44
Web: www.parkassociati.it

Set up in 2000 by Filippo Pagliani and Michele Rossi, Park Associati is an architectural design studio which works on a large scale. Investigation into the dynamics and flows of urban spaces form the basis of a wide range of projects for the tertiary, manufacturing and residential sectors. Equally important for the studio is the planning of interiors by paying close attention to both craftsmanship and design. This adopts a transversal approach to the planning process and a vision that seeks to interrelate cultures and specializations, experiments with new technologies and is sensitive to the issues of sustainability and energy saving.

Such an approach opens up new possibilities in terms of scope and vision, and on each occasion gives rise to original compositional landscapes where architecture, place, technology and materials combine to create different spatial forms.

The projects on urban and architectonic scales underline the aptitude of Park Associati to research solutions that can combine local identity and technological innovation, this is confirmed by the recent awards in competitions and following realized projects. The projects of Park Associati have been shown in several exhibitions and published in Italian and international magazines.

Paul Le Quernec & Michel Grasso

Add: 50 rue de la course - 67000 strasbourg
Tel: 09 50 15 55 11 Fax: 09 55 15 55 11
Web: www.michelgrasso.fr

Paul Le Quernec: After a scientific bachelor's degree and one year study in the field of mathematics, Paul Le Quernec has been admitted by competitive examination to ENSAIS (Ecole Nationale Supérieure des Arts & Industries de Strasbourg – National Grad School of Art and Industry of Strasbourg). After the graduation, his work has consisted of small sized private assignments, small sized architectural open competitions.

Michel Grasso: Michel Grasso is 31 years old. He has created his own company in 2006, just after his degree performed at ENSAIS (Ecole Nationale Supérieure des Arts & Industries de Strasbourg – National Grad School of Art and Industry of Strasbourg).

His first projects were some houses, a wine bar on a barge in Strasbourg, a transient hostel for the Centre Pompidou in Metz and some collaborations as subcontractor for other architects on public buildings.

He met Paul Le Quernec through an internship at his office in 2004, during his studies. They became very good friends and continued to work together for some architectural projects, design or graphics.

Paulo David Arquitecto

Add: Rua da Carreira nº 73 5º, 9000-042 Funchal, Portugal
Tel: +351 291 281 840
Web: www.paulodavidarquitecto.com

Located in the core of Funchal's city, Paulo David Arquitecto was founded on the last level of a 20th Century building. The configuration of its interior space, spread in open space, offers to the architects who inhabit the space several views towards the city. These views are framed as slides in a sequence of windows that flood the working space with natural light.

Embodying a philosophy of project design in working teams, the intentional reduced number of collaborators use and move freely within the interior space of the studio, from working table to working table. This attitude allows a greater cooperation and dialogue among the architects, where everyone participates in every project.

PlanB Arquitectos

Add:Crr 33 # 5G 13 Apto 301 Medellin, Colombia
Tel: +57 – 4 – 3119151; Fax:
Web: www.planbarquitectura.com

PlanB Arquitectos is an architectural office that defines the work through a practice in which equal status is given to dialogue, drawing, travel, layout, construction, etc. and which are handled continuously, professional or academic situations, publication of books, college classes or construction of buildings. Plan:B Arquitectos trusts in working collaborative, to make of it a statement on the architecture and understands the practice and the architectural project as open situations, interim agreements, not imposed phenomena embedded in eco-social networks, either local or worldwide.

Since 2000 to 2005, this working group was led by architects Felipe Mesa and Alejandro Bernal, from 2006 until 2010 was led by Felipe Mesa, and is currently led by Felipe Mesa in partnership with Federico Mesa. The Plan:B Arquitectos work is generated primarily through participation in architectural competitions, and collaborating with other professionals in those projects is constant and diverse.

PRODUCTORA

Add: Insurgentes Sur 348, piso 9, Colonia Roma Sur, Delegación Cuauhtémoc, CP 06700 Mexico City, Mexico
Tel: +52 (55) 55 84 12 78
Web: www.productora-df.com.mx

PRODUCTORA is an architecture practice founded in 2006 in Mexico City that brings together architects from several countries: Abel Perles (1972, Argentina), Carlos Bedoya (1973, Mexico), Víctor Jaime (1978, Mexico) and Wonne Ickx (1974, Belgium). PRODUCTORA develops its ideas by means of intuitive explorations, rather than on the basis of an established methodology. The name PRODUCTORA emerges from the conviction that the design process advances through a continuous production of material to be evaluated.

PRODUCTORA is developing a variety of projects in Mexico and abroad (Asia, South America) ranging from single family dwellings to office or public buildings. The office has been teaching and lecturing on national and international events. They were selected for the Young Architect's Forum organized by the Architectural League in New York (2007) and presented their work on the 2nd architectural Biennale in Beijing (2006) and the Venice Biennale in Italy (2008).

PRODUCTORA was one of the architecture studios chosen to build a villa for the Ordos100 Project in Inner Mongolia (China/2008). They won the International Competition for the CAF Headquarters in Caracas (Venezuela) in collaboration with Lucio Muniain (2008). They were asked to build an installation in the National Museum of Art in Beijing (China/2009) and were showcase at the Victoria & Albert Museum in London (2010). Recently, PRODUCTORA and Ruth Estevez opened LIGA, a platform for Latin-American Architecture in Mexico City (2011).

Projektarbeitsgemeinschaft Behnisch Architekten Pohl Architekten

Add: Rotebühlstraße 163A, 70197 Stuttgart, Germany
Tel: +49 (0)711 60772-0
Web: www.behnisch.com

Projektarbeitsgemeinschaft Behnisch Architekten Pohl Architekten was founded in 1989 and works out of three offices – Stuttgart, Munich, and Boston. These offices are directed by Stefan Behnisch and his partners in varying combinations. The partners are Robert Hösle, Robert Matthew Noblett and Stefan Rappold.

Projektarbeitsgemeinschaft Behnisch Architekten Pohl Architekten realized such signature projects as the Institute for Forestry and Nature Research in Wageningen, the Netherlands (1998); the LEED Platinum Genzyme Center in Cambridge, MA (2004); as well as the new Unilever Headquarters (2009) and the Marco Polo Tower in Hamburg's HafenCity (2010). The most recently completed major construction project is a new administrative building for the World Intellectual Property Organization in Geneva (2011) which is adjacent to the Organization's main headquarters building. Over the years the practice has established an international reputation as a firm that combines design excellence with advanced expertise in sustainability, and the firm's innovative work has been awarded widely.

Ross Barney ArchitectsS

Add: 10 West Hubbard Street, Chicago, Illinois 60610, USA
Tel: 312.832.0600; Fax: 312.832.0601
Web: www.r-barc.com

Ross Barney Architects strives to improve the built environment. r_barc believes that design should be symbolic of the higher purposes of public building capturing a contemporary vision of today's society. They enjoy an international reputation for work primarily in the field of institutional and public buildings that include park buildings, libraries, public utilities, government, transportation buildings, and elementary schools.

RTA-Office

Add: Passatge de Domingo 2, entlo 1ª (Rambla Catalunya - Pg. de Gracia) 08007 Barcelona, Spain
Tel: +34 93 467 49 77; Fax: +34 93 467 49 77
Web: http://www.rta-office.com

RTA-Office (Real Time Architecture Office) is a branch of Santiago Parramon architectural office, which since 1991 has its headquarters in Barcelona (SPAIN). Since the development of their work recognized at the local level, the architectural office begins around the year 1999 to show its global claim. First in prestigious publications around the world, then in conferences in the USA and Europe and later in competitions, projects and buildings, his work started to be recognized internationally.

They believe their team is only one, working in different locations and with a great capacity of response to projects from different countries. This is their future: think globally, understand the contemporary condition and act on it. To live a global architectural experience.

Salto AB

Add: Salto AB OÜ, Kalaranna 6, 10415 Tallinn, Estonia
Tel: +372 682 5222
Web: www.salto.ee

Maarja Kask (b 1979), Ralf Lõoke (b 1978) and Karli Luik (b 1977) founded Salto AB in Tallinn, Estonia in 2004. It is an office devoted to architectural practices from interior design and art projects to landscape architecture and urban planning projects. In recent years Salto AB has gained prizes in over 40 architectural competitions in Estonia and abroad. Salto AB is a Tallinn-based practice established in 2004, renowned for their works that blur the line between different levels of architecture; a small team of people means that the partners are involved in each project and as seen in the press, the approach has brought them a fair share of recognition with some good news.

SAOTA - Stefan Antoni Olmesdahl Truen Architects

Add: 109 Hatfield Street, Gardens, Cape Town, South Africa
Tel: +27 21 468 4400; Fax: +27 21 461 5408
Web: www.saota.com

SAOTA - Stefan Antoni Olmesdahl Truen Architects is driven by the dynamic combination of partners Stefan Antoni, Philip Olmesdahl and Greg Truen who share a potent vision easily distinguished in their buildings and an innovative and dedicated approach to the execution of projects internationally, nationally and locally. Projects range from large scale commercial and institutional to individual high-end homes. Inspired by the challenges and the opportunities of site, context, brief and budget they strive to create distinctive, memorable and yet timeless buildings optimizing the Clients' quality of lifestyle and return on investment. The company has received numerous awards and commendations from some of the most respected institutions worldwide.

Over the past decade Antoni Associates, the Interior Design & Décor studio of the iconic architectural firm SAOTA has become known for creating some of the most exclusive interiors in South Africa as well as international locations including London, Paris, Moscow, New York, Dubai & Geneva. This dynamic and innovative practice conceptualizes and creates contemporary interior spaces and bespoke décor for the full spectrum of interior design briefs which includes domestic, hospitality, retail, corporate and leisure sectors.

Led by Mark Rielly and partners Vanessa Weissenstein & Adam Court, together with Associates Ashleigh Gilmour, Jon Case & Michele Rhoda and a dedicated and skilled team of designers and decorators, Antoni Associates prides itself on its dedication to cutting-edge contemporary design, sound technical knowledge and up to the moment computer skills and a design finesse that combined is unique in South Africa. The synergy of these attributes combined with carefully orchestrated logistics allows them to stay ahead of the market. Their interiors meet the international standard of being modern, luxurious and seductive while at the same time remaining understated and timeless and in tune with the delights of quality living demanded by their discerning clients.

Saunders Architecture

Add: VESTRE TORGGATE 22, 5015 BERGEN, NORWAY
Tel: +47 55 36 85 06
Email: post@saunders.no
Web: http://www.saunders.no

Bringing together dynamic buildings and material experimentation with traditional methods of craft, Bergen-based Saunders Architecture has worked on cultural and residential projects right across Norway, as well as England, Denmark, Italy, Sweden and Canada.

Led by a strong contemporary design sensibility, the studio believes that architecture must play an important role in creating place, using form, materials and texture to help evoke and shape memory and human interaction. The office operates within existing nature as well as manmade contexts, with examples ranging from an award - nominated dramatic viewpoint structure set amidst a rich protected landscape to several newly-built houses within more traditional suburban settings.

Saunders Architecture was founded by the Canadian architect Todd Saunders in 1998. Saunders has lived and worked in Bergen since 1996, following his studies at the Nova Scotia College of Art and Design in Halifax and McGill University in Montreal. He continues to combine teaching with practice and has been a part-time teacher at the Bergen Architecture School since 2001. Saunders has also lectured and taught at schools in Scandinavia, the UK and Canada and was a visiting professor at The University of Quebec in Montreal.

Studio B architects

Add: 501 rio grande place, suite 104, aspen, co 81611, USA
Tel: 970 920 9428
Web: www.studiobarchitects.net

Rooted in a strong modernist aesthetic, Studio B Architects fresh, minimalist design, elegant site integration and seamless interplay among architecture, interiors and nature have garnered increasing national attention since the firm's founding in Aspen, Colorado in 1991. With more than 45 prestigious design awards received over the last decade, and recognition by the American Institute of Architects/ Colorado as the organization's 2009 Firm of the Year, Studio B Architects, influence and growing portfolio of modernist work have continued to draw attention far beyond the Rocky Mountain West.

Studio B Architects brings a rare level of artistry to their notable designs elegantly accomplished through an unforgettable, egoless, engaging experience that catalyzes new clients to seek them out and former patrons to work with them again.

Tank Architectes

Add: 33 Rue de la Justice 59000 Lille, France
Tel: 00 33 (0) 3 28 14 03 59; Fax: 00 33 (0) 3 28 14 00 47
Web: www.tank.fr

Tank Architectes was created in 2005 by Olivier Camus and Lyderic Veauvy, both architects graduated from the Superior Institute of Architecture of St LUC Tournai, and taught in 1st and 2nd year workshops at the Faculty of Architecture, Architectural Engineering and Urbanism LOCI of Catholic University of Louvain La Neuve, Belgium, on the site of Tournai.

Tank is a team of architects working as hard for sense and contents as for achievements. The process is based on research and brainstorming in order to create sensitive and poetic buildings. The projects of Tank cover their whole built environment : the High School Levi Strauss in Lille, crèche and nursery in Roubaix, high school Léonard de Vinci in St germain en Laye (associated to Colboc & Franzen Associés), hive of small enterprises La Tossée in Tourcoing, media libraries in Proville and La Madeleine and housing projects in Lille, Arras, Lens or Aubervilliers.

Tank has been rewarded by the 3rd Trophy of Nouveaux Albums des Jeunes NAJA and nominated at the Grand Prix de l'Equerre d'Argent 2009 for the rehabilitation of the Minoterie, and 2010 for the High School Levi Strauss in Lille.

The Jerde Partnership, Inc

Los Angeles, Shanghai, Hong Kong, Seoul, and Berlin

Add: 913 Ocean Front Walk, Venice, Los Angeles
California 90291, USA
Tel: +310 399 1987, Fax: +310 392 1316
Web: www.jerde.com

The Jerde Partnership is a visionary architecture and urban planning firm that designs unique places that deliver memorable experiences and attract millions of people every day. Nearly 1 billion people visit Jerde-designed Places every year.

As the firm that pioneered "placemaking", Jerde has created projects throughout the world that provide lasting social, cultural and economic values and promote further investment and revitalization. For over 30 years, they have partnered with developers, city planners and local officials throughout the world to achieve some astounding results.

Their journey began in 1977, when founder Jon Jerde, FAIA, broke from conventional architects who focus on advancing architectural forms, to advance the creation of memorable places where people can gather and experience a sense of community. That singular, founding vision, which they call Jerde Placemaking, continues to inform their work today.

Since its founding, Jerde has grown to a multi-disciplinary, international design studio based in Los Angeles, with offices in Amsterdam, Hong Kong, Shanghai, Seoul, and Dubai. Approximately one-third of their designers were born outside of the United States in diverse places such as Australia, Belgium, El Salvador, China, Korea, Latvia, Russia, Singapore, Uruguay, the United Kingdom, and more.

Jerde's design talent possesses individual and collective passions about what they do, and they work closely together bringing a diversity of cultural backgrounds and ideas to continually evolve the global application of Jerde Placemaking.

They currently offer clients conceptual and design services that integrate pre-development and leasing strategies, planning, architecture design, landscape design, and interiors.

Their international portfolio has expanded beyond its retail and entertainment roots to include hotels, casinos and resorts; residential complexes; office and commercial facilities; transit-oriented mixed-use hubs; major urban districts; waterfronts; town centers; community plans; and visionary master plans.

To date, over 100 Jerde-designed places have opened in cities throughout the world, such as Budapest, Hong Kong, Las Vegas, Los Angeles, Osaka, Rotterdam, Seoul, Shanghai, Tokyo, Istanbul, Warsaw, Dubai and others. Projects are currently under construction in Las Vegas, Mexico City, Korea, Macao, Dubai, Russia and Los Angeles. Wherever they are built, Jerde Places are widely embraced by the public and ultimately transform the economic and social landscape of neighborhoods, cities and regions. They integrate public life, shops, parks, restaurants, entertainment, housing and nature into one place and set the bar for the way urban mixed-use projects are created.

Topos Atelier de Arquitectura, Lda

Add: Rua Andrade Corvo, 242 – 3º, Sala 301,
4700-204 Braga, Portugal
Tel: +351 253 272 187; Fax +351 253 267 296
Web: www.toposatelier.com

Topos Atelier de Arquitectura, Lda was founded in 1986 by Jean Pierre Porcher and Margarida Oliveira and has developed projects spanning the entire architectural and land specification spectrum. Albino Freitas integrated the Atelier in 1992.

From the moment of its conception, each project is accompanied by a team of skilled architects and engineers who ensure that every requirement is met to specification throughout each stage of development, maintaining permanent contact with the client. Throughout the process of finalising a building contract with the developers, each client receives support from the studio and every project undertaken is accompanied from start to finish.

Topos is currently working on various programmes including equipment, housing and the management of public spaces. Some of their projects and developments were recently published and featured introductions by Eduardo Souto Moura, Alain Gunst, Gonçalo Byrne and Manuel Graça Dias.

Wahana Cipta Selaras

Add: Jl Ciputat Raya No. 351, Kebayoran Lama Utara,
Jakarta Selatan, Indonesia
Tel: +62-21 72793419/+62-21 72793409; Fax: +62-21 72793417
Web: www.wahanaarchitects.com

Wahana Cipta Selaras (WCS) is one of the leading and well-known architectural firm serving clients in business building, industrial, residential, interior and contractor in Indonesia. Their general practice has developed many successful projects. In WCS, the core values are providing clients with responsive design, technological expertise and exceptional service. The architects can coordinate your projects from its initial inception trough design, community, governmental approvals, bidding and construction. The firm helps integrating different design solutions and incorporate sustainable design principles as their responsibility to each and every project they created.

Zerafa Studio LLC.

Add: 17 Vestry Street 2nd Floor, New York NY 10013, USA
Tel: 212 966 7083; Fax: 212 966 7086
Web: www.zerafaarchitecture.com

Zerafa Studio LLC. is a multi-disciplinary architectural design firm based in New York City. Founded in 2005 by Jason Zerafa, the firm has a fundamental belief that each project offers them the opportunity to develop innovative architectural design solutions that can make a positive and lasting contribution to the built environment. They strongly believe in a rigorous and holistic design process that encourages the direct participation of their clients and specialty consultants in a collaborative work environment.

The firm is fully committed to creating sophisticated modern architecture that is sensitive to the culture of their individual clients and the specific context of each project. It is their responsibility to develop cost effective design solutions to address complex technical and environmental challenges and to fully support and realize the programmatic and functional needs of each project.

They believe that a strong architectural solution should embody and clearly illustrate a series of fundamental ideas inherent to the spirit of the project. These ideas should ultimately be used to test and inform design decisions throughout the design process. Their professional commitment to their clients and the profession is paramount, and they initiate each project with the understanding that innovative design adds value to their projects.

图书在版编目(CIP)数据

一筑一景．文化·教育·住宅 / 度本图书 编．－武汉：华中科技大学出版社，2012.6
ISBN 978-7-5609-8061-4

Ⅰ．①一… Ⅱ．①度… Ⅲ．①建筑设计－作品集－世界－现代 Ⅳ．①TU206

中国版本图书馆CIP数据核字(2012)第113049号

一筑一景　文化·教育·住宅　度本图书　编

出版发行：华中科技大学出版社（中国·武汉）
地　　址：武汉市武昌珞喻路1037号（邮编：430074）
出 版 人：阮海洪

责任编辑：刘锐桢　责任监印：秦英
责任校对：杨　睿　装帧设计：Dopress Books

印　刷：利丰雅高印刷（深圳）有限公司
开　本：965 mm × 1270 mm　1/16
印　张：20
字　数：160千字
版　次：2012年8月第1版 第1次印刷
定　价：288.00元

投稿热线：(010)64155588-8000 hzjztg@163.com
本书若有印装质量问题，请向出版社营销中心调换
全国免费服务热线：400-6679-118　竭诚为您服务